Spectral Theory of Multivalued Linear Operators

Spectral Theory of Multivalued Linear Operators

Aymen Ammar, PhD
Aref Jeribi, PhD

APPLE ACADEMIC PRESS

ublished [2022]

nic Press Inc.
ɔd Circle, NE,
32905 USA

ɾe Road, Burlington,
Canada

CRC Press
6000 Broken Sound Parkway NW,
Suite 300, Boca Raton, FL 33487-2742 USA

2 Park Square, Milton Park,
Abingdon, Oxon, OX14 4RN UK

Academic Press, Inc.

ic Press exclusively co-publishes with CRC Press, an imprint of Taylor & Francis Group, LLC

nd Archives Canada Cataloguing in Publication

tral theory of multivalued linear operators / Aymen Ammar, PhD, Aref Jeribi, PhD.
nmar, Aymen, author. | Jeribi, Aref, author.
ɪ: Includes bibliographical references and index.
Canadiana (print) 20210157038 | Canadiana (ebook) 20210157194 | ISBN 9781771889667 (hardcover) |
ʹ781774639382 (softcover) | ISBN 9781003131120 (ebook)
CSH: Linear operators. | LCSH: Spectral theory (Mathematics)
ion: LCC QA329.2 .A46 2022 | DDC 515/.7246—dc23

f Congress Cataloging-in-Publication Data

nmar, Aymen, author. | Jeribi, Aref, author.
ɪ: tral theory of multivalued linear operators / Aymen Ammar, Aref Jeribi.
ɪ: Burlington ON ; Palm Bay, Florida : Apple Academic Press, [2022] | Includes bibliographical references
łex. | Summary: "The concept of multivalued linear operators-or linear relations, the one of the most exciting
luential fields of research in modern mathematics. Applications of this theory can be found in economic theory,
perative games, artificial intelligence, medicine, and more. This new book, Spectral Theory of Multivalued
Operators, focuses on the theory of multivalued linear operators, responding to the lack of resources exclusively
, with the spectral theory of multivalued linear operators. The subject of this book is the study of linear relations
al or complex Banach spaces. The main purposes are the definitions and characterization of different kinds of
and extending the notions of spectra that are considered for the usual one single-valued operator bounded or
ınded. The volume introduces the theory of pseudospectra of multivalued linear operators. The main topics
demicompact linear relations, essential spectra of linear relation, pseudospectra, and essential pseudospectra of
elations. The volume will be very useful for researchers since it represents not only a collection of a previously
ʒeneous material but is also an innovation through several extensions. Beginning graduate students who wish to
ıe field of spectral theory of multivalued linear operators will benefit from the material covered, and an expert
will also find some of the results interesting enough to be sources of inspiration. Prerequisites for the book are
ic courses in classical real and complex analysis and some knowledge of basic functional analysis. In fact,
ory constitutes a harmonious mixture of analysis (pure and applied), topology, and geometry"-- Provided by
er.

LCCN 2021010465 (print) | LCCN 2021010466 (ebook) | ISBN 9781771889667 (hardcover) | ISBN
ʹ4639382 (paperback) | ISBN 9781003131120 (ebook)
CSH: Linear operators. | Spectral theory (Mathematics)
ion: LCC QA329.2 .A395 2022 (print) | LCC QA329.2 (ebook) | DDC 515/.7246--dc23
available at https://lccn.loc.gov/2021010465
record available at https://lccn.loc.gov/2021010466

77188-966-7 (hbk)
77463-938-2 (pbk)
ɔ0313-112-0 (ebk)

Dedication

To
the memory of my mother Ayada,
the memory of my father Khalifa,
my wife Nihel, and my children Nourcen and Ayen,
my brother Houcem, my sisters Imen, Ines, Dorsaf, and Awatef,
all members of my extended family ...

—Aymen Ammar, PhD

To
my mother Sania, my father Ali,
my wife Fadoua, my children Adam and Rahma,
my brothers Sofien and Mohamed Amin,
my sister Elhem,
and all members of my extended family...

—Aref Jeribi, PhD

Contents

About the Authors

Aymen Ammar, PhD

Associate Professor, Department of Mathematics, University of Sfax, Tunisia

Aymen Ammar, PhD, is currently working as an Associate Professor in the Department of Mathematics, Faculty of Sciences of Sfax at the University of Sfax, Tunisia. He has published many articles in international journals. His areas of interest include spectral theory, matrice operators, transport theory, and linear relations.

Aref Jeribi, PhD

Professor, Department of Mathematics, University of Sfax, Tunisia

Aref Jeribi, PhD, is a Professor in the Department of Mathematics at the University of Sfax, Tunisia. He is the author of the book Spectral Theory and Applications of Linear Operators and Block Operator Matrices (Springer-Verlag, New-York, 2015), co-author of the book *Nonlinear Functional Analysis in Banach Spaces and Banach Algebras: Fixed Point Theory under Weak Topology for Nonlinear Operators* and *Block Operator Matrices with Applications* (Taylor-Francis, 2015), the author of the book *Denseness, Bases, and Frames in Banach Spaces and Applications* (De Gruyter, Berlin, 2018), the author of the book *Linear Operators and Their Essential Pseudospectra* (Apple Academic Press, CRC Press, Oakville, Boca Raton, 2018), and the co-author of the book *Analyse Numérique Matricielle, Méthodes et Algorithmes, Exercices et Problèmes Corrigés* (Références Sciences. Paris: Ellipses, 2020). He has published many journal articles in international journals. His areas of interest include spectral theory, matrice operators, transport theory, Gribov operator, Bargmann space, fixed point theory, Riesz basis, and linear relations.

Preface

This book is focused on the theory of multivalued linear operators. To some extent, it is a sequel of the authors' recent work on linear operators on linear relation as well as their spectral theory. There are not many books exclusively dealing with the spectral theory of multivalued linear operators, and the authors wish to add to the scarce literature on the subject. A minimum necessary background material has been gathered, which will allow relatively friendly access to the book. Beginning graduate students who wish to enter the field of spectral theory of multivalued linear operators should benefit from the material covered, but an expert reader might also find some of the results interesting enough to be sources of inspiration. Prerequisites for the book are the basic courses in classical real and complex analysis and basic functional analysis knowledge. In fact, this theory constitutes a harmonious mixture of analysis (pure and applied), topology, and geometry.

The concept of multivalued linear operators or linear relations is one of the most exciting and influential fields of research in modern mathematics. Applications of this theory can be found in economic theory, non-cooperative games, artificial intelligence, medicine, and existence of solutions for differential inclusions, the theory of a multivalued operator, assigning to the points of some set X a subset of another set Y, has arisen naturally by refining the classical concept of a multivalued function. The concept of a multivalued linear operator appeared in the literature a few decades ago, not only through the need of considering adjoints (conjugates) of non-densely defined linear differential operators, but also through the necessity of considering the inverses of certain operators used; for example, in the study of some Cauchy problems associated with parabolic type equations in Banach spaces. Let us, for example, the abstract degenerate equation in a Banach space. It is generally true that the degenerate linear evolution equations with respect to the time derivative are rewritten into non-degenerate equations in such a way using multivalued linear operators. This fact then leads us naturally to consider a problem of generalizing the well-developed results concerning the ordinary linear evolution equations with univalent coefficient operators to those with multivalued operators and to handling the degenerate equations using analogous techniques to non-degenerate ones.

The subject of this book is the study of linear relations over real or complex Banach spaces. The primary purposes are the definitions and characterization of different kinds of spectra, someone extending the notions of spectra considered for the usual one single-valued operators bounded or not bounded. One of the objectives of this book is to introduce the theory of pseudospectra of multivalued linear operators. The main topics include demicompact linear relations, essential spectra of linear relation, pseudospectra, and essential pseudospectra of linear relations.

We hope that this book will be handy for researchers, since it represents a collection of a previously heterogeneous material and innovation through several extensions; of course, a single book can't cover such a huge research field. In making personal choices for material inclusion, we tried to give useful complementary references in this research area, hence probably neglecting some relevant works. We would be very grateful to receive any comments from readers and researchers, providing us with some information concerning some missing references.

We would like to mention that the thesis work results by my students Bilel Boukettaya, Houcem Daoud, Bilel Saadaoui, Slim Fakhfakh, Nawrez Lazrag, and Ameni Bouchekoua, the obtained results have helped us in writing this book.

Sfax, Tunisia   *Aymen Ammar, PhD*
April 2021   *Aref Jeribi, PhD*

Symbol Description

The most frequently used notations, symbols, and abbreviations are listed below:

$\mathbb{N}$ — The set of natural numbers.

$\mathbb{R}, \mathbb{C}$ — The fields of real and complex numbers, respectively.

$\mathbb{R}^n$ — The n-dimensional real space.

$\inf(A)$ — The infimum of the set A.

$\sup(A)$ — The supremum of the set A.

$\widetilde{T}$ — The completion of T.

B_r — The ball with radius r.

B_X — The unit ball in X.

$\mathrm{dist}(x, y)$ — The distance between x and y.

X^*, X' — The dual of X.

$\dim(X)$ — The dimension of the space X.

$\|x\|$ — The norm of x.

$(X, \|\cdot\|)$ — The linear normed space X.

$(x_n)_n$ — The sequence $(x_n)_n$.

$G(T)$ — The graph of T.

$\mathcal{C}(\Omega, \mathbb{R})$ — The space of real continuous functions on Ω.

L_p — The L_p space.

$\rho(T)$ — The resolvent set of T.

$\sigma(T)$ — The spectrum of T.

$\sigma_\varepsilon(T)$ — The pseudospectrum of T.

$\Sigma_\varepsilon(T)$ — The ε-pseudospectrum of T.

$T_{|M}$ — The restriction of T to M.

$T_n \xrightarrow{g} T$ — The sequence $(T_n)_n$ converge in the generalized sense to T.

$\mathrm{asc}(T)$ — The ascent of T.

$\mathrm{des}(T)$ — The descent of T.

$\mathfrak{R}_c(T)$ — The singular chain manifold of T.

$\mathcal{W}(X, Y)$ — The set of weakly compact linear operators from X into Y.

$\mathcal{L}(X, Y)$ — The space of linear operators from X into Y.

$\mathcal{R}(U, V)$ — The collection of relation from U into V.

$L\mathcal{R}(X, Y)$ — The collection of linear relations from X into Y.

$B\mathcal{R}(X, Y)$ — The collection of bounded linear relations from X into Y.

$C\mathcal{R}(X, Y)$ — The class of all closed linear relations from X into Y.

$BC\mathcal{R}(X, Y)$ — The class of closed everywhere defined linear relations from X into Y.

$\mathcal{KR}(X,Y)$ The families of all compact linear relations from X into Y.

$\mathcal{DC}(X,Y)$ The families of all demicompact linear relations from X into Y.

$N(T)$ The null space of T.

$R(T)$ The range space of T.

$\mathcal{PRK}(X,Y)$ The families of all precompact linear relations from X into Y.

$\mathcal{SSR}(X,Y)$ The set of strictly singular linear relations from X into Y.

J_M^X injection operator from M into X.

$x_n \to x$ x_n converges strongly to x.

$\overline{M}$ The closure of the set M.

$co(\cdot)$ The convex hull.

$\overline{co}(\cdot)$ The closed convex hull.

$\mathcal{PR}(\Phi_+(X,Y))$ The set of upper semi-Fredholm perturbations of linear relations from X into Y.

$\mathcal{PR}(\Phi_-(X,Y))$ The set of lower semi-Fredholm perturbations of linear relations from X into Y.

$\mathcal{PR}(\mathcal{A}_\alpha(X,Y))$ The set of α-Atkinson perturbations linear relations from X into Y.

$\mathcal{PR}(\mathcal{A}_\beta(X,Y))$ The set of β-Atkinson perturbations linear relations from X into Y.

$\mathcal{F}(X,Y)$ The set of Fredholm linear relations from X into Y.

$\mathcal{F}_+(X,Y)$ The set of upper semi-Fredholm linear relations from X into Y.

$\mathcal{F}_-(X,Y)$ The set of lower semi-Fredholm linear relations from X into Y.

$\Phi(X,Y)$ The set of Fredholm closed linear relations from X into Y.

$\Phi_+(X,Y)$ The set of upper semi-Fredholm closed linear relations from X into Y.

$\Phi_-(X,Y)$ The set of lower semi-Fredholm closed linear relations from X into Y.

$\mathcal{A}_\alpha(X,Y)$ The class of α-Atkinson multivalued linear operators from X into Y.

$\mathcal{A}_\beta(X,Y)$ The class of β-Atkinson multivalued linear operators from X into Y.

$\mathcal{W}^l(X,Y)$ The set of left Weyl linear relations from X into Y.

$\mathcal{W}^r(X,Y)$ The set of right Weyl linear relations from X into Y.

$\mathcal{B}(X,Y)$ The set of Browder linear relations from X into Y.

Chapter 1

Introduction

The spectral theory of linear operators is an important part of functional analysis which has application in several areas of modern mathematical analysis and physics; for instance, in differential and integral equation as well as quantum theory. The spectral theory of linear operators in a Banach space is one of the major advances in mathematics of the 20th century. It was initiated by T. Kato in his famous book (see Ref. [67]) which contains the most basic ideas for the spectral theory of linear operators. Another important field was stimulated by problems from mathematics physics is the concept of multivalued linear operators. The concept of a linear relation, (or multivalued linear operator) in a linear space which generalizes the notion of a (single valued) linear operator to that of a multivalued operator. This notion was introduced by R. Arens who gave a scientific treatment in Ref. [42] and by E. A. Coddington (see Refs. [48, 49]). Since that time, it will be helpful in different areas and it has been studied in diverse specific contexts, (see Ref. [6]). At present, the study of linear relations in Banach or Hilbert spaces is of significance since it has applications in many problems in physics and other areas of applied mathematical. A. Favini and A. Yagi, in his paper Ref. [57], introduced the idea of multivalued linear operators as an approach toward the degenerate linear evolution equations with respect to the time derivative. In Ref. [56, Sect. 5] he handled a time homogeneous equation

$$\begin{cases} \dfrac{dMv}{dt} + Lv &= f(t), \quad 0 \leq t \leq T, \\ \lim_{t \to 0^+} Mv(t) &= u_0, \end{cases} \tag{1.1}$$

in X and indicated that the operator defined by

$$Z_0(t) = \frac{1}{2\pi i} \int_\Gamma e^{-\lambda t} (\lambda M - L)^{-1} d\lambda, \quad t > 0,$$

plays a role like a fundamental solution under the assumption that the bounded inverse $(\lambda M - L)^{-1}$, called M-modified resolvent of L, exists for any $\lambda \in \Sigma := \{\lambda \in \mathbb{C} : |arg(\lambda)| \geq w\}$, $0 < w < \frac{\pi}{2}$, and $(\lambda M - L)^{-1}$ satisfies

1

$$\|L(\lambda M - L)^{-1}\| \leq Const, \quad \lambda \in \Sigma.$$

To understand this result intuitively, let us rewrite (1.1) into a non degenerate form by putting $u(t) = Mv(t)$ (such a change of unknown function was already seen in R. E. Showalter [87], in which M is a nonlinear operator). It then turns out that

$$\begin{cases} \dfrac{du}{dt} + Au & \ni \quad f(t), \quad 0 \leq t \leq T, \\ u(0) & = \quad u_0, \end{cases}$$

where $A = LM^{-1}$ and M^{-1} is the inverse of M. Of course M^{-1} is no longer defined as a univalent operator, but conserves its linearity; such an operator is called a multivalued linear operator. This fact then leads us naturally to consider a problem of generalizing the well developed results concerning the ordinary linear evolution equations with univalent coefficient operators to those with multivalued operators and to in tend handling the degenerate equations by means of analogous techniques to non-degenerate ones. Consider an initial boundary value problem for an elliptic-parabolic equation of the type

$$\begin{cases} \dfrac{\partial}{\partial t}(m(x)u(x,t)) & = \quad \nabla \bullet (a(x)\nabla u(x,t)) + \displaystyle\sum_{i=1}^{n} a_i(x)\dfrac{\partial(m(x)u(x,t))}{\partial x_i} \\ & \quad\quad +c_0(x)u(x,t) + f(x,t) \quad (x,t) \in \Omega \times (0,\tau], \\ u(x,t) & = \quad 0 \quad (x,t) \in \partial\Omega \times (0,\tau], \\ m(x)u(x,0) & = \quad v_0(x), \quad\quad x \in \Omega, \end{cases}$$

$$(1.2)$$

where Ω is a bounded domain in $\mathbb{R}^n$ of class C^2, $n = 2,3$; ∇ denotes the gradient vector with respect to x variable; $m(x) \geq 0$ on $\overline{\Omega}$ and $m \in C^2(\Omega)$; $a(x)$, $a_i(x)$, $c_0(x)$ are real-valued smooth functions on $\overline{\Omega}$; $a(x) \geq \delta > 0$ and $c_0(x) \leq 0$ for all $x \in \overline{\Omega}$. Take the space $X := L^2(\Omega)$, the operators $Mu := m(\cdot)u$ with domain $\mathcal{D}(M) := X$, and $Lu = \nabla \bullet (a(\cdot)\nabla u) + c_0(\cdot)$ with domain $\mathcal{D}(L) := H^2(\Omega) \times H_0^1(\Omega)$. Introduce the new unknown function $v := Mu$ and, in X, consider the operators $A := LM^{-1}$ with domain $\mathcal{D}(A) := M(\mathcal{D}(L))$, and $Bv := \displaystyle\sum_{i=1}^{n} a_i(x)\dfrac{\partial v}{\partial x_i}$ with domain $\mathcal{D}(B) = H_0^1(\Omega)$. Note that A is a multivalued linear operator. Then, problem (1.2) is reduced to the multivalued Cauchy problem (inclusion)

$$\begin{cases} \dfrac{\partial v}{\partial t} & \in \quad Av + Bv + f(t), \quad t \in (0, \tau] \\ v(0) & = \quad v_0, \end{cases}$$

where $f(t) := f(t, .)$ and $v_0 := v_0(\cdot)$. The class of univalent linear operators is unstable under the operations closure, inverse and adjoint. This is not the case if we consider the more general class of multivalued linear operators. So, it is interesting to extend some results of spectral theory to the multivalued linear case as well as obtaining perturbation theorems for multivalued linear operators. As in the case of linear operators, the spectral theory of linear relations, including the associated analytic functional calculus, is an important tool for studying various properties of these objects and for deriving some of their applications.

One of the objectives of this book is the study of the perturbation theory of multivalued linear operators, in order to establish the existence results of the second kind multivalued operator equations, hence allowing to describe the spectrum, and essential spectra of a linear relation. Fredholm theory and perturbation results are also widely investigated. Recall that the perturbation problems are among the main topics in both pure and applied mathematics. The perturbation theory of operators, from which many valuable results have been obtained has been extensively studied. In 1907, F. Riesz introduced and studied the theory of compact operators. This theory forms a part of the classic core of functional analysis and operator theory. A fundamental result proved by F. Riesz is that, if T is a compact operator defined on a Banach space, then $I - T$ is a Fredholm operator with index 0. This contribution came with some us stability results of the essential spectra. But this is wrong, if T is a compact multivalued linear operator. It is well-known that if T is a closed Fredholm linear operator between Banach spaces, then $T + S$ has the same property, whenever S is a compact linear operator (respectively, bounded with sufficiently small norm). Among the works in this direction, we can state, for example, I. C. Gohberg, A. S. Markus, and I. A. Feldman [58], A. Lebow and M. Schechter [77], E. Albrecht and F.-H. Vasilescu [7], and C.-G. Ambrozie and F.-H. Vasilescu [15]. In 1998, R. W. Cross [50], has investigated the question of stability in the sets of upper semi-Fredholm and lower semi-Fredholm (see Theorems 2.16.1 and 2.16.3).

Considerable attention has also been devoted by T. Álvarez, R. W. Cross and D. Wilcox [13] introduced the class of α- and β-Atkinson linear relations in normed spaces. Specifically these authors gave important characterization theorems of such multivalued linear operators in terms of the existence of left and right generalized inverses. After that, T. Álvarez and D. Wilcox [14] continued their study of α-and β-Atkinson linear relations. They inspect essentially, these characterization theorems which will be used in conjunction with perturbation theorems for Fredholm linear relations in order to establish a perturbation theory for Atkinson type relations. To have further details we may refer to Ref. [14].

The concept of demicompactness was introduced into functional analysis by W. V. Petryshyn [83], in order to discuss fixed points and it has been studied in a large number of papers (see for example, Refs. [43, 82]). In 2012, W. Chaker, A. Jeribi and B. Krichen achieved some results on Fredholm and upper semi-Fredholm operators involving demicompact operators [47]. Since much attention was paid to this notation, several research papers, used it (see Refs. [65, 71]). In 2018, A. Ammar, H. Daoud and A. Jeribi, extended the concept of demicompact and k-set-contraction of linear operators to multivalued linear operators and developed some properties (see Ref. [24]). In Refs. [39, 40], A. Ammar, A. Jeribi and B. Saadaoui have firstly obtained more results in the spirit of those obtained in Ref. [24]. Secondly, they applied these new results in order to establish some perturbation results of multivalued linear operator. Recently, A. Ammar, S. Fakhfakh and A. Jeribi introduced in Ref. [28] concept of relatively demicompactness with respect to multivalued linear operator as a generalization of the class of demicompact linear relation introduced by A. Ammar, H. Daoud and A. Jeribi [24] and established some new results in Fredholm theory.

The notion of essentially semi regularity operators amongst the various concepts of regularity originated by the classical treatment of perturbation theory owed to T. Kato and was studied by many authors; for instance, we cite Refs. [4, 5, 18, 65, 80, 95]. We remark that all the above authors considered only the case of bounded linear operators. It is the purpose of this section to consider these class of essentially semi regularity in the more general setting of linear relations in Hilbert spaces. Many properties of essentially semi regularity for the case of linear operators remain valid in the context of linear relations, sometimes under supplementary conditions. In Ref. [21], A. Ammar, B. Boukettaya and A. Jeribi studied some properties of the semi regular and essentially semi regular, of a multivalued linear operator.

There are many ways to define the essential spectrum of a closed densely defined linear operator on a Banach space. Indeed, several definitions may be found in the literature. Particularly, various notions of essential spectrum appear in applications of the spectral theory (see for example, Refs. [4,5,65]). Five of these definitions were studied by D. E. Edmunds and W. D. Evans [55]. In 1998, R. W. Cross [50] introduced a concept of the essential spectrum of a multivalued linear operator on a complex normed space in terms of the nullity and the deficiency of its complete closure, and he showed its stability under relative compact perturbation with certain additional conditions. Subsequently, in 2012, T. Álvarez has introduced several essential spectra of a linear relation on a normed space. We investigate the closedness and the emptiness of such essential spectra. As an application we prove two results, the first of which characterizes the class of quotient indecomposable normed spaces in terms of upper semi-Fredholm, lower semi-Fredholm and strictly cosingular linear relations, and it gives conditions under which a linear relation on a complex quotient indecomposable normed space is a strictly cosingular perturbation of multiple identities. In 2014, D. Wilcox has came with five distinct essential spectra of linear relations on Banach spaces in terms of semi-Fredholm properties, and has showed their stability under relative compact perturbations with some additional conditions and under compact perturbations. More recently, T. Álvarez, A. Ammar and A. Jeribi [10, 11] have made a detailed treatment of some subsets of the essential spectra of a matrix of multivalued linear operator. What is notable is that the investigation of the essential spectra of a linear relation present a guide to the study of operators on normed spaces since every continuous operator between normed spaces is the inverse of an injective upper semi-Fredholm relation. In addition, the class of all bounded Fredholm operators on Banach spaces coincides with the class of the inverses of closed Fredholm linear relations which are both surjective and injective. In Refs. [16,17], A. Ammar characterized some essential spectra of a closed linear relation in terms of certain linear relations type α- and β-Atkinson. This study leads us to generalize some well-known results for operators and to extend some results of small perturbations given by T. Álvarez and D. Wilcox [14]. The investigation of the notion of the essential spectra of linear relations has important applications to several problems of operator theory. For instance, we cite Refs. [44] and [85]. Many spectral and basic properties about essential spectra were given in Refs. [1,11,17,26–28,37,61–66,93]. Besides, in Ref. [36], A. Ammar, A. Jeribi and N. Lazrag found a relationship between the essential spectra of $(T_n)_n$ and essential spectra of T, where $(T_n)_n$ is a sequence of closed linear relation converging in the generalized sense to T. In order to study this convergence, they seek a counterpart of the gap between two closed linear relations as defined by T. Kato [67].

Several mathematical and physical problems lead to operator pencils, (operator-valued functions of a complex argument). Recently, the spectral theory of operator pencils attracted the attention of many mathematicians (see Ref. [2]). In this book, we give a characterization of the essential spectrum of the closed pencil linear relation in order to extend many known results in the literature. In fact, we give the characterization of the S-essential spectra of the sum of two multivalued linear operators and we study the stability of the S-essential approximate point spectrum and the S-essential defect spectrum of closed and closable linear relations under relatively precompact perturbations on a Banach space. On another level, we study the stability of the S-essential approximate point spectrum but under assumptions different from those adopted above (see Refs. [10, 29, 41]).

An important direction was established by J. M. Varah [90] which investigated the notion of pseudospectra in the domain of linear operators and has been employed by other authors, such as, H. J. Landau [76], L. N. Trefethen [88, 89], D. Hinrichsen and A. J. Pritchard [60] and E. B. Davies [53]. More precisely, L. N. Trefethen developed this notion for matrices and operators and applied it to several interesting problems, and to study interesting problems in mathematical physics. The concept of pseudospectra appeared as a result of the realization that various properties of highly non-self-adjoint operators are related. We refer the reader to L. N. Trefethen for the definitions of the pseudospectra and the ε-pseudospectra of the closed linear operator T, respectively, by

$$\sigma_\varepsilon(T) \;=\; \sigma(T) \bigcup \left\{ \lambda \in \mathbb{C} \text{ such that } \|(\lambda - T)^{-1}\| > \frac{1}{\varepsilon} \right\},$$

$$\Sigma_\varepsilon(T) \;=\; \sigma(T) \bigcup \left\{ \lambda \in \mathbb{C} \text{ such that } \|(\lambda - T)^{-1}\| \geq \frac{1}{\varepsilon} \right\},$$

where $\sigma(.)$ is the spectrum of T and by convention $\|(\lambda - T)^{-1}\| = +\infty$ if and only if, $\lambda \in \sigma(T)$. Importance characterization of pseudospectra given by

$$\bigcup_{\|S\| < \varepsilon} \sigma(T + S) = \sigma_\varepsilon(T).$$

In Ref. [86], E. Shargorodsky has proved that the ε-pseudospectra of a bounded linear operator is not equal to the union of the spectra of all perturbed operators with perturbations that have norms less than ε,

$$\bigcup_{\|S\|<\varepsilon} \sigma(T+S) \subsetneq \Sigma_\varepsilon(T).$$

In Ref. [46], F. C. Chaitin-Chetelin and A. Harrabi have shown if the resolvent norm of closed linear operator T cannot be constant in an open set, then $\Sigma_\varepsilon(T) = \overline{\sigma_\varepsilon(T)}$. In Ref. [94], M. P. H. Wolff has proved that there exists $n_0 \in \mathbb{N}$ such that for all $n \geq n_0$, $\Sigma_{\varepsilon_1}(T) \subset \Sigma_{\varepsilon_2}(T_n)$, where $0 \leq \varepsilon_1 < \varepsilon_2$ and $(T_n)_n$ is a sequence of bounded linear operators on a Banach space which converges with respect to the operator norm to the operator T.

Inspired by the notion of pseudospectra, A. Ammar and A. Jeribi in their works [20, 31–33], thought to extend these results for the essential spectra of closed, densely defined, and linear operators on a Banach space. They declared the new concept of the essential pseudospectra of closed, densely defined, and linear operators on a Banach space. In 2016, A. Ammar, H. Daoud and A. Jeribi have extended the concept of pseudospectra to the linear relation and they have defined it by

$$\sigma_\varepsilon(T) = \sigma(T)\bigcup\left\{\lambda \in \mathbb{C} \text{ such that } \|(\lambda - \widetilde{T})^{-1}\| > \frac{1}{\varepsilon}\right\},$$

where $\widetilde{T}$ is the completion of T. Moreover, they have investigated some properties and the characterization of the pseudospectra and the essential pseudospectra of linear relations (see Refs. [22,23]). Subsequently, in 2017, A. Ammar, A. Jeribi and B. Saadaoui introduced and gave a characterization of the essential pseudospectra of linear relations on a Banach space (see Ref. [38]). In 2019, A. Ammar, A. Bouchekoua and A. Jeribi [19] have extended the notion of the ε-pseudospectra to linear relations and have introduced the new concept of the essential ε-pseudospectra of linear relations by

$$\Sigma_\varepsilon(T) := \sigma(T)\bigcup\left\{\lambda \in \mathbb{C} \text{ such that } \|(\lambda - \widetilde{T})^{-1}\| \geq \frac{1}{\varepsilon}\right\},$$

Going on to a brief indication of the contents of the individual chapters, we mention, first of all that our book consists of four chapters.

Chapter 2 summarises a selection of known properties of linear relations acting on normed spaces. Many results from the theory of linear relations in normed spaces due to R. W. Cross [50] are recalled. In particular, results concerning the closure, the adjoint, the nullity, the deficiency, the norm and the minimum modulus of a linear relation and some small perturbation results are presented. We also introduce the concept of quasi-Fredholm, demicompact, measure of noncompactness, semi regular, pseudospectra and essential pseudospectra, precompact, compact and strictly singular linear relations.

In Chapter 3, we develop the closedness of linear relations, which include both single valued and multivalued linear operators. Several characterizations for closability, bundedness, and relative boundedness from linear relations are given in terms of their induced linear operators. Subsequently, we study the semi regular and essentially semi regular linear relation. The notion of essentially semi regularity operators amongst the various concepts of regularity originated by the classical treatment of perturbation theory owed to Kato and was studied by many authors, for instance, we cite Refs. [4, 5, 18, 80, 95]. We remark that all the above authors considered only the case of bounded linear operators. It is the purpose of this chapter to consider these class of essentially semi regularity in the more general setting of linear relations in Hilbert spaces. Many properties of essentially semi regularity for the case of linear operators remain be valid in the context of linear relations, sometimes under supplementary conditions. We also address the concept of demicompactness and k-set-contactive linear operators on multivalued linear operators and we develop some properties.

In Chapter 4, we emphasize the strong connection between the spectral theory of closed linear relations and that of some closed linear operators. As a matter of fact, we develop a spectral theory for a certain class of linear operators, obtaining as consequences most of the main spectral properties of linear relations. This chapter deals with the elegant interaction we give a characterization of the essential spectrum of the closed pencil linear relation in order to extend many known results in the literature. In fact, we give the characterization of the S-essential spectra of the sum of two multivalued linear operators and we study the stability of the S-essential approximate point spectrum and the S-essential defect spectrum of closed and closable linear relations under relatively precompact perturbations on a Banach space. On another level, we study the stability of the S-essential approximate point spectrum but under assumptions different from those adopted above. Finally, we focus on the study of the pseudospectra and the essential pseudospectra of linear relations. We start by giving the definition and we investigate the characterization and some properties of these pseudospectra. We end this chapter by gives some new results related to the pseudospectra and the essential pseudospectra of linear relations. We start by studying the stability of these pseudospectra and some characterization.

We do hope that this book will be very useful for researchers, since it represents not only a collection of a previously heterogeneous material, but also an innovation through several extensions. Of course, it is impossible for a single book to cover such a huge field of research. In making personal choices for inclusion of material, we tried to give useful complementary references in this research area. Hence, probably neglecting some relevant works. We would be very grateful to receive any comments from readers and researchers, providing us with some information concerning some missing references.

Chapter 2

Fundamentals

In this chapter, the concept and basic properties of a linear relation defined in a normed space are developed. We collect some results of the theory of linear relations needed in the sequel, in the attempt of making our work as selfcontained as possible. Before beginning let us recall some basic definitions following the notation and terminology of the book.

2.1 Banach Space

2.1.1 Direct Sum

Let M and N be vector subspaces of a vector space X over $\mathbb{K}$ ($\mathbb{R}$ or $\mathbb{C}$), the sum $M + N$ of M and N is given by

$$M + N = \{m + n \text{ such that } m \in M \text{ and } n \in N\},$$

and it is the smallest subspace of X which contains M and N. It is obvious that the sum $M + N$ of two linear subspaces M and N of a vector space X is again a linear subspace. If $M \bigcap N = \{0\}$, then this sum is called the direct sum of M and N, and will be denoted by $M \bigoplus N$. In this case, for every $z = x + y$ in $M + N$, the components x, y are uniquely determined. A subspace $M \subset X$ is said to be complemented, if there exists a closed subspace $N \subset X$, such that $X = M \bigoplus N$. If $X = M \bigoplus N$, then N is called an algebraic complement of M. A particularly important class of endomorphisms are the so-called projections. If $X = M \bigoplus N$ and $x = y + z$, with $x \in M$ and $y \in N$, define $P : X \longrightarrow M$ by $Px := y$. The linear map P projects X onto M along N. Clearly, $I - P$ projects X onto N along M and, we have

$$P(X) = M,$$

$$(I - P)(X) = N,$$

with $P^2 = P$, i.e., P is an idempotent operator. Let $M \subset X$ be a closed subspace. Define an equivalence relation

$$x \, R \, y \text{ if and only if, } y - x \in M.$$

Denote the equivalence class of an element $x \in X$ under this equivalence relation by

$$[x] = x + M = \{x + m \text{ such that } m \in M\},$$

and denote the quotient space by X/M or $\dfrac{X}{M}$

$$X/M = \frac{X}{M} = \{x + M \text{ such that } x \in X\}.$$

If X is a linear space, then the dimension of X is denoted by $\dim X$.

Theorem 2.1.1 [85] *If M is a subspace of a vector space X, then*

$$\dim X = \dim(X/M) + \dim(M). \qquad \Diamond$$

Definition 2.1.1 *Two linear spaces X_1 and X_2 are said to be isomorphic, denoted by $X_1 \cong X_2$, if there exists a one-to-one linear mapping from X_1 into X_2.* $\qquad \Diamond$

We would like to mention that the isomorphic linear spaces have the same dimension.

Theorem 2.1.2 (Second isomorphism theorem) [68, Theorem 2.28] *Let M and N be vector subspaces of a vector space X over $\mathbb{K}$ ($\mathbb{R}$ or $\mathbb{C}$). Then,*

$$\frac{M+N}{N} \cong \frac{M}{M \bigcap N},$$

where $\cong$ is a canonical isometry. $\qquad \Diamond$

For subspaces M, N of X we write $M \subset_e N$ (M is essentially contained in N) if there exists a finite dimensional subspace $F \subset X$ such that $M \subset N + F$. In Ref. [80], V. Muller proved that $M \subset_e N$, if and only if,

$$\dim \left(\frac{M}{M \bigcap N} \right) < \infty$$

if and only if,

$$\dim \left(\frac{M+N}{N} \right) < \infty$$

if and only if, there is a finite-dimensional subspace $F \subset M$ such that $M \subset N + F$.

Lemma 2.1.1 [59, Lemma 4.2.8] *Let M be a closed subspace having finite deficiency in X. Then,*

(i) For any subspace V of X, there exists a finite-dimensional subspace N contained in V such that $\overline{V} = (\overline{V} \cap M) \oplus N$.
(ii) If V is dense in X, then $V \cap M$ is dense in M. ◇

2.1.2 Distance Function

Definition 2.1.2 *A metric space is a pair (X, dist) consisting of a set X and a function*

$$\mathrm{dist} : X \times X \longrightarrow \mathbb{R}_+$$

that satisfy the following axioms.

(i) $\mathrm{dist}(x, y) = 0$ for all x, $y \in X$, if and only if, $x = y$.
(ii) $\mathrm{dist}(x, y) = \mathrm{dist}(y, x)$ for all x, $y \in X$.
(iii) $\mathrm{dist}(x, z) \leq \mathrm{dist}(x, y) + \mathrm{dist}(y, z)$ for all x, y, $z \in X$.

A function $\mathrm{dist} : X \times X \longrightarrow \mathbb{R}_+$ that satisfies these axioms is called a distance function and the inequality in (iii) is called the triangle inequality. A subset $U \subset X$ of a metric space (X, dist) is called open if, for every $x \in U$, there exists a constant $\varepsilon > 0$ such that the open ball

$$B_\varepsilon(x) = \{y \in X \text{ such that } \mathrm{dist}(x, y) < \varepsilon\}$$

is included in U. ◇

For a subset $\Omega \subset \mathbb{C}$ we set as usual

$$\mathrm{dist}(\lambda, \Omega) = \inf\{|z - \lambda| \text{ such that } z \in \Omega\},$$

and note that if Ω is compact, then the infimum is attained for some point in Ω. Recall that a Cauchy sequence in a metric space (X, dist) is a sequence $(x_n)_n$ with the property that, for every $\varepsilon > 0$, there exists an $n_0 \in \mathbb{N}$, such that any two integers $n, m \geq n_0$ satisfy the inequality $\mathrm{dist}(x_n, x_m) < \varepsilon$. Recall also that a metric space (X, dist) is called complete if every Cauchy sequence in X converges.

2.1.3 Normed Vector Space

The most important metric spaces in the field of functional analysis are the normed vector spaces.

Definition 2.1.3 *A normed vector space is a pair* $(X, \|\cdot\|)$ *consisting of a real vector space* X *and a function* $X \longrightarrow \mathbb{R}$, $x \longrightarrow \|\cdot\|$ *satisfying the following:*

(i) $\|x\| \geq 0$ *for all* $x \in X$, *with equality if and only if,* $x = 0$.
(ii) $\|\lambda x\| = |\lambda| \|x\|$ *for all* $x \in X$ *and* $\lambda \in \mathbb{R}$.
(iii) $\|x + y\| \leq \|x\| + \|y\|$ *for all* $x, y \in X$. $\diamond$

The Cartesian product $X \times Y$ is a normed space with the usual definition of addition and multiplication by scalars, and its norm is defined by

$$\|(x, y)\|_p = (\|x\|^p + \|y\|^p)^{\frac{1}{p}}, \ x \in X, \ y \in Y, \ p \geq 1.$$

Let U and V be nonempty subsets of a normed vector space X. We define the distance between U and V by the formula

$$\operatorname{dist}(U, V) = \inf \{\|u - v\| \text{ such that } u \in U, \ v \in V\}.$$

We will write $\operatorname{dist}(x, V)$, or $\operatorname{dist}(V, x)$ for the distance between $\{x\}$ and V. We shall assume that the following geometric property of normed space (see Ref. [59]).

Lemma 2.1.2 *Let* X *be a normed vector space. If* M *and* N *are subspaces of* X *with* $\dim N < \dim M$. *Then, there exists* $m \in M$ *and* $m \neq 0$ *such that* $\|m\| = \operatorname{dist}(m, N)$. $\diamond$

Definition 2.1.4 *Let* $(X, \|\cdot\|)$ *be a real normed vector space and* $M \subset X$ *be a closed subspace. For* $x \in X$, *we define*

$$\|[x]\|_{X/M} = \inf_{m \in M} \|x - m\|. \tag{2.1}$$

The quotient space X/M *is a real vector space and the formula (2.1) defines a norm function on* X/M. $\diamond$

Definition 2.1.5 *For a given closed linear subspace* E *of a normed space* X, *the natural quotient map is defined as follows:*

$$\begin{aligned} Q_E^X : X &\longrightarrow X/E \\ x &\longmapsto [x]. \end{aligned}$$

$\diamond$

The following lemma is from L. E. Labuschagne [75].

Lemma 2.1.3 *Let M be a closed subspace of a normed vector space E and let N be a subspace of E containing M. Then,*

(i) N is closed if and only if, N/M is closed.
(ii) $E/N \cong (E/M)\big/(N/M)$ and $Q_N = Q_{N/M}Q_M$, where $\cong$ is a canonical isometry. ◇

Lemma 2.1.4 *[50, Lemma 4.5.2] Let M be a closed space of a normed vector space X. If N is a closed subspace of X containing M, then*

$$Q_N^X = Q_{N/M}^{X/M} Q_M^X.$$ ◇

Lemma 2.1.5 *(i) Let D be a subset of a normed space X contains 0. Then, for every bounded sequence $(x_n)_n$ in X, $(Q_D x_n)_n$ is bounded.*
(ii) If $(x_n)_n$ in X is a convergent sequence, then $(Q_D x_n)_n$ is also a convergent sequence. ◇

Proof. *(i)* Let D be a closed linear subspace of X and let $(x_n)_n$ be a bounded sequence in X. Then, for all $n \in \mathbb{N}$ there exists a real number $M > 0$ such $\|x_n\| \leq M$. It is obvious that

$$
\begin{aligned}
\|Q_D x_n\| &= \operatorname{dist}(x_n, D) \\
&\leq \operatorname{dist}(x_n, 0) \\
&= \|x_n\| \\
&\leq M.
\end{aligned}
$$

Then, $(Q_D x_n)_n$ is bounded. *(ii)* Let $(x_n)_n$ be a sequence converges to a limit x, by continuity of Q_D, $(Q_D x_n)_n$ converges to $Q_D x$. Q.E.D.

Theorem 2.1.3 *Let D be a linear subspace of a normed space X such that $\dim D < \infty$. If $(x_n)_n$ is a bounded sequence in X such that $(Q_D x_n)_n$ is a convergent sequence, then $(x_n)_n$ has a convergent subsequence.* ◇

Proof. Let D be a linear subspace of X such that $\dim D < \infty$, let $(x_n)_n$ in X be a bounded sequence such that $(Q_D x_n)_n$ is a convergent sequence. Then, there exists $x \in X$ such that $(Q_D x_n)_n$ converges to $Q_D x$. It is clear, since $\dim D < \infty$, that there exists $d_0 \in D$ such that

$$
\begin{aligned}
\|Q_D x\| &= \operatorname{dist}(x, D) \\
&= \|x - d_0\| \\
&\leq \operatorname{dist}(x, 0) \\
&= \|x\| \tag{2.2}
\end{aligned}
$$

and there exists $d_n \in D$ such that

$$
\begin{aligned}
\|Q_D x_n - Q_D x\| &= \|Q_D(x_n - x)\| \\
&= \operatorname{dist}(x_n - x, D) \\
&= \|x_n - x - d_n\|.
\end{aligned}
\tag{2.3}
$$

Since $(x_n)_n$ is bounded, then there exists $M > 0$ such that $\|x_n\| \leq M$. Let $\varepsilon > 0$. Since $(Q_D x_n)_n$ converge to $Q_D x$, then there exists $N_1 \in \mathbb{N}$ such that for all $n \geq N_1$, we have

$$
\|Q_D x_n - Q_D x\| < \frac{\varepsilon}{2}.
$$

Hence, in view of (2.3), we have

$$
\|x_n - x - d_n\| < \frac{\varepsilon}{2}.
$$

Moreover,

$$
\|Q_D x\| \leq \|Q_D x_n - Q_D x\| + \|Q_D x_n\|
$$

and, by using (2.2), we have

$$
\begin{aligned}
\|x - d_0\| &\leq \frac{\varepsilon}{2} + \|x_n\| \\
&\leq \frac{\varepsilon}{2} + M.
\end{aligned}
$$

Thus,

$$
\begin{aligned}
\|d_n\| &\leq \|x_n - x - d_n\| + \|x_n - x\| \\
&\leq \frac{\varepsilon}{2} + \|x_n\| + \|x - d_0\| + \|d_0\| \\
&\leq \frac{\varepsilon}{2} + M + \frac{\varepsilon}{2} + M + \|d_0\| \\
&\leq \varepsilon + 2M + \|d_0\|.
\end{aligned}
$$

Hence, $(d_n)_n$ is bounded. Since $\dim D < \infty$, then $(d_n)_n$ has a subsequence convergent $(d_{\varphi(n)})_n$ converges to $d \in D$. So, there exists $N_2 \in \mathbb{N}$, such that for all $n \geq N_2$, we have $\|d_{\varphi(n)} - d\| < \frac{\varepsilon}{2}$. This implies that

$$
\begin{aligned}
\|x_{\varphi(n)} - x - d\| &= \|x_{\varphi(n)} - x - d_{\varphi(n)} + d_{\varphi(n)} - d\| \\
&\leq \|x_{\varphi(n)} - x - d_{\varphi(n)}\| + \|d_{\varphi(n)} - d\| \\
&< \varepsilon
\end{aligned}
$$

for all $n \geq \max(N_1, N_2)$. This completes the proof.   Q.E.D.

2.1.4 Banach Space

Let $(X, \|\cdot\|)$ be a normed vector space. Then, the formula

$$\operatorname{dist}(x, y) = \|x - y\|$$

for $x, y \in X$ defines a distance function on X. The space X is called a Banach space if the metric space (X, dist) is complete, i.e., if every Cauchy sequence in X converges. Obviously, if X and Y are Banach spaces, then $X \times Y$ is also Banach space.

The following examples illustrate the definition. We will study many of these examples in greater detail later on, so we do not present proofs here.

Example 2.1.1 *For $1 \leq p < \infty$, we define the p-norm on $\mathbb{R}^n$ (or $\mathbb{C}^n$) by*

$$\|(x_1, x_2, \ldots, x_n)\|_p := (|x_1|^p + |x_2|^p + \ldots + |x_n|^p)^{\frac{1}{p}}.$$

For $p = \infty$, we define the sup-norm or maximum norm, by

$$\|(x_1, x_2, \ldots, x_n)\|_\infty := \max\{|x_1|, |x_2|, \ldots, |x_n|\}.$$

Then, $\mathbb{R}^n$ equipped with the p-norm is a finite-dimensional Banach space for $1 \leq p \leq \infty$.

Example 2.1.2 *The space $C([a, b])$ of continuous, real-valued (or complex-valued) functions on $[a, b]$ with the sup-norm is a Banach space. More generally, the space $C(K)$ of continuous functions on a compact metric space K equipped with the sup-norm is a Banach space.*

Example 2.1.3 *The space $C^k([a, b])$ of k-times continuously differentiable functions on $[a, b]$ is not a Banach space with respect to the sup-norm $\|\cdot\|_\infty$ for $k \geq 1$, since the uniform limit of continuously differentiable functions need not be differentiable. We define the C^k-norm by*

$$\|f\|_{C^k} := \|f\|_\infty + \|f'\|_\infty + \ldots + \|f^{(k)}\|_\infty.$$

Then, $C^k([a, b])$ is a Banach space with respect to the C^k-norm. The convergence with respect to the C^k-norm is uniform convergence of functions and their first k derivatives.

Example 2.1.4 *For $1 \leq p < \infty$ the set of p-summable sequences of real numbers is denoted by*

$$l_p(\mathbb{N}) = \left\{ (x_j)_j \text{ such that } x_j \in \mathbb{R} \text{ and } \sum_{j=0}^{+\infty} |x_j|^p < \infty \right\}.$$

This is a Banach space with the norm

$$\|x\|_p = \left(\sum_{j=0}^{+\infty} |x_j|^p \right)^{\frac{1}{p}}, \quad x \in l_p(\mathbb{N}).$$

Lemma 2.1.6 [67, Lemma 3.1.12] *Let X be a Banach space, and let M be a closed linear subspace of X with $M \subsetneq X$. Then, for any $\varepsilon > 0$, there exists $x \in X$ with $\|x\| = 1$ such that $1 - \varepsilon < dist(x, M)$. We can even achieve $dist(x, M) = 1$ if $\dim M < \infty$.* ◇

Remark 2.1.1 *If X is a Banach space, $X = M \bigoplus N$ and the projection P is continuous, then M is said to be complemented and N is said to be a topological complement of M. Note that each complemented subspace is closed, but the converse is not true, for instance c_0, the Banach space of all sequences which converge to 0, is a not complemented closed subspace of l_∞, where l_∞ denotes the Banach space of all bounded sequences, (see Ref. [81]).* ◇

Definition 2.1.6 *Let M be a vector subspace of a Banach space X. M is called paracomplet subspace of X, if M is a Banach space and the injection map of M into X is continuous.* ◇

Proposition 2.1.1 [72, Proposition 2.1.1] *Let M and N be two paracomplet subspaces of X such that $M + N$ and $M \cap N$ are closed on X. Then, M and N are two closed subspaces of X.* ◇

2.2   Relations on Sets

Let U and V be arbitrary non-empty sets. A relation T from U into V is a mapping defined on a non-empty subset $\mathcal{D}(T)$ of U, called the domain of T, which takes on values in $2^V \backslash \emptyset$ (the collection of non-empty subset of Y). If T maps the points of its domain to singletons, then T is said to be a single valued operator, or function. The collection of relations as defined above will be denoted by $\mathcal{R}(U, V)$ and as useful we write $\mathcal{R}(U) := \mathcal{R}(U, U)$. Examples of relations are functions, inverse of functions, adjoints of linear operators, partial order relations, equivalent relations and convex processes. For $u \in U$, $u \notin \mathcal{D}(T)$, we define $Tu = \emptyset$. With this convention, we get

$$\mathcal{D}(T) = \{u \in U \text{ such that } Tu \neq \emptyset\}.$$

Let $T \in \mathcal{R}(U, V)$, $M \subset U$ and $\emptyset \neq N \subset V$. The graph of T is a subset of $U \times V$ defined by

$$G(T) = \{(u, v) \in U \times V \ \text{ such that } u \in \mathcal{D}(T) \ \text{ and } v \in Tu\}.$$

A relation in $\mathcal{R}(U, V)$ is uniquely determined by its graph and conversely any non-empty subset of $U \times V$ uniquely determines a relation. The identity relation defined on a nonempty subset E of U is denoted by I_E (or simply I) when E is understood, it is the relation in $\mathcal{R}(U)$ whose graph is

$$G(I_E) = \{(e, e) \ \text{ such that } \ e \in E\}.$$

Definition 2.2.1 *Let M be a subspace of a normed vector space X, the operator J_M^X is called injection operator if it denote the natural injection map of M into X, i.e., $J_M^X \in \mathcal{R}(M, X)$, $\mathcal{D}(J_M^X) = M$ and $J_M^X(m) = m$ for $m \in M$.* $\diamond$

Let $\widetilde{X}$ be the completion of the normed space X. The completion of T, denoted $\widetilde{T}$ is defined in terms of her corresponding graph by

$$G(\widetilde{T}) = \widetilde{G(T)} \subset \widetilde{X} \times \widetilde{Y}. \tag{2.4}$$

We write

$$T(M) = \bigcup \left\{ Tm \ \text{ such that } m \in M \bigcap \mathcal{D}(T) \right\},$$

called the image of M, with $R(T) = T(\mathcal{D}(T))$ called the range of T. The subspace $T^{-1}(0)$ is called the null space (or kernel) of T, and is denoted by $N(T)$. The inverse of T is the relation T^{-1} given by

$$G(T^{-1}) = \{(v, u) \in V \times U \ \text{ such that } (u, v) \in G(T)\}.$$

We write

$$T^{-1}(N) := \left\{ u \in \mathcal{D}(T) \ \text{ such that } Tu \bigcap N \neq \emptyset \right\}.$$

In particular, for $v \in R(T)$,

$$T^{-1}(v) = \{u \in \mathcal{D}(T) \ \text{ such that } v \in Tu\}.$$

If $R(T) = Y$, then T is called surjective. We shall be using both $N(T)$ and $T^{-1}(0)$ throughout the sequel. If T^{-1} is single valued, then T is called injective.

Proposition 2.2.1 *Let $T \in \mathcal{R}(U)$. If T is injective, then for all u_1, $u_2 \in \mathcal{D}(T)$ such that $T(u_1) = T(u_2)$, we have $u_1 = u_2$.* $\diamond$

Proof. Let $v \in T(u_1) = T(u_2)$, then $(u_1, v) \in G(T)$ and $(u_2, v) \in G(T)$. Hence, $(v, u_1) \in G(T^{-1})$ and $(v, u_2) \in G(T^{-1})$. Since T^{-1} is single valued, we infer that $u_1 = u_2$. Q.E.D.

2.3   Linear Relations

2.3.1   The Algebra of Linear Relations

Definition 2.3.1 *Let X and Y be two vector spaces over the field $\mathbb{K}$ ($\mathbb{R}$ or $\mathbb{C}$). A multivalued linear operator (or a linear relation) is a mapping $T \subset X \times Y$ which goes from a subspace $\mathcal{D}(T) \subset X$ called the domain of T, into the collection of nonempty subsets of Y such that*

$$T(\alpha_1 x_1 + \alpha_2 x_2) = \alpha_1 T(x_1) + \alpha_2 T(x_2)$$

for all nonzero scalars α_1, α_2 and $x_1, x_2 \in \mathcal{D}(T)$. For $x \in X \backslash \mathcal{D}(T)$, we define $Tx = \emptyset$. With this convention, we get

$$\mathcal{D}(T) = \{x \in X \text{ such that } Tx \neq \emptyset\}. \qquad \diamond$$

The collection of linear relations as defined above will be denoted by $LR(X, Y)$ and as useful we write $LR(X) := LR(X, X)$.

Definition 2.3.2 *Let X and Y be two normed spaces. If $T \in LR(X, Y)$ maps the points in its domain to singletons, then T is said to be a single valued (or simply an operator).* $\qquad \diamond$

Remark 2.3.1 *(i) A linear relation $T \in LR(X, Y)$ is uniquely determined and identified with its graph, $G(T)$, which is defined by*

$$G(T) = \{(x, y) \in X \times Y \text{ such that } x \in \mathcal{D}(T), \ y \in Tx\}.$$

(ii) T is a linear relation if and only if, $G(T)$ is a linear subspace of $X \times Y$.
(iii) $T(0) = \{y \text{ such that } (0, y) \in G(T)\}$.
(iv) The zero relation is denoted by O_X, which is defined by

$$G(O_X) = \{(x, 0) \text{ such that } x \in X\}.$$

(v) The identity relation on a subspace M of X is denoted by I_M (or simply I) when M is understood, it is the linear relation whose graph is

$$G(I_M) = \{(m, m) \text{ such that } m \in M\}. \qquad \diamond$$

Proposition 2.3.1 *Let X and Y be two vector spaces over the field $\mathbb{K}$ ($\mathbb{R}$ or $\mathbb{C}$), and $T \in R(X, Y)$. Then, $T \in LR(X, Y)$ if and only if, for all x_1, $x_2 \in \mathcal{D}(T)$ and for all α_1, $\alpha_2 \in \mathbb{K}$, we have*

$$\alpha_1 T(x_1) + \alpha_2 T(x_2) \subset T(\alpha_1 x_1 + \alpha_2 x_2). \qquad (2.5)$$

◇

Proof. Let $T \in LR(X,Y)$, x_1, $x_2 \in \mathcal{D}(T)$ and α_1, $\alpha_2 \in \mathbb{K}$, then in view of Remark 2.3.1 (ii), we have $(\alpha_1 x_1 + \alpha_2 x_2, \alpha_1 y_1 + \alpha_2 y_2) \in G(T)$ for all $y_1 \in Tx_1$ and $y_2 \in Tx_2$. Hence, $\alpha_1 y_1 + \alpha_2 y_2 \in T(\alpha_1 x_1 + \alpha_2 x_2)$. So, (2.5) holds. Conversely, let T satisfy (2.5) and let (x_1, y_1), $(x_2, y_2) \in G(T)$. Then, for α_1, $\alpha_2 \in \mathbb{K}$, we have

$$\alpha_1 y_1 + \alpha_2 y_2 \in \alpha_1 T(x_1) + \alpha_2 T(x_2) \subset T(\alpha_1 x_1 + \alpha_2 x_2).$$

Then, $\alpha_1(x_1, y_1) + \alpha_2(x_2, y_2) \in G(T)$. Therefore, $G(T)$ is a linear subspace of $X \times Y$. Consequently, T is a linear relation by Remark 2.3.1 (ii).   Q.E.D.

2.3.2   Inverse of Linear Relation

Let $T \in LR(X,Y)$. The inverse of T is the linear relation T^{-1} defined by

$$G(T^{-1}) = \{(y, x) \in Y \times X \text{ such that } (x, y) \in G(T)\}.$$

If $M \subset Y$, then the inverse image of M under T is defined by

$$T^{-1}(M) := \left\{ x \in \mathcal{D}(T) \text{ such that } Tx \bigcap M \neq \emptyset \right\}.$$

Hence,

$$T^{-1}(y) = \{x \in \mathcal{D}(T) \text{ such that } y \in Tx\}.$$

Example 2.3.1 *Let* $X = l_p(\mathbb{N})$ *be the space of sequences* $x : \mathbb{N} \longrightarrow \mathbb{C}$ *summable with a power* $p \in [1, \infty)$ *with the standard norm. Consider* T *the left shift single valued operator defined by*

$$T(x(n)) = x(n+1), \quad n \geq 1 \quad and \quad x \in X.$$

Then, T^{-1} *is the linear relation defined by*

$$G(T^{-1}) = \{(x, y) \in X \times X \text{ such that } x(n) = y(n-1), n \geq 2\}.$$

Definition 2.3.3 *Let* $T \in LR(X,Y)$. *Then, the range* $R(T)$ *of* T *is defined by*

$$R(T) = \bigcup_{x \in \mathcal{D}(T)} Tx.$$

The subspace $T^{-1}(0)$ *is called the null space (or kernel) of* T, *and is denoted by* $N(T)$.

◇

If $R(T) = Y$, then we say that T is surjective, and if $M \subset X$, then the image of M under T is defined to be the set

$$T(M) = \bigcup_{x \in M \cap \mathcal{D}(T)} Tx.$$

If T^{-1} is single valued, then T is called injective, i.e., $T^{-1}(0) = \{0\}$.

Proposition 2.3.2 [50, Subsection 1.1.3 (9), p. 3] *If $T \in LR(X, Y)$, then*

(i) T is injective if and only if, $T^{-1}T = I_{\mathcal{D}(T)}$.
(ii) T is single valued if and only if, $TT^{-1} = I_{R(T)}$.   ◇

Remark 2.3.2 *(i) $R(T) = \mathcal{D}(T^{-1})$ and $\mathcal{D}(T) = R(T^{-1})$.*
(ii) $R(T) = \{y \text{ such that } (x, y) \in G(T)\}$.
(iii) $N(T) = \{x \in \mathcal{D}(T) \text{ such that } (x, 0) \in G(T)\}$.
(iv) The assertion (i) of Theorem 2.3.1, is equivalent to say that

$$N(T) = \{x \in \mathcal{D}(T) \text{ such that } Tx = T(0)\}.$$

(v) If T is a linear operator such that $N(T) \neq \{0\}$, then T^{-1} is a linear relation ($T^{-1}(0) = N(T) \neq \{0\}$).   ◇

Example 2.3.2 *Let T be an ordinary differential operator defined by*

$$T : C^n([a, b]) \longrightarrow C([a, b])$$

of the kind

$$(Tx(t)) = x^{(n)}(t) + a_1(t)x^{(n-1)}(t) + \cdots + a_n(t)x(t),$$

where $a_k \in C([a, b])$, $k = 1, \ldots, n$ that acts in the Banach space $C([a, b])$ of bounded continuous complex functions on $[a, b] \subset \mathbb{R}$ and has a finite-dimensional kernel $N(T)$ of dimension $n \geq 1$. Therefore, $T^{-1} \in LR(C([a, b]))$ and if $Tx = x'$, then

$$T^{-1}x = \int x(t)dt, \quad x \in C([a, b]).$$ ◇

Theorem 2.3.1 [50, Proposition 1.2.8 and Corollary 1.2.11] *Let $T \in LR(X, Y)$. Then,*

(i) for $x \in \mathcal{D}(T)$, we have the following equivalence

$$y \in Tx \text{ if and only if, } Tx = y + T(0).$$

In particular,

$$0 \in Tx \ \text{if and only if,} \ Tx = T(0).$$

(ii) For $x_1, x_2 \in \mathcal{D}(T)$, we have the equivalence

$$Tx_1 \bigcap Tx_2 \neq \emptyset \ \text{if and only if,} \ Tx_1 = Tx_2.$$

(iii) $T^{-1}Tx = x + T^{-1}(0)$ for all $x \in \mathcal{D}(T)$.
(iv) $TT^{-1}y = y + T(0)$ for all $y \in R(T)$.
(v) $T(0)$ and $T^{-1}(0)$ are linear subspaces. ◇

Remark 2.3.3 *By using (iii) and (iv) of Theorem 2.3.1, we have $TT^{-1}(0) = T(0)$ and $T^{-1}T(0) = T^{-1}(0)$.* ◇

2.3.3 Sum and Product of Linear Relations

Definition 2.3.4 *Let $T, S \in LR(X, Y)$. Then, the linear relation $S + T$ is defined by $(S + T)x := Sx + Tx$, for $x \in X$.* ◇

Remark 2.3.4 *Let $T, S \in LR(X, Y)$. As an immediate consequence of the Definition 2.3.4, we have the following properties*

(i) $\mathcal{D}(T + S) = \mathcal{D}(T) \bigcap \mathcal{D}(S)$.
(ii) $T + S = S + T$.
(iii) The algebraic sum $T + S$ is also a linear relation defined by

$$\begin{aligned} G(T + S) &= \{(x, y + z) \ \text{such that} \ (x, y) \in G(T), \ (x, z) \in G(S)\} \\ &= \{(x, y) \ \text{such that} \ y = u + v \ \text{with} \ (x, u) \in G(T), \ (x, v) \in G(S)\}. \end{aligned}$$

In particular, for all $\lambda \in \mathbb{K}$, we have

$$G(\lambda - T) = \{(x, \lambda x - y) \ \text{such that} \ (x, y) \in G(T)\}.$$

(iv) Let $T, S, R \in LR(X, Y)$. Then, $T + (R + S) = (T + R) + S$. ◇

Proposition 2.3.3 *Let $T, S \in LR(X, Y)$. If $G(T) \subset G(S)$, then $G(T) \subset G(T + S)$.* ◇

Proof. To prove this, let us observe that

$$\begin{aligned} G(T) &= \{(x, y) \in X \times Y \ \text{such that} \ x \in \mathcal{D}(T) \subset \mathcal{D}(S) \ \text{and} \ y \in Tx \subset Sx\} \\ &\subset \{(x, y) \in X \times Y \ \text{such that} \ x \in \mathcal{D}(T) \bigcap \mathcal{D}(S) \ \text{and} \ y \in Tx + Sx\} \\ &= \{(x, y) \in X \times Y \ \text{such that} \ x \in \mathcal{D}(T + S) \ \text{and} \ y \in (T + S)x\} \\ &= G(T + S), \end{aligned}$$

which completes the proof. Q.E.D.

Proposition 2.3.4 *Let* $S, T \in LR(X, Y)$. *If* $S(0) \subset T(0)$ *and* $\mathcal{D}(T) \subset \mathcal{D}(S)$, *then* $T - S + S = T$. $\diamondsuit$

Proof. Let $(x, y) \in G(T + S - S)$, then $x \in \mathcal{D}(T - S + S) = \mathcal{D}(T)$ and $y \in (T - S + S)x$. So, $y \in Tx + S(0)$. Using the fact $S(0) \subset T(0)$, it follows that $Tx + S(0) \subset Tx + T(0)$ which yields $y \in Tx$ and $x \in \mathcal{D}(T)$, that is $(x, y) \in G(T)$. Therefore, $G(T + S - S) \subset G(T)$. Conversely, let $(x, y) \in G(T)$, then $x \in \mathcal{D}(T)$ and $Tx = y + T(0)$. Since $S(0) \subset T(0)$ and $\mathcal{D}(T) \subset \mathcal{D}(S)$, we have $(T + S - S)(0) = T(0)$ and $\mathcal{D}(T - S + S) = \mathcal{D}(T)$. This implies that $x \in \mathcal{D}(T - S + S)$ and $(T + S - S)x = y + (T + S - S)(0)$. Thus, in view of Theorem 2.3.1 (i), we have $y \in (T + S - S)x$. Hence $(x, y) \in G(T + S - S)$. Consequently, $G(T) \subset G(T + S - S)$. This completes the proof. Q.E.D.

Definition 2.3.5 *Let* $T \in LR(X, Y)$ *and* $S \in LR(Y, Z)$. *If* $R(T) \bigcap \mathcal{D}(S) \neq \emptyset$, *then their product* ST *is also a linear relation defined by*

$$STx = S(Tx), \quad x \in X.$$

For $\alpha \in \mathbb{K}$, *we define the relation* αT *by*

$$(\alpha T)x = \alpha(Tx), \quad x \in X. \qquad \diamondsuit$$

Remark 2.3.5 *Let* X, Y *and* Z *be linear spaces and let* $T \in LR(X, Y)$, $S \in LR(Y, Z)$. *If* $R(T) \bigcap \mathcal{D}(S) \neq \emptyset$, *then*

(i) $G(ST) = \{(x, z) \in X \times Z$ *such that* $(x, y) \in G(T)$ *and* $(y, z) \in G(S)$ *for some* $y \in Y\}$, *and*

$$
\begin{aligned}
\mathcal{D}(ST) &= \{x \in X \text{ such that } STx \neq \emptyset\} \\
&= \{x \in X \text{ such that } Tx \bigcap \mathcal{D}(S) \neq \emptyset\} \\
&= T^{-1}(\mathcal{D}(S)).
\end{aligned}
$$

In particular, for all $\alpha \in \mathbb{K} \backslash \{0\}$, *we have*

$$\mathcal{D}(\alpha T) = \mathcal{D}(T),$$

and

$$G(\alpha T) = \{(\alpha x, y) \text{ such that } (x, y) \in G(T)\}.$$

(ii) *If* $X = Y$, *we also define the product* $STx = \bigcup\{Sy : y \in \mathcal{D}(S) \cap Tx\}$ *with* $\mathcal{D}(ST) = \{x \in \mathcal{D}(T) : \mathcal{D}(S) \cap Tx \neq \emptyset\}$. $\diamondsuit$

Theorem 2.3.2 *Let X, Y and Z be linear spaces and let $T \in LR(X, Y)$, $S \in LR(Y, Z)$. If $R(T) \bigcap \mathcal{D}(S) \neq \emptyset$, then*

(i) $N(T) \subset N(ST) \subset \mathcal{D}(ST) \subset \mathcal{D}(T)$, and
(ii) $S(0) \subset ST(0) \subset R(ST) \subset R(S)$. $\diamondsuit$

Proof. *(i)* Let $x \in N(T)$, then $(x, 0) \in G(T)$. Since $(0, 0) \in G(S)$, it follows that $(x, 0) \in G(ST)$. Hence, $x \in N(ST)$, which shows that $N(T) \subset N(ST)$. Clearly, $N(ST) \subset \mathcal{D}(ST)$. Let $x \in \mathcal{D}(ST)$, then $(x, z) \in G(ST)$ for some $z \in Z$. Hence, $(x, y) \in G(T)$ and $(y, z) \in G(S)$ for some $y \in Y$, which shows that $x \in \mathcal{D}(T)$. So, $\mathcal{D}(ST) \subset \mathcal{D}(T)$.

(ii) Let $z \in S(0)$, then $(0, z) \in G(S)$. Since $(0, 0) \in G(T)$, it follows that $(0, z) \in G(ST)$. Hence, $z \in ST(0)$, which shows that $S(0) \subset ST(0)$. Clearly, $ST(0) \subset R(ST)$. Let $z \in R(ST)$, then $(x, z) \in G(ST)$ for some $x \in X$. Then, $(x, y) \in G(T)$ and $(y, z) \in G(S)$ for some $y \in Y$. Thus, $z \in R(S)$. Hence, $R(ST) \subset R(S)$. Q.E.D.

Lemma 2.3.1 [50, Lemma 1.6.8] *Let $T \in LR(X)$ and M, $N \subset X$ such that $X = M + N$, $M \bigcap N = \{0\}$, $\mathcal{D}(T) = X$, and $N(T) \subset N$. Then, $R(T) = T(M) + T(N)$ and $T(M) \bigcap T(N) = T(0)$.* $\diamondsuit$

Proposition 2.3.5 [50, Proposition 1.3.1] *Let $T \in LR(X)$, $\alpha \in \mathbb{K} \backslash \{0\}$, and let M, $N \subset X$. Then,*

(i) $T(\alpha M) = \alpha T(M)$,
(ii) $T(M + N) \supset T(M) + T(N)$,
(iii) if $M \subset \mathcal{D}(T)$ or $N \subset \mathcal{D}(T)$, then $T(M + N) = T(M) + T(N)$. Moreover, if $M \bigcap N = \{0\}$, then $T(M \bigcap N) = T(0) = T(M) \bigcap T(N)$,
(iv) $TT^{-1}(M) = M \bigcap R(T) + T(0)$, for all $M \subset X$, and
(v) $T^{-1}T(M) = M \bigcap \mathcal{D}(T) + N(T)$, for all $M \subset X$. $\diamondsuit$

2.3.4 Restrictions and Extensions of Linear Relations

Definition 2.3.6 *(i) Let $T \in LR(X, Y)$, and M be a subspace of X such that $M \bigcap \mathcal{D}(T) \neq \emptyset$. Then, the restriction of T to M, denoted by $T_{|M}$ is defined by*

$$\begin{cases} T_{|M} \in LR(X, Y) \\ \mathcal{D}(T_{|M}) = \mathcal{D}(T) \bigcap M \\ (T_{|M})m = Tm \text{ for all } m \in M. \end{cases}$$

(*ii*) Let $T, S \in LR(X, Y)$. The relation S is said an extension of T if $S_{|D(T)} = T$. We write $T \subset S$. $\diamond$

Remark 2.3.6 (*i*) Let $T \in LR(X, Y)$, and M be a subspace of X such that $M \bigcap D(T) \neq \emptyset$. Then,

(i_1) $G(T_{|M}) = \{(x, y) \in G(T) \text{ such that } x \in M\} = G(T) \bigcap (M \times Y)$.

(i_2) $T_{|M} = T_{|D(T) \cap M}$.

(*ii*) By using Definition 2.2.1, we deduce that

$$T_{|M} = T J_M^X (J_M^X)^{-1}.$$

Hence, $I_M = J_M^X (J_M^X)^{-1}$.

(*iii*) If T is an extension of S^{-1}, then it is not implies T^{-1} is an extension of S. $\diamond$

Example 2.3.3 Let $X = \mathbb{R}^3$ and consider the relation T defined by

$$\left\{ T \begin{pmatrix} x_1 \\ x_2 \\ x_3 \end{pmatrix} = \begin{pmatrix} x_1 \\ x_2 \\ 0 \end{pmatrix} + \left\{ \begin{pmatrix} 0 \\ 0 \\ x_3 \end{pmatrix} \text{ such that } x_3 \in \mathbb{R} \right\}, \\ D(T) = \mathbb{R}^3. \right.$$

Let S be the single valued map $\begin{pmatrix} 1 & 0 & 0 \\ 0 & 1 & 0 \\ 0 & 0 & 0 \end{pmatrix}$. Writing $x = \begin{pmatrix} x_1 \\ x_2 \\ x_3 \end{pmatrix}$ and

$y = \begin{pmatrix} y_1 \\ y_2 \\ y_3 \end{pmatrix}$ for arbitrary vectors in $\mathbb{R}^3$. Then, the graph of S is defined by

$$G(S) = \left\{ \left(\begin{pmatrix} x_1 \\ x_2 \\ x_3 \end{pmatrix}, \begin{pmatrix} y_1 \\ y_2 \\ y_3 \end{pmatrix} \right) \text{ such that } \begin{pmatrix} x_1 \\ x_2 \\ 0 \end{pmatrix} = \begin{pmatrix} y_1 \\ y_2 \\ y_3 \end{pmatrix} \right\}.$$

The relation S^{-1} is defined by

$$S^{-1} \begin{pmatrix} x_1 \\ x_2 \\ 0 \end{pmatrix} = \begin{pmatrix} x_1 \\ x_2 \\ 0 \end{pmatrix} + S^{-1}(0),$$

and

$$\begin{cases} S^{-1}(0) = \left\{ \begin{pmatrix} 0 \\ 0 \\ x_3 \end{pmatrix} \text{ such that } x_3 \in \mathbb{R} \right\} \text{ is a linear subspace of } \mathbb{R}^3, \\ \mathcal{D}(S^{-1}) = \left\{ \begin{pmatrix} x_1 \\ x_2 \\ 0 \end{pmatrix} \text{ such that } (x_1, x_2) \in \mathbb{R}^2 \right\}. \end{cases}$$

Then, T is an extension of S^{-1}, and by using Remark 2.3.6 (ii), we have

$$TI_{R(A)} = TJ_{R(A)}J_{R(A)}^{-1} = S^{-1}.$$

Hence,

$$S = J_{R(A)}J_{R(A)}^{-1}T^{-1}.$$

However, $T^{-1} \neq S$.

Proposition 2.3.6 (i) If $G(S) \subset G(T)$, then T is an extension of S if and only if, $S(0) = T(0)$.
(ii) If $G(S) \subset G(T)$, $N(T) \subset N(S)$, and $R(T) \subset R(S)$, then $T = S$.
(iii) If $\mathcal{D}(T) = \mathcal{D}(S)$ and $T(0) = S(0)$, then $T = S$ or the graphs of T and S are incomparable.
(iv) $(R + S)T \subset RT + ST$ with equality if T is single valued.
(v) $TR + TS = T(R + S)$ if $\mathcal{D}(T)$ is the whole space.
(vi) $T(R + S)$ is an extension of $TR + TS$. $\diamond$

Remark 2.3.7 (i) If S is an extension of T, then $G(T) \subset G(S)$.
(ii) Let $R = -S$, where S is single valued and nonzero, and let $G(T) = X \times Y$ where $Y \neq \{0\}$. Then,

$$(R + S)T(0) = (S - S)(Y) = \bigcup_{y \in \mathcal{D}(S)} (S - S)(y) = \{0\}$$

and

$$(RT+ST)(0) = (ST-ST)(0) = ST(0)-ST(0) = S(Y)-S(Y) = S(Y) \neq \{0\}.$$

This shows that $RT + ST$ is not in general an extension of $(R + S)T$.

(iii) Let R, S, $T \in L\mathcal{R}(X)$. If $RT = TR$ and $ST = TS$, then

$$G((R + S)T) \subset G(T(R + S)). \tag{2.6}$$

The equality fails in (2.6), even if R, S, T are all single valued, as seen by considering the following example: $S = -R = I$, and T is a linear operator with $\mathcal{D}(T) \neq X$. $\diamond$

We construct an example making $T(R + S)$ in Proposition 2.3.6 (vi) a proper extension of $TR + TS$.

Example 2.3.4 *Let R, S, $T \in LR(\mathbb{R}^2)$ be defined by*

$$R = I$$

$$G(S) = \left\{ \left(\begin{pmatrix} x \\ 0 \end{pmatrix}, \ \begin{pmatrix} w \\ x \end{pmatrix} \right) \ for \ x, \ w \in \mathbb{R} \right\}$$

$$T = I_{|M}, \quad where \ \ M = \left\{ \begin{pmatrix} 0 \\ y \end{pmatrix} \ for \ y \in \mathbb{R} \right\}.$$

Then, $\mathcal{D}(TR) = \mathcal{D}(T)$ and

$$\mathcal{D}(TS) = \left\{ \begin{pmatrix} x \\ 0 \end{pmatrix} \ for \ x \in \mathbb{R} \right\}.$$

Hence, $\mathcal{D}(TR + TS) = \{0\}$. Now,

$$G(R + S) = \left\{ \left(\begin{pmatrix} x \\ 0 \end{pmatrix}, \ \begin{pmatrix} z \\ x \end{pmatrix} \right) \ for \ x, \ z \in \mathbb{R} \right\}.$$

Thus,

$$\mathcal{D}(T(R + S)) = (R + S)^{-1}(\mathcal{D}(T)) = \left\{ \begin{pmatrix} x \\ 0 \end{pmatrix} \ for \ x \in \mathbb{R} \right\}.$$

Therefore, $TR + TS \neq T(R + S)$. In this example S is the inverse of the single valued map $\begin{pmatrix} 0 & 1 \\ 0 & 0 \end{pmatrix}$. $\Diamond$

2.4 Index and Co-Index of Linear Relations

Definition 2.4.1 *Let $T \in LR(X, Y)$, we define*

$$\alpha(T) = \dim(N(T)),$$

$$\beta(T) = \dim(Y/R(T)),$$

and

$$\overline{\beta}(T) = \dim(Y/\overline{R(T)}).$$

If $\alpha(T)$ and $\beta(T)$ are finite, then the index of T is defined by

$$i(T) = \alpha(T) - \beta(T).$$

The index of T^{-1} is called co-index of T and is denoted by $ci(T)$. Then,

$$ci(T) = i(T^{-1}) = \alpha(T^{-1}) - \beta(T^{-1}) = \dim(T(0)) - \mathrm{codim}(\mathcal{D}(T)).$$

If T is a single valued everywhere defined, then $ci(T) = 0$. ◇

Remark 2.4.1 *For a given closed linear subspace E of a normed vector space X, we have*

$$i(Q_E^X) = \dim E.$$

In fact, it is obvious that $N(Q_E^X) = E$ and $R(Q_E^X) = X/E$. Hence, $\alpha(Q_E^X) = \dim(E)$ and $\beta(Q_E^X) = 0$. ◇

Lemma 2.4.1 *Let $A, B \in LR(X, Y)$ such that $G(A) \subset G(B)$. Then,*

(i) $N(A) \subset N(B)$ and $\alpha(A) \le \alpha(B)$.
(ii) $R(A) \subset R(B)$ and $\beta(B) \le \beta(A)$.
(iii) $i(A) \le i(B)$. ◇

Proof. *(i)* To prove this, let us observe that

$$
\begin{aligned}
N(A) &= \{x \in \mathcal{D}(A) \text{ such that } (x, 0) \in G(A)\} \\
&\subset \{x \in \mathcal{D}(A) \text{ such that } (x, 0) \in G(B)\} \\
&= N(B_{|\mathcal{D}(A)}) \\
&\subset N(B).
\end{aligned}
$$

So, $\alpha(A) \le \alpha(B)$.

(ii) Let $y \in R(A)$. Then, there exists $x \in \mathcal{D}(A)$ such that $y \in Ax$. By using $G(A) \subset G(B)$, we have $y \in Bx$. Hence, $y \in R(B_{|\mathcal{D}(A)})$. So, $y \in R(B)$. Thus, $Y/_{R(B)} \subset Y/_{R(A)}$ and hence $\beta(B) \le \beta(A)$.

(iii) $i(A) = \alpha(A) - \beta(A) \le \alpha(B) - \beta(B) = i(B)$. Q.E.D.

Proposition 2.4.1 *Let $T, S \in LR(X, Y)$. Then,*

(i) If $N(S) = N(T)$, $R(S) = R(T)$, and S is an extension of T, then $S = T$.
(ii) If S be an extension of T, and $i(T) = i(S)$, then $T = S$. ◇

Proof. (i) Let $(x, y) \in G(S)$. Then, $y \in Sx \subset R(S) = R(T)$. So, there exists $x_1 \in \mathcal{D}(T)$ such that $y \in Tx_1 = Sx_1$. It follows that $Sx = Sx_1$ which means that $x - x_1 \in N(S) = N(T) \subset \mathcal{D}(T)$. Therefore, $x = (x - x_1) + x_1 \in \mathcal{D}(T)$. Thus, $(x, y) \in G(T)$.

(ii) To show that $T = S$, it suffices to prove that $S \subset T$ (equivalently $G(S) \subset G(T)$). In view of Lemma 2.4.1 (i) and (ii), it is easy to see that $\alpha(T) \leq \alpha(S)$ and $\beta(S) \leq \alpha(T)$. Since $i(T) = i(S)$, then $R(T) = R(S)$ and $N(T) = N(S)$. Hence, by using (i), we have $T = S$.   Q.E.D.

Lemma 2.4.2 [50, Corollary 1.6.12] *(Index of product) Let X, Y, Z be three vector spaces, $T \in LR(X, Y)$, $S \in LR(Y, Z)$ and $\mathcal{D}(S) = Y$. Suppose that T and S have finite indices. Then, ST has a finite index and,*

$$
\begin{aligned}
i(ST) &= i(S) + i(T) - \dim\left(T(0) \cap N(S)\right) \\
&= i(S) + i(T) - \dim\left(T(0) \cap S^{-1}(0)\right).
\end{aligned}
$$ ◇

Example 2.4.1 *Let $X = Y = Z = \mathbb{R}^2$ and $S = \begin{pmatrix} 0 & 0 \\ 0 & 1 \end{pmatrix}$. Then, the graph of S is defined by*

$$
G(S) = \left\{ \left(\begin{pmatrix} x_1 \\ x_2 \end{pmatrix}, \begin{pmatrix} y_1 \\ y_2 \end{pmatrix} \right) : \begin{pmatrix} 0 \\ x_2 \end{pmatrix} = \begin{pmatrix} y_1 \\ y_2 \end{pmatrix} \right\}.
$$

Hence, S^{-1} is the linear relation defined by

$$
\begin{cases}
S^{-1} \begin{pmatrix} 0 \\ x_2 \end{pmatrix} = \begin{pmatrix} 0 \\ x_2 \end{pmatrix} + S^{-1}(0), \\[2mm]
\mathcal{D}(S^{-1}) = \left\{ \begin{pmatrix} 0 \\ x_2 \end{pmatrix} \text{ such that } x_2 \in \mathbb{R} \right\},
\end{cases}
$$

and

$$
S^{-1}(0) = \left\{ \begin{pmatrix} x_1 \\ 0 \end{pmatrix} \text{ such that } x_1 \in \mathbb{R} \right\}
$$

is a linear subspace of $\mathbb{R}^2$. Let $T = S^{-1}$, then $N(T) = \{0\} \times \{0\}$, $R(T) = \mathbb{R} \times \mathbb{R}$, $N(S) = \mathbb{R} \times \{0\}$, $N(ST) = \{0\} \times \{0\}$, and $R(ST) = \{0\} \times \mathbb{R}$. Hence, $\alpha(T) = 0$, $\beta(T) = 0$, $\alpha(S) = 1$, $\beta(S) = 1$, $\alpha(ST) = 0$, and $\beta(ST) = 1$. So, $i(T) = \alpha(T) - \beta(T) = 0$, and $i(S) = \alpha(S) - \beta(S) = 0$. Thus, $i(ST) = i(S) + i(T) - \dim(T(0) \cap N(S)) = -1$ and $\dim(T(0) \cap N(S)) = 1$.

We would like to mention that, A. Sandovici and H. de Snoo [84] study some proprieties of index formula for the product of unbounded linear relations.

2.4.1 Properties of the Quotient Map Q_T

We shall denote $Q^Y_{\overline{T(0)}}$ by Q_T, or simply Q when T is understood.

Theorem 2.4.1 *Let* $T, S \in LR(X,Y)$ *such that* $\overline{T(0)} + S(0)$ *is closed. Then,*

(i)

$$
\begin{aligned}
Q^Y_{T+S} &= Q^{Y/\overline{T(0)}}_{\frac{\overline{T(0)}+S(0)}{T(0)}} \; Q^Y_T \\
&= Q^{Y/\overline{S(0)}}_{\frac{\overline{T(0)}+S(0)}{S(0)}} \; Q^Y_S.
\end{aligned}
$$

(ii) $$\dim\left(\frac{\overline{T(0)}+S(0)}{T(0)}\right) = \dim S(0) - \dim\left(\overline{T(0)} \cap S(0)\right).$$

(iii) $$\dim\left(\frac{\overline{T(0)}+S(0)}{S(0)}\right) = \dim \overline{T(0)} - \dim\left(\overline{T(0)} \cap S(0)\right). \qquad \diamond$$

Proof. *(i)* Since $\overline{T(0)} + S(0)$ is closed, then

$$\overline{T(0) + S(0)} \subset \overline{T(0)} + S(0). \tag{2.7}$$

Moreover, since $\overline{T(0)} \subset \overline{T(0) + S(0)}$ and $S(0) \subset \overline{T(0) + S(0)}$, then

$$\overline{T(0)} + S(0) \subset \overline{T(0) + S(0)}. \tag{2.8}$$

From (2.7) and (2.8), it follows that $\overline{T(0)} + S(0) = \overline{T(0) + S(0)}$. Therefore,

$$Q^Y_{T+S} = Q^Y_{\overline{T(0)+S(0)}} = Q^Y_{\overline{T(0)}+S(0)}.$$

Since $\overline{T(0)} + S(0)$ is closed containing $\overline{T(0)}$, then by using Lemma 2.1.4, we obtain

$$Q^Y_{T+S} = Q^Y_{\overline{T(0)}+S(0)} = Q^{Y/\overline{T(0)}}_{\frac{\overline{T(0)}+S(0)}{T(0)}} \; Q^Y_{\overline{T(0)}}.$$

Hence,

$$Q^Y_{T+S} = Q^{Y/\overline{T(0)}}_{\frac{\overline{T(0)}+S(0)}{T(0)}} \; Q^Y_T.$$

In the same way, since $\overline{T(0)} + S(0)$ is closed containing $S(0)$, we obtain

$$Q^Y_{T+S} = Q^Y_{\overline{T(0)}+S(0)} = Q^{Y/\overline{S(0)}}_{\frac{\overline{T(0)}+S(0)}{S(0)}} \; Q^Y_{S(0)}.$$

So,

$$Q^Y_{T+S} = Q^{Y/\overline{S(0)}}_{\frac{\overline{T(0)}+S(0)}{S(0)}} \; Q^Y_S.$$

(*ii*) By using Theorem 2.1.2, we have

$$\frac{\overline{T(0)} + S(0)}{\overline{T(0)}} \simeq \frac{S(0)}{\overline{T(0)} \cap S(0)}. \tag{2.9}$$

It follows that

$$\dim\left(\frac{\overline{T(0)} + S(0)}{\overline{T(0)}}\right) = \dim\left(\frac{S(0)}{\overline{T(0)} \cap S(0)}\right)$$
$$= \dim(S(0)) - \dim\left(\overline{T(0)} \cap S(0)\right).$$

(*iii*) The proof may be achieved in a similar way as (*ii*). It is sufficient to replace (2.9) by

$$\frac{\overline{T(0)} + S(0)}{S(0)} \simeq \frac{\overline{T(0)}}{\overline{T(0)} \cap S(0)}. \qquad \text{Q.E.D.}$$

Corollary 2.4.1 *Let* T, $S \in LR(X,Y)$ *such that* $\dim S(0) < +\infty$. *Then*,

$$Q^Y_{T+S} = Q^{Y/\overline{T(0)}}_{\frac{\overline{T(0)}+S(0)}{\overline{T(0)}}} Q^Y_T,$$
$$= Q^{Y/S(0)}_{\frac{\overline{T(0)}+S(0)}{S(0)}} Q^Y_S,$$

and

$$\dim\left(\frac{\overline{T(0)} + S(0)}{\overline{T(0)}}\right) = \dim S(0) - \dim\left(\overline{T(0)} \cap S(0)\right). \qquad \diamond$$

Proposition 2.4.2 *Let* $T \in LR(X,Y)$. *Then,* $Q_T T$ *is single valued.* $\diamond$

Proof. Let $x \in \mathcal{D}(T)$ and $z_1, z_2 \in Q_T T x$, then, $z_1 - z_2 \in Q_T T x - Q_T T x = Q_T T(0) \subset Q_T \overline{T(0)} = \{0\}$. Hence $z_1 = z_2$.   Q.E.D.

2.5   Generalized Kernel and Range of Linear Relations

The integer powers of a multivalued linear operator $T \in LR(X)$ are recursively defined as follows: $T^0 = I$, and if T^{n-1} is defined, then we set

$$T^n x := (TT^{n-1})x := \bigcup \{Ty \text{ such that } y \in \mathcal{D}(T) \cap T^{n-1}x\},$$

with

$$\mathcal{D}(T^n) := \{x \in \mathcal{D}(T^{n-1}) \text{ such that } \mathcal{D}(T) \cap T^{n-1}x \neq \emptyset\}.$$

It is easy to see that

$$(T^{-1})^n = (T^n)^{-1}, \quad \text{for all } n \in \mathbb{Z}.$$

Remark 2.5.1 *Let T and S be two linear relations in a vector space such that T and S commute, that is, $TS = ST$ in the sense of the product of linear relations, then*

$$(\lambda - S)T = T(\lambda - S),$$

and

$$(\lambda - T)S = S(\lambda - T)$$

holds for all $\lambda \in \mathbb{C}$. Let $\mu \in \mathbb{K}$, by applying this property to $\mu - T$ instead and $\mu - S$, we have

$$(\lambda - S)(\mu - T) = (\mu - T)(\lambda - S)$$

and

$$(\lambda - T)(\mu - S) = (\mu - S)(\lambda - T).$$

Furthermore, it is clear that $T^n S^m = S^m T^n$ holds for all $n, m \in \mathbb{N}$. By using this equality with $\lambda - S$, $\mu - T$ and $\lambda - S$ instead of T and S, it follows that

$$(\lambda - S)^m(\mu - T)^n = (\mu - T)^n(\lambda - S)^m$$

and

$$(\lambda - T)^m(\mu - S)^n = (\mu - S)^n(\lambda - T)^m. \qquad \diamond$$

The following results can be found in Refs. [84, 85].

Lemma 2.5.1 *Let T be a linear relation in a vector space X. Then,*

(i) For all $\alpha, \beta \in \mathbb{K}$ and for all $m, n \in \mathbb{N}$, we have

(a) $\mathcal{D}((T - \alpha)^n(T - \beta)^m) = \mathcal{D}(T^{n+m})$.
(b) $(T - \alpha)^n(T - \beta)^m(0) = T^{n+m}(0)$.
(c) $(T - \alpha)^n(T - \beta)^m = (T - \beta)^m(T - \alpha)^n$.
(d) If $\alpha = \beta$, then $N(T - \alpha)^n \subset R(T - \beta)^m$.

(ii) If there exists $\lambda \in \mathbb{C}$ such that $(\lambda - T)^{-1}$ is bounded single valued, then $N(T^n) \cap T^m(0) = \{0\}$ for all $m, n \in \mathbb{N}$.

(iii) Let $n, p \in \mathbb{N}$. If there is $\eta \in \mathbb{C}$ such that $T - \eta$ is bijective, then

$$\mathcal{D}(T^n) + R(T^p) = X, \quad and$$

$$T^n(0) \bigcap N(T^p) = \{0\}$$

for all n, $p \in \mathbb{N}$. Further, if T has a finite index, then $i(T^n) = ni(T)$ and $i(T^{n+p}) = i(T^n) + i(T^p) = ni(T) + pi(T) = (n+p)i(T)$. $\diamondsuit$

Theorem 2.5.1 *Let T be a linear relation in a vector space X. Then, for all n, $m \in \mathbb{N}$*

$$\mathcal{D}(T^{n+m}) \subset \mathcal{D}(T^n), \quad R(T^{n+m}) \subset R(T^n), \tag{2.10}$$

$$N(T^{n+m}) \supset N(T^n), \quad T^{n+m}(0) \supset T^n(0), \tag{2.11}$$

and for all p, $k \in \mathbb{N}$

$$N(T^p) \subset \mathcal{D}(T^k), \quad T^p(0) \subset R(T^k). \tag{2.12}$$

$\diamondsuit$

Proof. Assume that $x \in \mathcal{D}(T^{n+m})$, so that $(x, y) \in G(T^{n+m})$ for some $y \in X$. Since $T^{n+m} = T^m T^n$, it follows that $(x, z) \in G(T^n)$ and $(z, y) \in G(T^m)$ for some $z \in X$, which shows that $x \in \mathcal{D}(T^n)$. Therefore, $\mathcal{D}(T^{n+m}) \subset \mathcal{D}(T^n)$, and the first inclusion in Eqn. (2.10) is proved. In order to prove the first inclusion in Eqn. (2.11) assume that $x \in N(T^n)$, so that $(x, 0) \in G(T^n)$. Since $(0, 0) \in G(T^m)$ as $G(T^m)$ is a subspace of $X \times X$, it follows that $(x, 0) \in G(T^{n+m})$, which shows that $x \in N(T^{n+m})$. Therefore, $N(T^{n+m}) \supset N(T^n)$. If $p \leq k$, then it follows from Eqn. (2.11) that $N(T^p) \subset N(T^k)$ and since $N(T^k) \subset \mathcal{D}(T^k)$, it follows that in this case, Eqn. (2.12) holds true. Assume now that $p > k$ and let $x \in N(T^p)$, so that $(x, 0) \in G(T^p) = G(T^{p-k}T^k)$. Thus, $(x, y) \in G(T^k)$ and $(y, 0) \in G(T^{p-k})$, which shows that $x \in \mathcal{D}(T^k)$. Therefore, the first inclusion in (2.12) is proved. The remaining inclusions in Eqs. (2.10)–(2.12) follow from the duality of T and T^{-1}. This completes the proof.   Q.E.D.

Corollary 2.5.1 *(i) Let $T \in LR(X)$. Then, the kernels and the ranges of the iterates T^n, $n \in \mathbb{N}$ form two increasing and decreasing chains, respectively, i.e., the chain of kernels*

$$\{0\} = N(T^0) \subset N(T) \subset N(T^2) \subset \cdots$$

and the chain of ranges

$$X = R(T^0) \supset R(T) \supset R(T^2) \supset \cdots.$$

(ii) $T^n(0) \subset T^{n+1}(0)$, for all $n \in \mathbb{N}$. $\diamondsuit$

Definition 2.5.1 *Let T be a linear relation in X and let*

$$\Delta(T) = \left\{ n \in \mathbb{N} \text{ such that } R(T^n) \bigcap N(T) = R(T^m) \bigcap N(T) \text{ for all } m \geq n \right\}.$$

The degree $\eta(T)$ of T, is defined as

$$\eta(T) = \begin{cases} \inf \Delta(T) & \text{if } \Delta(T) \neq \emptyset \\ +\infty & \text{if } \Delta(T) = \emptyset. \end{cases} \qquad \diamond$$

Lemma 2.5.2 [74, Lemmas 2.5, 2.7 and Corollary 2.6] *Let T be a linear relation in X. Then,*

(i) $d \in \Delta(T)$ if and only if, $N(T^m) \subset N(T^d) + R(T^n)$ if and only if, $R(T^d) \bigcap N(T^n) \subset R(T^m) \bigcap N(T^n)$ for all $m, n \in \mathbb{N}$.

(ii) $\eta(T) = 0$ if and only if, $N(T^m) \subset R(T^n)$ for all $m, n \in \mathbb{N}$ if, and only if, $N(T^n) \subset R(T)$ for all $n \in \mathbb{N}$ if and only if, $N(T) \subset R(T^m)$ for all $m \in \mathbb{N}$. $\diamond$

Lemma 2.5.3 *Let $m \in \mathbb{N}$. Then, $\eta(T^m) \leq \eta(T) \leq m\eta(T^m)$.* $\diamond$

Proof. If $\eta(T) = \infty$, the inequality $\eta(T^m) \leq \eta(T)$ is obvious. Let $d := \eta(T) < \infty$. Then, by virtue of Lemma 2.5.2 (i), we have

$$R(T^d) \bigcap N(T^m) \subset R(T^{mr}) \bigcap N(T^m),$$

whenever $r \geq d$. So, for all $r \geq d$,

$$N(T^m) \bigcap R(T^{md}) = N(T^m) \bigcap R(T^{mr}).$$

So, $d \in \Delta(T^m)$. Hence,

$$\eta(T^m) \leq \eta(T).$$

If $\eta(T^m) = \infty$, then the inequality $\eta(T) \leq m\eta(T^m)$ is clear. Assume that $r := \eta(T^m) < \infty$ and let $n \geq r$. Then,

$$\begin{aligned} N(T) \bigcap R(T^{mr}) &= N(T) \bigcap N(T^m) \bigcap R(T^{mr}) \\ &= N(T) \bigcap N(T^m) \bigcap R(T^{mn}) \\ &= N(T) \bigcap R(T^{mn}). \end{aligned}$$

Let $p \in \mathbb{N}$ such that $p \geq mr$ and we write $p = mk + s$ for some nonnegative integers k and s such that $k + 1 \geq r$ and $0 \leq s \leq m - 1$. Then,

$$\begin{aligned} N(T) \bigcap R(T^{mr}) &= N(T) \bigcap R(T^{m(k+1)}) \\ &\subset N(T) \bigcap R(T^p) \\ &\subset N(T) \bigcap R(T^{mr}). \end{aligned}$$

Therefore, $N(T) \cap R(T^{mr}) = N(T) \cap R(T^p)$ for all $p \geq mr$. So, $mr \in \Delta(T)$. Hence,

$$\eta(T) \leq m\eta(T^m).$$

This completes the proof.   Q.E.D.

Lemma 2.5.4 *Let T be a relation in a linear space X and let $i, k \in \mathbb{N} \cup \{0\}$. Then,*

$$\frac{N(T^{i+k})}{(N(T^i) + R(T^k)) \cap N(T^{i+k})} \cong \frac{N(T^k) \cap R(T^i)}{N(T^k) \cap R(T^{i+k})},$$

and

$$\frac{T^{i+k}(0)}{(T^i(0) + \mathcal{D}(T^k)) \cap T^{i+k}(0)} \cong \frac{T^k(0) \cap \mathcal{D}(T^i)}{T^k(0) \cap \mathcal{D}(T^{i+k})}. \qquad \diamond$$

Proposition 2.5.1 *Let T be a relation in a linear space X. Then,*

(i) If $N(T^k) = N(T^{k+1})$ for some $k \in \mathbb{N}$, then $N(T^n) = N(T^k)$ for all nonnegative integers $n \geq k$.
(ii) If $T^k(0) = T^{k+1}(0)$ for some $k \in \mathbb{N}$, then $T^n(0) = T^k(0)$ for all nonnegative integers $n \geq k$. $\diamond$

Proof. *(i)* Assume that $N(T^{n+1}) = N(T^n)$. It will be shown that $N(T^{n+2}) = N(T^{n+1})$, and then, the statement will follow by induction. Clearly, (2.11) shows that $N(T^{n+1}) \subset N(T^{n+2})$. So, only the converse inclusion remains to be proved. Let $x \in N(T^{n+2})$, then $(x, 0) \in G(T^{n+2}) = G(T^{n+1}T)$. Thus, $(x, y) \in G(T)$ and $(y, 0) \in G(T^{n+1})$ for some $y \in X$. Now, $y \in N(T^{n+1}) = N(T^n)$ by the induction hypothesis, which shows that $(y, 0) \in G(T^n)$. Therefore, $(x, 0) \in G(T^{n+1})$. So, $x \in N(T^{n+1})$, which implies *(i)*.

The statement *(ii)* follows from the statement in *(i)* due to the duality of T and T^{-1}.   Q.E.D.

Theorem 2.5.2 *Let T be a linear relation in a vector space X. Then,*

(i) If $\mathcal{D}(T^k) = \mathcal{D}(T^{k+1})$ for some $k \in \mathbb{N}$, then $\mathcal{D}(T^n) = \mathcal{D}(T^k)$ for all nonnegative integers $n \geq k$.
(ii) If $R(T^k) = R(T^{k+1})$ for some $k \in \mathbb{N}$, then $R(T^n) = R(T^k)$ for all nonnegative integers $n \geq k$. $\diamond$

Proof. *(i)* Assume that $\mathcal{D}(T^n) = \mathcal{D}(T^{n+1})$. It suffices to show that $\mathcal{D}(T^{n+1}) = \mathcal{D}(T^{n+2})$. Clearly, (2.10) shows that $\mathcal{D}(T^{n+2}) \subset \mathcal{D}(T^{n+1})$. So, only the converse inclusion remains to be proved. Assume that $x \in \mathcal{D}(T^{n+1})$, then $(x, y) \in G(T^{n+1}) = G(T^n T)$. Hence, $(x, z) \in G(T)$ and $(z, y) \in G(T^n)$ for some $z \in X$. Now, $z \in \mathcal{D}(T^n) = \mathcal{D}(T^{n+1})$ implies that $(z, u) \in G(T^{n+1})$ for some $u \in X$. This leads to $(x, u) \in G(T^{n+2})$. So, $x \in \mathcal{D}(T^{n+2})$. Hence, $\mathcal{D}(T^{n+1}) \subset \mathcal{D}(T^{n+2})$.

The statement (ii) follows from the statement in (i) due to the duality of T and T^{-1}.

Q.E.D.

Corollary 2.5.2 *Let T be a linear relation in a vector space X. Then,*

(i) $\mathcal{D}(T) = X$ if and only if, $\mathcal{D}(T^p) = X$ for some (for all) $p \in \mathbb{N}$,
(ii) $R(T) = X$ if and only if, $R(T^p) = X$ for some (for all) $p \in \mathbb{N}$. ◇

Proof. (i) If $\mathcal{D}(T) = X$, then it follows from Theorem 2.5.2 that $\mathcal{D}(T^p) = X$ for all $p \in \mathbb{N}$. Conversely, if $\mathcal{D}(T^p) = X$ for some $p \in \mathbb{N}$, then due to (2.10), it is clear to see that $\mathcal{D}(T^i) = X$ for $1 \leq i \leq p$.

The proof of (ii) is analogous.

Q.E.D.

An important role is played by certain root manifolds of a relation T in a linear space X. The root manifold (the generalized kernel) $\mathfrak{R}_0(T)$ is defined by

$$\mathfrak{R}_0(T) := N^\infty(T) = \bigcup_{i=1}^\infty N(T^i). \tag{2.13}$$

Similarly, the root manifold (the generalized range) $R^\infty(T)$, is defined by

$$R^\infty(T) := \bigcap_{i=1}^\infty R(T^i) = \bigcup_{i=1}^\infty T^i(0). \tag{2.14}$$

Clearly the root manifolds $\mathfrak{R}_0(T)$ and $R^\infty(T)$ are subspaces of $\mathcal{D}(T) \subset X$ and $R(T) \subset X$, respectively.

Lemma 2.5.5 [9, Lemma 20] *Let T be a linear relation in a vector space X. Then,*

(i) If $\lambda \in \mathbb{K} \backslash \{0\}$, then $N(\lambda - T) \subset R^\infty(T)$.
(ii) If λ, $\mu \in \mathbb{K}$ are distinct, then $N(\lambda - T)^n \subset R^\infty(\mu - T)$ for all $n \in \mathbb{N}$.
(iii) If there exists $d \in \mathbb{N}$ such that $N(T) \bigcap R(T^d) = N(T) \bigcap R(T^{n+d})$ for all $n \in \mathbb{N}$, then $T(\mathcal{D}(T) \bigcap R^\infty(T)) = R^\infty(T)$.
(iv) If $N(T) \subset R^\infty(T)$ or $\alpha(T) < \infty$, then $T(\mathcal{D}(T) \bigcap R^\infty(T)) = R^\infty(T)$. ◇

2.5.1   Ascent and Descent and Singular Chain of Linear Relations

Likewise, the statements in Theorem 2.5.1 and Proposition 2.5.1 lead to the introduction of the ascent and descent of linear relation.

Definition 2.5.2 *Let $T \in LR(X, Y)$. The ascent and the descent of T are defined as follows*

$$asc(T) = \inf \left\{ p \in \mathbb{N} \text{ such that } N(T^p) = N(T^{p+1}) \right\}$$

and

$$des(T) = \inf \left\{ p \in \mathbb{N} \text{ such that } R(T^p) = R(T^{p+1}) \right\},$$

respectively, whenever these minima exist. If no such numbers exist the ascent and descent of T are defined to be ∞. ◇

Let $T \in LR(X)$. The singular chain manifold $\mathfrak{R}_c(T)$ is defined as the intersection of the root manifolds $\mathfrak{R}_0(T)$ in (2.13) and $\mathfrak{R}_\infty(T)$ in (2.14):

$$\mathfrak{R}_c(T) = \mathfrak{R}_0(T) \bigcap R^\infty(T) = \left(\bigcup_{n=1}^{\infty} N(T^n) \right) \bigcap \left(\bigcup_{n=1}^{\infty} T^n(0) \right).$$

The linear space $\mathfrak{R}_c(T)$ is non-trivial if and only if, there exists a number $s \in \mathbb{N}$ and elements $x_i \in X$, $1 \leq i \leq s$, not all equal to zero, such that

$$(0, x_1), (x_1, x_2), \cdots, (x_{s-1}, x_s), (x_s, 0) \in G(T).$$

We say that T has trivial singular chain if

$$\mathfrak{R}_c(T) = \{0\}.$$

Example 2.5.1 *Let $X = span\{e_1, e_2\}$ with e_1 and e_2 are linearly independent and define the relation T by*

$$T = span\left\{ \{0, e_1\}, \{e_1, 0\}, \{e_2, 0\} \right\} \text{ and } S = span\left\{ \{e_1, e_2\}, \{e_2, e_1\} \right\}.$$

Then, $S^2 = I$ and $\mathfrak{R}_c(S) = \{0\}$. The nullity and defect of T are given by $\alpha(T) = 2$ and $\beta(T) = 1$. Moreover, $T^2 = T$ and the ascent and descent of T are given by $asc(T) = 1$ and $des(T) = 1$. So, $\mathfrak{R}_c(T) \neq \{0\}$.

Proposition 2.5.2 *Let T be a relation in a linear space X. If one of the following conditions*

$$\mathcal{D}(T^r) \bigcap T(0) = \{0\} \quad \text{or} \quad R(T^r) \bigcap N(T) = \{0\}$$

is satisfied for some $r \in \mathbb{N}$, then $\mathfrak{R}_c(T) = \{0\}$. ◇

Proof. Assume that $\mathcal{D}(T^r) \bigcap T(0) = \{0\}$. If $r = 0$, then $T(0) = \{0\}$ and hence $\mathfrak{R}_c(T) = \{0\}$. Now, let $r \in \mathbb{N}^*$. If $\mathfrak{R}_c(T) \neq \{0\}$, then T has a non-trivial singular chain of the form

$$(0, x_1), (x_1, x_2), (x_2, x_3), \ldots, (x_{n-1}, x_n), (x_n, 0)$$

for some non-zero vectors $x_i \in X$, $1 \leq i \leq n$. Clearly, $x_1 \in T(0)$ and $x_1 \in N(T^n) \subset \mathcal{D}(T^r)$ by (2.12). Therefore,

$$x_1 \in \mathcal{D}(T^r) \bigcap T(0) = \{0\},$$

which shows that $x_1 = 0$. This contradiction implies that $\mathfrak{R}_c(T) = \{0\}$. The argument for the other case is completely similar.　　　　　　　Q.E.D.

Lemma 2.5.6 (*i*) [85, Lemmas 4.4 and 5.1] *Let T be a linear relation in a vector space X with $\mathfrak{R}_c(T) = \{0\}$ and let i, $k \in \mathbb{N}$. Then,*

$$\frac{N(T^{i+k})}{N(T^i)} \cong N(T^k) \bigcap R(T^i).$$

(*ii*) *Let T be a linear relation in a vector space X and let $r \in \mathbb{N}$. Then,*

$$
\begin{aligned}
\dim\left(\frac{N(T^r) + R(T)}{R(T)}\right) &= \dim\left(\frac{N(T^r)}{N(T^r) \bigcap R(T)}\right) \\
&\leq \dim\left(\frac{\mathcal{D}(T^r) + R(T)}{R(T)}\right) \\
&= \dim\left(\frac{\mathcal{D}(T^r)}{\mathcal{D}(T^r) \bigcap R(T)}\right).
\end{aligned}
$$
◇

2.5.2　Norm of a Linear Relation

Let X and Y be two normed spaces.

Definition 2.5.3 *Let $T \in LR(X, Y)$. We define the norm of Tx and T by*

$$\|Tx\| = \|Q_T Tx\| \quad (x \notin \mathcal{D}(T)),$$

and

$$\|T\| = \|Q_T T\|.$$
◇

We note $\|T\|$ is not a true "norm" function since $\|T\| = 0$ does not imply that $T = 0$. For example, the linear relation $T \in LR(X, Y)$ defined as $G(T) = X \times Y$ has a zero norm.

Lemma 2.5.7 *Let* $T \in LR(X, Y)$. *Then,*

(i) For $x \in \mathcal{D}(T)$,

$$
\begin{aligned}
\|Tx\| &= \operatorname{dist}(y, T(0)) \text{ for all } y \in Tx \\
&= \inf_{y \in Tx} \|y\| \\
&= \operatorname{dist}(y + T(0), 0) \text{ for all } y \in Tx \\
&= \operatorname{dist}(Tx, 0) \\
&= \operatorname{dist}(Tx, T(0)).
\end{aligned}
$$

(ii) $\|T\| = \sup\limits_{x \in B_{\mathcal{D}(T)}} \|Tx\|$, *where* $B_{\mathcal{D}(T)} = \{x \in \mathcal{D}(T) \text{ such that } \|x\| \leq 1\}$. $\diamondsuit$

Proof. (*i*) The first equality follows from definition of $\|Tx\|$ and Theorem 2.3.1. The second equality follows from the definition and properties of the norm on $Y/\overline{T(0)}$. The rest are obvious.

(*ii*) $\|T\| = \|Q_T T\| = \sup\limits_{x \in B_{\mathcal{D}(T)}} \|Q_T Tx\| = \sup\limits_{x \in B_{\mathcal{D}(T)}} \|Tx\|$.   Q.E.D.

Proposition 2.5.3 *Let* X *be a normed space and* $T \in LR(X)$. *Then,* $\|T\| = 0$ *if and only if,* $T(B_{\mathcal{D}(T)}) \subset \overline{T(0)}$. $\diamondsuit$

Proof. If $\|T\| = 0$, then by using Lemma 2.5.7, we have $\|Tx\| = 0$, for all $x \in B_{\mathcal{D}(T)}$. Hence, for all $x \in B_{\mathcal{D}(T)}$ and $y \in Tx$, we have $\operatorname{dist}(y, T(0)) = 0$. So, for all $x \in B_{\mathcal{D}(T)}$ and $y \in Tx$, we have $\operatorname{dist}(y, \overline{T(0)}) = 0$. This implies that for all $x \in B_{\mathcal{D}(T)}$, $Tx \subset \overline{T(0)}$. Thus, $T(B_{\mathcal{D}(T)}) \subset \overline{T(0)}$. Conversely, by using both the same above reasoning and Lemma 2.5.7, we find the result. Q.E.D.

Proposition 2.5.4 *For* T *and* $S \in LR(X, Y)$, *we have*

(i) $\|Tx + Sx\| \leq \|Tx\| + \|Sx\|$, *for* $x \in \mathcal{D}(T) \cap \mathcal{D}(S)$.
(ii) $\|\alpha Tx\| = |\alpha| \|Tx\|$, *for* $\alpha \in \mathbb{K}$ *and* $x \in \mathcal{D}(T)$.
(iii) If $S(0) \subset \overline{T(0)}$ *and* $\mathcal{D}(T) \subset \mathcal{D}(S)$, *then* $Q_T S$ *is a single valued linear operator and* $\|Q_T S\| \leq \|S\|$. $\diamondsuit$

Proof. (*i*) Let $x \in \mathcal{D}(T) \cap \mathcal{D}(S)$, $s \in Sx$ and $t \in Tx$, then $s + t \in Sx + Tx = (S + T)x$. Hence, by using Lemma 2.5.7 (*i*), we have

$$
\begin{aligned}
\|Tx + Sx\| &= \operatorname{dist}(t + s, (T + S)(0)) \\
&\leq \operatorname{dist}(t, T(0) + S(0)) + \operatorname{dist}(s, T(0) + S(0)) \\
&\leq \operatorname{dist}(t, T(0)) + \operatorname{dist}(s, S(0)) \\
&= \|Tx\| + \|Sx\|.
\end{aligned}
$$

(ii) The case $\alpha = 0$ is obviously. If $\alpha \neq 0$, then for all $x \in \mathcal{D}(T)$

$$\begin{aligned}
\|\alpha Tx\| &= \|Q_T(\alpha T)(x)\| \text{ (since } \alpha T(0) = T(0) \text{ for } \alpha \neq 0)\\
&= \|\alpha Q_T T(x)\|\\
&= |\alpha| \|Q_T T(x)\|.
\end{aligned}$$

(iii) Since $S(0) \subset \overline{T(0)}$, then $Q_T(S(0)) \subset Q_T(\overline{T(0)}) = \{0\}$. Consequently, $Q_T S$ is a single valued operator, and $\|Q_T Sx\| = \text{dist}(Sx, T(0)) = \text{dist}(Sx, \overline{T(0)}) \leq \text{dist}(Sx, S(0)) = \|Sx\| = \|Q_S Sx\|$. This completes the proof. $\hfill$ Q.E.D.

Proposition 2.5.5 [50, Corollary 2.3.13] *Let* $T \in LR(X, Y)$ *and* $S \in LR(Y, Z)$. *Then,*

(i)
$$\|ST\| \leq \|S\| \, \|I_{\mathcal{D}(S)}T\| \ (\infty.0 \text{ excluded})$$

with $\|ST\| = 0$ *whenever* $\|S\| = 0$ *even if* $\|I_{\mathcal{D}(S)}T\| = \infty$.
(ii) *Moreover, if* $T(0) \subset \mathcal{D}(S)$, *then* $\|ST\| \leq \|S\| \, \|T\|$. $\hfill \diamond$

Proposition 2.5.6 *Let* $T \in LR(X, Y)$. *Then,*

(i) $\|T\| < \infty$ *if and only if, there exists* $\lambda > 0$ *such that*

$$T(B_{\mathcal{D}(T)}) \subset \lambda B_{R(T)} + T(0). \tag{2.15}$$

(ii) *If* $\|T\| < \infty$, *then*

$$\|T\| = \inf_{\lambda > 0} \left\{ \lambda \text{ such that } T(B_{\mathcal{D}(T)}) \subset \lambda B_{R(T)} + T(0) \right\}. \tag{2.16}$$

$\hfill \diamond$

Proof. (i) Suppose $\|T\| < \infty$. By using Lemma 2.5.7 (ii), we have for $x \in B_{\mathcal{D}(T)}$ and $y \in Tx$, there exists $m \in T(0)$ such that for all $\varepsilon > 0$,

$$\|y - m\| < \|T\| + \varepsilon,$$

that is, $y - m \in (\|T\| + \varepsilon)B_{R(T)}$. So,

$$y \in (\|T\| + \varepsilon)B_{R(T)} + T(0). \tag{2.17}$$

Conversely, suppose (2.15) holds. Let $x \in B_{\mathcal{D}(T)}$ and choose $y \in Tx$. Then, $y = \lambda y_1 + m$, where $\|y_1\| \leq 1$ and $m \in T(0)$. Thus, $\|y - m\| \leq \lambda$, in particular, $\text{dist}(y, T(0)) \leq \lambda$. It follows from Lemma 2.5.7 (i) that $\|T\| \leq \lambda < \infty$.

(*ii*) Suppose $\|T\| < \infty$. If $\|T\| = 0$, then by using Proposition 2.5.3, we have $T(B_{\mathcal{D}(T)}) \subset \overline{T(0)}$. So, (2.16) holds. Suppose $\|T\| > 0$. It follows from (2.17) that

$$\|T\| \geq \inf_{\lambda > 0} \left\{ \lambda \text{ such that } T(B_{\mathcal{D}(T)}) \subset \lambda B_{R(T)} + T(0) \right\}.$$

Let $\alpha \in]0, \|T\|[$ and choose $x \in B_{\mathcal{D}(T)}$, $y \in Tx$ such that

$$\alpha < \text{dist}(y, T(0)). \tag{2.18}$$

If $T(B_{\mathcal{D}(T)}) \subset \alpha B_{R(T)} + T(0)$, then there exist $y_1 \in B_{R(T)}$ and $m \in T(0)$ such that $y = \alpha y_1 + m$. So, $\|y - m\| \leq \alpha$, which contradicts (2.18). Thus,

$$\alpha < \inf_{\lambda > 0} \left\{ \lambda \text{ such that } T(B_{\mathcal{D}(T)}) \subset \lambda B_{R(T)} + T(0) \right\}. \qquad \text{Q.E.D.}$$

Definition 2.5.4 *Let $T \in LR(X, Y)$. The T graph norm $\| \cdot \|_T$ is defined on $\mathcal{D}(T)$ by*

$$\|x\|_T = \|x\| + \|Tx\|, \text{ for } x \in \mathcal{D}(T).$$

We denote by X_T the vector space $(\mathcal{D}(T), \| \cdot \|_T)$ and $G_T \in LR(X_T, X)$ the identity injection (or the graph operator) of X_T into X, i.e., $\mathcal{D}(G_T) = X_T$, and $G_T x = x$ for all $x \in X_T$. G_T is called the graph operator of T. ◇

Remark 2.5.2 (*i*) *If $T = \widetilde{T}$, then X_T and $T(0)$ are complete.*

(*ii*) *Let X be a complete space. If we denote by $(G_T)^{-1}$ the inverse of G_T, then $(G_T)^{-1}$ and G_T are bijective everywhere defined. So, $R(TG_T) = TG_T(\mathcal{D}(TG_T)) = TG_T(G_T)^{-1}(\mathcal{D}(T)) = R(T)$ and $\alpha(TG_T) = \alpha(T)$.* ◇

Theorem 2.5.3 *Let S, $T \in LR(X, Y)$ such that $\mathcal{D}(S) = \mathcal{D}(T)$. Then, $\|SG_T\| = \frac{\|S\|}{1 + \|T\|}$.* ◇

Proof. We have

$$\begin{aligned} \|SG_T\| &= \sup_{x \in X_T} \frac{\|SG_T x\|}{\|x\|_T} \\ &= \sup_{x \in \mathcal{D}(T) \setminus \{0\}} \frac{\|Sx\|}{\|x\| + \|Tx\|} \\ &= \sup_{x \in \mathcal{D}(T) \setminus \{0\}} \frac{\|Sx\|}{\left(1 + \frac{\|Tx\|}{\|x\|}\right) \|x\|} \\ &= \frac{\|S\|}{1 + \|T\|} \quad (\text{as } \mathcal{D}(S) = \mathcal{D}(T)). \end{aligned}$$

This completes the proof. Q.E.D.

Corollary 2.5.3 *Let $T \in LR(X, Y)$. Then, $\|TG_T\| = \frac{\|T\|}{1 + \|T\|}$.* ◇

2.5.3 Continuity and Openness of a Linear Relation

Definition 2.5.5 (*i*) *Let T be an arbitrary linear relation from one topological space X to another Y. Then, T is said to be continuous if for each neighbourhood $\mathcal{V}$ in $R(T)$ the image $T^{-1}(\mathcal{V})$ is a neighbourhood in $\mathcal{D}(T)$.*
(*ii*) *T is called open if whenever $\mathcal{V}$ is a neighbourhood in $\mathcal{D}(T)$, the image $T(\mathcal{V})$ is a neighbourhood of $R(T)$.*
(*iii*) *If $\mathcal{D}(T) = X$ and T is a continuous linear relation, then we shall say that T is bounded.* ◇

The collection of bounded linear relations as defined above will be denoted by $BR(X, Y)$ and as useful we write $BR(X) := BR(X, X)$ and the collection of bounded linear operators as defined above will be denoted by $\mathcal{L}(X, Y)$ and as useful we write $\mathcal{L}(X) := \mathcal{L}(X, X)$.

Remark 2.5.3 (*i*) *We note that T is continuous if and only if, T^{-1} is open.*
(*ii*) *Let M be a subspace of a normed vector space X, then the injection operator J_M^X is continuous.* ◇

Theorem 2.5.4 [50, Theorem 3.4.2] *(Closed graph and open mapping theorem) Let $T \in LR(X, Y)$. Then,*

(*i*) *$\widetilde{T}$ is continuous if and only if, $\mathcal{D}(\widetilde{T})$ is closed, where $\widetilde{T}$ is given in (2.4).*
(*ii*) *$\widetilde{T}$ is open if and only if, $R(\widetilde{T})$ is closed.* ◇

Theorem 2.5.5 *Let $T \in LR(X, Y)$. Then,*

(*i*) *T is continuous if and only if, $\|T\| < \infty$.*
(*ii*) *If $\dim \mathcal{D}(T) < \infty$, then T is continuous.* ◇

Proof. (*i*) Suppose T is continuous. Since $T(0) + B_Y$ is a neighbourhood of $T(0)$, it follows that $T^{-1}(T(0) + B_Y) = T^{-1}(B_{R(T)})$ is a neighbourhood of 0. So, there is $\lambda > 0$ such that $\lambda B_{\mathcal{D}(T)} \subset T^{-1}(B_{R(T)})$. Thus, $\lambda T(B_{\mathcal{D}(T)}) \subset B_{R(T)} + T(0) = TT^{-1}(B_{R(T)})$. Using Proposition 2.5.6 (*i*), we have $\|T\| < \infty$. Conversely, suppose $\|T\| < \infty$. Let $x \in \mathcal{D}(T)$, $y \in Tx$ and let V be a nontrivial closed ball in $R(T)$ with center y. Then, $V_0 = V \backslash \{y\} = \alpha B_{R(T)}$ for some $\alpha > 0$. Using Proposition 2.5.6 (*i*), there exists $\lambda > 0$ such that $T(B_{\mathcal{D}(T)}) \subset \lambda B_{R(T)} + T(0)$. It follows that $B_{\mathcal{D}(T)} + T^{-1}(0) \subset \lambda T^{-1}(B_{R(T)}) = \alpha^{-1}\lambda T^{-1}(V_0)$ and $\lambda^{-1}\alpha B_{\mathcal{D}(T)} + T^{-1}(0) \subset T^{-1}(V_0) = T^{-1}(V - y)$. Hence, $\lambda^{-1}\alpha B_{\mathcal{D}(T)} + T^{-1}y \subset T^{-1}(V) - T^{-1}y + T^{-1}y = T^{-1}(V)$. So, $\lambda^{-1}\alpha B_{\mathcal{D}(T)} + T^{-1}y$ is a neighbourhood of x in $\mathcal{D}(T)$. Let W be a neighbourhood of Tx, let $U \subset W$

be an open set containing $y \in Tx$ and let $V \subset U$ be a non-trivial closed ball with center y. Then, $T^{-1}(W)$ is neighbourhood of x. Hence, T is continuous.

(ii) If $\dim \mathcal{D}(T) < \infty$, then $Q_T T$ is a continuous single valued linear operator, that is, $\|Q_T T\| < \infty$. On the other hand, $\|T\| = \|Q_T T\|$. Now, the result follows from (i).   Q.E.D.

Corollary 2.5.4 *Let X and Y be normed spaces, and $T \in LR(X,Y)$. If T is continuous, then $\|T(x)\| \leq \|T\| \|x\|$ for all $x \in \mathcal{D}(T)$.*

Proposition 2.5.7 *Let X and Y be normed spaces, and T, $S \in LR(X,Y)$ such that $\mathcal{D}(T) \subset \mathcal{D}(S)$ and $\overline{S(0)} \subset \overline{T(0)}$. Then,*

$$\|(T-S)x\| \geq \|Tx\| - \|Sx\|, \quad x \in \mathcal{D}(T). \tag{2.19}$$

Further, if T and S are bounded, then $T - S$ is bounded, and

$$\|T - S\| \geq \|T\| - \|S\|. \tag{2.20}$$

In addition, if either S is bounded and T is unbounded or T is bounded, $S_{|\mathcal{D}(T)}$ is unbounded, and $\overline{S(0)} = \overline{T(0)}$, then $T - S$ is unbounded.   ◇

Proof. We can easily check that $\overline{T(0)} = \overline{(T-S)(0)}$ since $\overline{S(0)} \subset \overline{T(0)}$. So, by (i) of Lemma 2.5.7, we have for any $x \in \mathcal{D}(T)$, and any given $y_1 \in Tx$ and $y_2 \in Sx$,

$$
\begin{aligned}
\|(T-S)x\| &= \operatorname{dist}(y_1 - y_2, \overline{(T-S)(0)}) \\
&= \operatorname{dist}(y_1 - y_2, \overline{T(0)}) \\
&\geq \operatorname{dist}(y_1, \overline{T(0)}) - \operatorname{dist}(y_2, \overline{T(0)}) \\
&\geq \operatorname{dist}(y_1, \overline{T(0)}) - \operatorname{dist}(y_2, \overline{S(0)}) \\
&= \|Tx\| - \|Sx\|, \tag{2.21}
\end{aligned}
$$

which yields that (2.19) holds. Further, suppose that T and S are bounded. Then, by using Lemma 2.5.7 (ii) and Theorem 2.5.5, we have $T-S$ is bounded. It follows from (2.21) that $\|Tx\| \leq \|(T-S)x\| + \|Sx\|$, for all $x \in \mathcal{D}(T)$ with $\|x\| \leq 1$. Hence, by (ii) of Lemma 2.5.7, we have $\|T\| \leq \|(T-S)\| + \|S\|$. Consequently, (2.20) holds. In addition, suppose that S is bounded and T is unbounded. It can be easily verified that $T - S$ is unbounded by (2.21), Lemma 2.5.7 and Corollary 2.5.4. Finally, suppose that T is bounded, $S_{|\mathcal{D}(T)}$ is unbounded, and $\overline{S(0)} = \overline{T(0)}$. It follows from (2.21) that

$$
\begin{aligned}
\|(T-S)x\| &\geq \operatorname{dist}(y_2, \overline{S(0)}) - \operatorname{dist}(y_1, \overline{T(0)}) \\
&= \|Sx\| - \|Tx\|,
\end{aligned}
$$

which, together with Lemma 2.5.7, implies that $T - S$ is unbounded. Q.E.D.

Remark 2.5.4 (*i*) *The conditions* $\mathcal{D}(T) \subset \mathcal{D}(S)$ *and* $\overline{S(0)} \subset \overline{T(0)}$, *in Proposition 2.5.7, are necessary. In fact, let* $X \neq \{0\}$, $T = I_X$, *and* $G(S) = X \times X$. *Then, for all* $x \neq 0$, *we have* $\|Tx - Sx\| = 0$ *and* $\|Tx\| - \|Sx\| = \|x\|$. *So,* $\|Tx\| - \|Sx\| \not\leq \|Tx - Sx\|$.
(*ii*) *Let* X *and* Y *be normed spaces, and* $S, T \in LR(X, Y)$ *satisfy that* $\mathcal{D}(S) \subset \mathcal{D}(T)$. *If* S *is bounded,* T *is unbounded,* $\overline{T(0)} \subset \overline{S(0)}$, *and* $\overline{T(0)} \neq \overline{S(0)}$, *then* $S - T$ *may be bounded or unbounded. For example, let* $X = Y = l_2$, *and*

$$T(x) = (nx_n)_{n \geq 1}, \ x = (x_n)_{n \geq 1} \subset \mathcal{D}(T),$$

where $\mathcal{D}(T) = \{x = (x_n)_{n \geq 1} \subset l_2 : (nx_n)_{n \geq 1} \subset l_2\}$. *Then,* T *is unbounded single valued. Let* $\mathcal{D}(S_1) = \mathcal{D}(S_2) = \mathcal{D}(T)$, *and*

$$S_1(x) = l_2, \ S_2(x) = x + S_2(0), \ x \in \mathcal{D}(T),$$

where $S_2(0) = \mathrm{span}\{e_1\}$ *with* $e_1(1) = 1$ *and* $e_1(n) = 0$ *for* $n \geq 2$. *It is evident that* $S_1(0) = l_2$, S_1 *is bounded with bound* $\|S_1\| = 0$, *and* S_2 *is bounded with bound* $\|S_2\| = 1$. *In addition* $\overline{T(0)} = \{0\} \subset \overline{S_i(0)}$ *and* $\overline{T(0)} \neq \overline{S_i(0)}$ *for* $i = 1, 2$. *Further, we get that for any* $x \in \mathcal{D}(T)$,

$$S_1(x) - T(x) = l_2, \ S_2(x) - T(x) = ((1 - n)x_n)_{n \geq 1} + S_2(0),$$

which implies that $S_1 - T$ *is bounded with bound* $\|S_1 - T\| = 0$ *and* $S_2 - T$ *is unbounded.* ◇

2.5.4 Selections of Linear Relation

We define the linear relation T_G associated with T by

$$\begin{cases} T_G \in LR(X, X \times Y), \\ \mathcal{D}(T_G) = \mathcal{D}(T), \text{ and} \\ T_G x = \{(x, y) \in X \times Y \text{ such that } y \in Tx\}. \end{cases} \quad (2.22)$$

Remark 2.5.5 *If* T *is continuous, then* T_G *is continuous. In fact,*

$$\begin{aligned} \|T_G x\| &= \inf\{\|x\| + \|y\| \text{ such that } y \in Tx\} \\ &= \|x\| + \|Tx\| \\ &\leq \|x\| + \|T\| \|x\| \quad \text{as Corollary 2.5.4} \\ &= (1 + \|T\|)\|x\|. \end{aligned}$$

◇

Definition 2.5.6 *Let $T \in LR(X, Y)$. A linear operator S is called a selection (or single valued part) of T if $T = S + T - T$ and $\mathcal{D}(T) = \mathcal{D}(S)$.* ◇

If S is a selection of T, then for all $x \in \mathcal{D}(T)$, we have $Tx = Sx + Tx - Tx = Sx + T(0)$. So, $G(T) = G(S) + \{0\} \times T(0)$.

Example 2.5.2 *Let X be denote the vector space $C[a, b]$ of all continuous valued functions defined on some given compact interval $[a, b]$ in $\mathbb{R}$. We define T by*

$$T : X \longrightarrow X$$
$$f \longrightarrow \int f(t)dt,$$

then $T \in LR(X)$ and $T(0)$ is the one dimensional subspace consisting of the constant functions on $[a, b]$. Clearly, for $x \in [a, b]$, the function

$$Af = \int_a^x f(t)dt$$

is a selection of T.

Proposition 2.5.8 [50, Proposition 2.4.2] *Let $T \in LR(X, Y)$. If T is continuous, and $T(0)$ is the kernel of a continuous projection P defined in $R(T)$, then PT is a continuous selection of T.* ◇

Theorem 2.5.6 *Let T and S be two linear relations on a vector space X and let A and B be two selections of T and S, respectively. Then,*
(i) $A + B$ is a selection of $T + S$.
(ii) If $R(T) \subset \mathcal{D}(S)$, then BA is a selection of ST.
(iii) If T be an everywhere defined linear relation on X and A be a selection of T. Then, for all integer $n \in \mathbb{N}$, A^n is a selection of T^n. ◇

Proof. (i) Let A be a selection of T and B be a selection of S, then

$$\mathcal{D}(T + S) = \mathcal{D}(T) \bigcap \mathcal{D}(S) = \mathcal{D}(A) \bigcap \mathcal{D}(B) = \mathcal{D}(A + B).$$

It is easy to prove that

$$T + S = A + T - T + B + S - S = A + B + (T + S) - (T + S).$$

(ii) Since $R(T) \subset \mathcal{D}(S)$, then $\mathcal{D}(ST) = \mathcal{D}(T)$. Since B is a selection of S, it follows that $\mathcal{D}(B) = \mathcal{D}(S)$. Hence, $R(A) \subset R(T) \subset \mathcal{D}(B)$ and then $\mathcal{D}(AB) = \mathcal{D}(ST)$. As B is a selection of S, it follows that

$$BTx \subset STx \quad \text{for all} \quad x \in \mathcal{D}(T). \tag{2.23}$$

As A is a selection of T, it follows that

$$BAx \subset BTx \quad for \ all \quad x \in \mathcal{D}(T). \tag{2.24}$$

Hence, the result follows from (2.23) in combination with (2.24).

(iii) Since A is a selection of T, then A is single valued. Hence, A^n is an operator for all $n \in \mathbb{N}$. It is enough to prove that, for all $x \in X$ and $n \in \mathbb{N}$,

$$A^n x \subset T^n x. \tag{2.25}$$

To do this, proceed by induction on $n \in \mathbb{N}$. The case $n = 1$ is obvious (A is a selection of T). Let $n \geq 2$ and suppose that (2.25) holds true. A is a selection of T, then it is clear that

$$AT^n x \subset T^{n+1} x \text{ for all } x \in X. \tag{2.26}$$

On the other hand, it follows from the induction hypothesis, (2.25), and (2.26), that

$$A^{n+1} x \subset AT^n x. \tag{2.27}$$

Finally, combining (2.25) and (2.27), we get the result. Q.E.D.

2.6 Relatively Boundedness of Linear Relations

Definition 2.6.1 *Let X and Y be two normed spaces and S, T be linear relations from X into Y and from X into Z, respectively. The linear relation S is called relatively bounded with respect to T (or T-bounded) if $\mathcal{D}(S) \supset \mathcal{D}(T)$ and there exist constants a, b for which the inequality*

$$\|Sx\| \leq a\|x\| + b\|Tx\| \tag{2.28}$$

holds for all $x \in \mathcal{D}(T)$. The infimum δ of all b such that the inequality (2.28) holds for some $a \geq 0$ is called the relative bound of S with respect to T (or T-bound of S). ◇

Remark 2.6.1 *Let S, $T \in L\mathcal{R}(X,Y)$. Then,*

(i) *If S is 0-bounded, then S is bounded.*

(ii) The inequality (2.28) is equivalent to

$$\|Sx\|^2 \le a_1^2\|x\|^2 + b_1^2\|Tx\|^2 \ , \quad x \in \mathcal{D}(T),$$

where $a_1 = (a^2 + ab)^{\frac{1}{2}}$, and $b_1 = (b^2 + ab)^{\frac{1}{2}}$.

(iii) S is T-bounded if and only if, S is $(\lambda - T)$-bounded for some $\lambda \in \mathbb{C}$.

(iv) S is T-bounded if and only if, $\mathcal{D}(T) \subset \mathcal{D}(S)$, and SG_T is bounded.

(v) S is T-bounded with T-bound δ if and only if, $Q_S S$ is $Q_T T$-bounded with $Q_T T$-bound δ. Indeed,

$$
\begin{aligned}
\|Q_T Sx\| \ &= \ \mathrm{dist}(\overline{T(0)}, Sx) \\
&\le \ \mathrm{dist}(S(0), Sx) \\
&= \ \|Sx\| \\
&\le \ a\|x\| + b\|Tx\| \\
&= \ a\|x\| + b\|Q_T Tx\|.
\end{aligned}
$$

$\diamond$

Lemma 2.6.1 *Let S, $T \in LR(X,Y)$ satisfy $S(0) \subset T(0)$ and $\mathcal{D}(T) \subset \mathcal{D}(S)$. If S is T-compact, then S is T-bounded.* $\diamond$

Proof. Suppose that S is not T-bounded. Then, assume without loss of generality that for each positive integer n, there exists an $x_n \in \mathcal{D}(T)$ such that $\|x_n\| + \|Tx_n\| = 1$, and $\|Sx_n\| > n$. This implies that $(x_n)_n$, and $(Tx_n)_n$ are bounded, and that S is T-compact. So, for all $y_n \in Sx_n$, we can extract a convergent subsequence of $(y_n)_n$, which is a contradiction.   Q.E.D.

Theorem 2.6.1 *Let A, B, and $S \in LR(X,Y)$ verifying $\overline{S(0)} \subset \overline{B(0)} \subset \overline{A(0)}$ and $\lambda \in \mathbb{C}$. If S is A-bounded with A-bound δ_1 and B is A-bounded with A-bound δ_2 such that $\delta_2 + |\lambda|\delta_1 < 1$. Then, $\|\cdot\|_A$ and $\|\cdot\|_{\lambda S - (A+B)}$ (see Definition 2.5.4) are equivalent.* $\diamond$

Proof. Since S is A-bounded with bound δ_1 and B is A-bounded with bound δ_2, then there exist non-negative constants a, b, a_1 and b_1 such that, for $x \in \mathcal{D}(A)$, $\|Sx\| \le a\|x\| + b\|Ax\|$ and $\|Bx\| \le a_1\|x\| + b_1\|Ax\|$. So, $-\|Bx\| \ge -a_1\|x\| - b_1\|Ax\|$. Thus,

$$\|Ax\| - \|Bx\| \ge -a_1\|x\| + (1 - b_1)\|Ax\|. \tag{2.29}$$

By using Proposition 2.5.7 and (2.29), we get

$$\|Ax + Bx\| \ge -a_1\|x\| + (1 - b_1)\|Ax\|. \tag{2.30}$$

On the other hand,

$$
\begin{aligned}
\|x\|_{\lambda S-(A+B)} &= \|x\| + \|(\lambda S - (A+B))x\| \\
&\geq \|x\| + \|(A+B)x\| - |\lambda|\|Sx\|, \quad (\text{since } \overline{S(0)} \subset \overline{(A+B)(0)}) \\
&\geq \|x\| - a_1\|x\| + (1-b_1)\|Ax\| - |\lambda|\|Sx\|, \quad (\text{as } (2.30)) \\
&\geq \|x\| - a_1\|x\| + (1-b_1)\|Ax\| - |\lambda|a\|x\| - b|\lambda|\|Ax\| \\
&\geq (1 - a_1 - |\lambda|a)\|x\| + (1 - b_1 - |\lambda|b)\|Ax\| \\
&\geq \min\left(1 - a_1 - |\lambda|a, 1 - b_1 - |\lambda|b\right)(\|x\| + \|Ax\|).
\end{aligned}
$$

Therefore,

$$
\|x\|_{\lambda S-(A+B)} \geq \min\left(1 - a_1 - |\lambda|a, 1 - b_1 - |\lambda|b\right)\|x\|_A.
$$

Hence, we obtain

$$
\begin{aligned}
\|x\|_{\lambda S-(A+B)} &= \|x\| + \|(\lambda S - (A+B))x\| \\
&\leq \|x\| + \|Ax\| + \|Bx\| + |\lambda|\|Sx\| \\
&\leq \|x\| + a_1\|x\| + b_1\|Ax\| + \|Ax\| + |\lambda|a\|x\| + b|\lambda|\|Ax\| \\
&\leq (1 + a_1 + |\lambda|a)\|x\| + (1 + b_1 + |\lambda|b)\|Ax\| \\
&\leq \max(1 + a_1 + |\lambda|a, 1 + b_1 + |\lambda|b)(\|x\| + \|Ax\|).
\end{aligned}
$$

Therefore,

$$
\|x\|_{\lambda S-(A+B)} \leq \max(1 + a_1 + |\lambda|a, 1 + b_1 + |\lambda|b)\|x\|_A,
$$

and we deduce that $\|\cdot\|_A$ and $\|\cdot\|_{\lambda S-(A+B)}$ are equivalent. Q.E.D.

Proposition 2.6.1 *If S is T-bounded with T-bound $\delta < 1$ and $S(0) \subset T(0)$, then S is $(T+S)$-bounded with $(T+S)$-bound $\leq \frac{\delta}{1-\delta}$.* ◇

Proof. First of all, it should be mentioned that the linear relation $T + S$ is well-defined as $\mathcal{D}(T + S) = \mathcal{D}(S) \bigcap \mathcal{D}(T) = \mathcal{D}(T)$ with $\mathcal{D}(T+S) \subset \mathcal{D}(S)$. Using the fact that S is T-bounded, it follows that there exist $a > 0, \delta \leq b < 1$, such that for all $x \in \mathcal{D}(T)$,

$$
\begin{aligned}
\|Sx\| &\leq a\|x\| + b\|Tx\| \\
&= a\|x\| + b\|Tx + Sx - Sx\| \quad (\text{as Proposition 2.3.4}) \\
&\leq a\|x\| + b\|Tx + Sx\| + b\|Sx\| \quad (\text{as Proposition 2.5.4}).
\end{aligned}
$$

Since $b < 1$, it follows that

$$
\|Sx\| \leq \left(\frac{a}{1-b}\right)\|x\| + \left(\frac{b}{1-b}\right)\|(T+S)x\|, \quad x \in \mathcal{D}(T). \tag{2.31}
$$

This completes the proof.   Q.E.D.

Proposition 2.6.2 *Let A, B, C and $T \in LR(X,Y)$. Then,*

(i) If A is B-bounded with B-bound δ_1, and B is C-bounded with C-bound δ_2, then A is C-bounded with C-bound $\delta_1\delta_2$.

(ii) If B is T-bounded with T-bound δ_1, and C is T-bounded with T-bound δ_2, then $A = B \pm C$ is T-bounded with T-bound $\delta_1 + \delta_2$.   ◇

Proof. (i) Since A is B-bounded, and B is C-bounded, then there exist a, b, c, $d \geq 0$, such that $\|Ax\| \leq a\|x\| + b\|Bx\|$ for all $x \in \mathcal{D}(B)$, and $\|Bx\| \leq c\|x\| + d\|Cx\|$ for all $x \in \mathcal{D}(C)$. It follows that, for all $x \in \mathcal{D}(C)$, $\|Ax\| \leq (a + bc)\|x\| + bd\|Cx\|$, and $\mathcal{D}(C) \subset \mathcal{D}(A)$.

(ii) Since B is T-bounded, and C is T-bounded, then there exist a, b, c, $d \geq 0$, such that $\|Bx\| \leq a\|x\| + b\|Tx\|$ for all $x \in \mathcal{D}(T)$, and $\|Cx\| \leq c\|x\| + d\|Tx\|$, for all $x \in \mathcal{D}(T)$. It follows that, for all $x \in \mathcal{D}(T)$, $\|Ax\| = \|(B \pm C)x\| \leq \|Bx\| + \|Cx\| \leq (a + c)\|x\| + (b + d)\|Tx\|$, and $\mathcal{D}(T) \subset \mathcal{D}(A)$. This completes the proof.   Q.E.D.

2.7   Closed and Closable Linear Relations

Since every linear relation T has a completion $\widetilde{T} \in LR(\widetilde{X}, \widetilde{Y})$, the question arises as to why non closed linear relations in incomplete spaces need be considered at all. The answer is simple: The sum of two closed linear relations is, in general, not closed or closable, likewise for the composition. Furthermore, continuity for $\widetilde{T}$ does not imply continuity for T, and the inverse of a closable linear relation (or operator) need not be closable.

2.7.1   Closed Linear Relations

Definition 2.7.1 *Let $T \in LR(X,Y)$. The relation T is called closed if its graph $G(T)$ is closed.*   ◇

We denote the class of all closed linear relations from X into Y by $CR(X,Y)$ and as useful we write $CR(X) := CR(X,X)$ and we denote by $BCR(X,Y) = \{T \in CR(X,Y)$ such that T is everywhere defined$\}$ and as useful we write $BCR(X) := BCR(X,X)$.

Proposition 2.7.1 [50, Proposition 2.5.3], [8, Lemma 5.3]

(*i*) *Let* $T \in LR(X,Y)$. *The following properties are equivalent*

    (i_1) T *is closed,*

    (i_2) T^{-1} *is closed, and*

    (i_3) $Q_T T$ *and* $T(0)$ *are closed.*

In particular, $N(T)$ *is closed, if* T *is closed.*

(*ii*) *Assume that* $T \in CR(X,Y)$. *Then,* $R(T)$ *is closed if and only if,* $R(Q_T T)$ *is closed.* ◇

Proposition 2.7.2 *Let* S, $T \in LR(X,Y)$ *such that* $S(0) \subset T(0)$ *and* $\overline{D(T)} \subset D(S)$. *If* T *is closed and* S *is continuous, then* $T + S$ *is closed.* ◇

Proof. <u>First case</u> : We shall first assume that T and S are single valued. Let $(x_n)_n \subset D(T + S) = D(T)$ such that $x_n \to x$ and $(T + S)x_n \to y$. Writing $Tx_n = (T + S)x_n - Sx_n$. Since S is continuous, then $Sx_n \to Sx$. Therefore, $Tx_n \to y - Sx$. Thus, since T is closed, then $Tx = y - Sx$ and $x \in D(T)$. Hence, $y = (S + T)x$ and $x \in D(S + T)$.

<u>Second case</u> : If T and S are linear relations, then since $S(0) \subset T(0)$, we have $(T + S)(0) = T(0) + S(0) = T(0)$. Hence, $(T + S)(0)$ is closed. On the other hand,

$$Q_{T+S}(T + S) = Q_T(T + S) = Q_T T + Q_T S \qquad (2.32)$$

and $\overline{S(0)} \subset \overline{T(0)} = T(0)$. So, by using Lemma 2.1.3, we infer that $Y/\overline{S(0)}\Big/R \equiv Y/T(0)$ and $Q_T = Q_R Q_S$, where $R = T(0)/\overline{S(0)}$. Thus, $Q_T S = Q_R Q_S S$. As S is continuous, then $Q_T S$ is a continuous single valued. Using (2.32), we obtain $Q_{T+S}(T + S)$ is closed. Again, applying Proposition 2.7.1, we infer that $T + S$ is closed. Q.E.D.

Remark 2.7.1 *If the hypotheses of Proposition 2.7.2 is satisfied, then by using Proposition 2.5.4 (iii), we can prove that* $Q_T S$ *is continuous.* ◇

Theorem 2.7.1 *Let* T, $S \in LR(X,Y)$ *such that* T *is closed and* S *is continuous with* $\dim S(0) < +\infty$ *and* $D(S) \supset \overline{D(T)}$. *Then,* $T + S$ *is closed.* ◇

Proof. First, since $D(Q_{T+S}) = Y$ and by using Proposition 2.3.6 $Q_{T+S}(T + S) = Q_{T+S}T + Q_{T+S}S$. Moreover, since T is closed, then by using Proposition 2.7.1, $T(0)$ is closed and $Q_T T$ is closed. Since $\dim S(0) < \infty$, then by using

Theorem 2.1.2, we have

$$\dim \left(\frac{T(0) + S(0)}{T(0)} \right) < \infty,$$

and by using Theorem 2.4.1 (i), we obtain that

$$Q_{T+S}T = Q^{Y/T(0)}_{\frac{T(0)+S(0)}{T(0)}} Q^Y_T T$$

is closed. On the other hand, $Q_{T+S}S$ is a continuous single valued linear operator. In fact, since $S(0) \subset (S+T)(0)$ and $\mathcal{D}(S+T) \subset \mathcal{D}(S)$ and by using Proposition 2.5.4 (iii), we have $\|Q_{T+S}S\| \leq \|S\|$. Finally, since $\mathcal{D}(Q_{T+S}S) = \mathcal{D}(S) \supset \overline{\mathcal{D}(Q_{T+S}T)} = \overline{\mathcal{D}(T)}$, then by using Proposition 2.7.2, we obtain that $Q_{T+S}(T + S) = Q_{T+S}T + Q_{T+S}S$ is closed. It is obvious that $(T + S)(0) = T(0) + S(0)$ is closed. Therefore, by Proposition 2.7.1 (i), $T+S$ is closed. This completes the proof.   Q.E.D.

Theorem 2.7.2 [50, Theorem 3.5.3 and Corollary 3.5.4] (i) *If* $T \in CR(X,Y)$ *and* $\mathcal{D}(T) = X$, *then* T *is bounded.*
(ii) *If* $T \in CR(X,Y)$ *and open, then* $R(T)$ *is closed.*
(iii) *If* $T \in CR(X,Y)$ *and* $R(T)$ *is closed, then* T *is open.*
(iv) *If* T *is continuous,* $\mathcal{D}(T)$ *and* $T(0)$ *are closed, then* T *is closed.*   ◇

Lemma 2.7.1 *Let* $T \in \mathcal{L}(X,Y)$ *and* $S \in CR(Y,Z)$. *Then,* $ST \in CR(X,Z)$.
                                                                                  ◇

Proof. Let $(x_n, y_n) \in G(ST)$ such that (x_n, y_n) converges to (x, y). Then, $y_n \in STx_n = S(Tx_n)$. This implies that $(Tx_n, y_n) \in G(S)$. Since T is an bounded operator, then (Tx_n, y_n) converges to (Tx, y). The fact that S is closed, then $(Tx, y) \in G(S)$. So, $y \in S(Tx) = STx$. Thus, $(x, y) \in G(ST)$. This completes the proof.   Q.E.D.

Remark 2.7.2 *If* X *and* Y *are two complete spaces and* $T \in CR(X,Y)$, *then the space* X_T *is also a Banach space and* $T(0)$ *is complete. Since* G_T *is a bounded operator, then it follows from Lemma 2.3.4 that* $TG_T \in CR(X_T,Y)$.
                                                                                  ◇

Proposition 2.7.3 [50, Proposition 6.5.2] *Let* X *be complete space, and let* $S, T \in CR(X)$ *be bijective. Then,* ST *has the same properties.*   ◇

Proposition 2.7.4 *Let X and Y be two Banach spaces. Then,*
(i) If $T \in LR(X,Y)$ has a continuous selection, $T(0)$ and $\mathcal{D}(T)$ are closed,
then $T \in CR(X,Y)$.
(ii) $T \in CR(X,Y)$, then T has a closed selection.   ◇

Proof. (i) Let S be a continuous selection of T. Then,

$$\|T\| = \|S + T - T\| \le \|S\| + \|T - T\| = \|S\| < \infty.$$

Hence, by using Theorem 2.5.5 (i) we have T is continuous. Since $\mathcal{D}(T)$ and $T(0)$ are closed, then by using Theorem 2.7.2 (iv), we have $T \in CR(X,Y)$.
(ii) If S is selection of T, then

$$G(T) = G(S) + (\{0\} \times T(0)). \tag{2.33}$$

It follows from (2.33) that

$$T_G x = S_G x + T_{G(0)} = S_G x + \{0\} \times T(0) \text{ for all } x \in \mathcal{D}(T),$$

where T_G (respectively, S_G) is the linear relation associated with T (respectively, S). Let P be any bounded projection defined on $G(T)$ with kernel $\{0\} \times T(0)$ such that $R(S_G) \bigcap (\{0\} \times T(0)) = \{0\}$. Then, for all $x \in \mathcal{D}(T)$, we have

$$
\begin{aligned}
PT_G x &= P(S_G x + T_G(0)) \\
&= P(S_G x) + P(T_G(0)) \\
&= PS_G x, \quad \left(\text{since } PT_G(0) = PP^{-1}(0) = 0\right) \\
&= PS_G x + S_G x - S_G x \\
&= S_G x + (P - I)S_G x.
\end{aligned}
$$

Since $P(P - I)S_G x = PS_G x - PS_G x = 0$, then

$$(P - I)S_G x \in R(S_G) \bigcap (\{0\} \times T(0)) = \{0\}.$$

Consequently, $PT_G x = S_G x$. By using Remark 2.7.2, we have $(\mathcal{D}(T), \|\cdot\|_T)$ is complete. Also,

$$\|x\|_S = \|S_G x\| = \|PT_G x\| \le \|P\|\|T_G x\| = \|P\|\|x\|_T. \tag{2.34}$$

We know that $\mathcal{D}(T) = \mathcal{D}(S)$, then $(\mathcal{D}(S), \|\cdot\|_T)$ is complete. This, together with (2.34) that $(\mathcal{D}(S), \|\cdot\|_S)$ is complete. Since S is single valued, then S is closed. This completes the proof.   Q.E.D.

2.7.2   Closable Linear Relations

The closure of a linear relation T, denoted $\overline{T}$, be defined in terms of their corresponding graphs in $X \times Y$ as follows

$$G(\overline{T}) = \overline{G(T)} \subset X \times Y.$$

Definition 2.7.2 *The linear relation T is said to be closable if $\overline{T}$ is an extension of T i.e., $Tx = \overline{T}x$ for all $x \in \mathcal{D}(T)$.* ◇

Remark 2.7.3 *It is easy to construct natural examples of non closable linear relations. Here is an example suggested by V. V. Shevchik (a discontinuous linear functional on a normed space). Let X be the space of continuous functions $C[0,1]$ with the norm*

$$\|x\| = \int_0^1 |x(t)| dt.$$

Fix $t \in [0,1]$ and define $f_t(x) = x(t)$. Then, $f_t(\cdot)$ is an unbounded linear functional on X. ◇

Proposition 2.7.5 [50, Proposition 2.5.1] *Let $T \in LR(X,Y)$, then $Q_T \overline{T} = \overline{Q_T T}$.* ◇

Lemma 2.7.2 [50, Exercise 5.19] *Let $T \in LR(X,Y)$. Then, $\overline{T(0)} \subset \overline{T}(0)$.* ◇

Lemma 2.7.3 [50, Definition 2.5.7] *Let $T \in LR(X,Y)$. Then, T is closable if and only if, $T(0) = \overline{T}(0)$ if and only if, $Q_T T$ is closable and $T(0)$ is closed. In particular, if T is continuous and $T(0)$ is closed, then T is closable.* ◇

Remark 2.7.4 *The condition $T(0)$ closed, in Lemma 2.7.3, is necessary. Indeed, let X, Y be two Banach spaces and F be a finite dimensional vector subspace of X. Let $T \in LR(X,Y)$ such that $T(0) \neq \overline{T(0)}$. Assume that $T_0 = T|_F$. On the one hand, since $\mathcal{D}(T_0) = \mathcal{D}(T) \cap F \subset F$, then we have $\dim(\mathcal{D}(T_0)) < +\infty$. Hence, by using Theorem 2.5.5 (ii) T_0 is a continuous linear relation. On the other hand, since $Tx = T_0 x$ for all $x \in \mathcal{D}(T) \cap F$ and $0 \in \mathcal{D}(T) \cap F$, then we deduce that $T_0(0) = T(0)$. Hence, $\overline{T_0(0)} = \overline{T(0)} \neq T(0)$. This implies that $\overline{T_0(0)} \neq T_0(0)$. Therefore, T_0 is not closable.* ◇

Proposition 2.7.6 [50, Definition 2.5.2(3)] *Let $T \in LR(X,Y)$. Then, $(\overline{T})^{-1} = \overline{T^{-1}}$.* ◇

Lemma 2.7.4 *Let E be a subspace of X. Then,*

$$\overline{I_E} = I_{\overline{E}}.$$ ◇

Proof. It suffices to show that $G(\overline{I_E}) = G(I_{\overline{E}})$. Let $(x, y) \in G(\overline{I_E})$, then there exists a sequence $(x_n, x_n) \in G(I_E)$ such that $(x_n, x_n) \to (x, y)$. Hence, $x = y \in \overline{E}$. Therefore, $(x, y) \in G(I_{\overline{E}})$. Conversely, if $(x, x) \in G(I_{\overline{E}})$, i.e., $x \in \overline{E}$, then there exists a sequence $(x_n)_n \in E$ converging to x. Hence, the sequence $(x_n, x_n) \in G(I_E)$ converges to (x, x). Therefore, $(x, x) \in G(\overline{I_E})$. Q.E.D.

Lemma 2.7.5 [67, Theorem 4.1.1 p. 190] *Let T and A be operators from X into Y, and let A be T-bounded with T-bound smaller than 1. Then, $S = T + A$ is closable if and only if, T is closable. In this case, the closure of T and S have the same domain. In particular S is closed if and only if, T is also closed.* ◇

2.8 Adjoint of Linear Relations

Let X be a normed vector space. We will denote by X^* the norm dual of X, i.e., the space of all continuous functionals x' defined on X, with norm

$$\|x'\| = \inf \{\lambda \in \mathbb{K} \text{ such that } |x'(x)| \leq \lambda \|x\| \text{ for all } x \in X\}.$$

A consequence of Hann-Banach theorem is the following:

Theorem 2.8.1 *Let X be a normed vector space. Then, for all $x \in X$ and $x \neq 0$, there exists $x' \in X^*$ (the adjoint of X) such that $\|x'\| = 1$ and $x'(x) = \|x\|$.* ◇

If M and N are subspaces of X, then

$$M^\perp = \{x' \in X^* \text{ such that } x'(x) = 0 \text{ for all } x \in M\}$$

and

$$N^\top = \{x \in X \text{ such that } x'(x) = 0 \text{ for all } x' \in N\},$$

where X^* is the dual of X. The adjoint T^* of T is defined by

$$G(T^*) = G(-T^{-1})^\perp \subset X^* \times Y^*,$$

where
$$\langle (y, x), (y', x') \rangle = \langle x, x' \rangle + \langle y, y' \rangle = x'(x) + y'(y). \qquad (2.35)$$

Thus, means that $(y', x') \in G(T^*)$ if and only if, $y'(y) - x'(x) = 0$ for all $(x, y) \in G(T)$. By using (2.35), we have $y'(y) = x'(x)$ for all $y \in Tx$, $x \in \mathcal{D}(T)$. Hence, $x' \in T^* y'$ if and only if, $y'(Tx) = x'(x)$ for all $x \in \mathcal{D}(T)$. So, we can characterize the adjoint as follows

$$G(T^*) = \{(y', x') \in Y^* \times X^* \text{ such that } x' \text{ is an extension of } y'T \}.$$

Remark 2.8.1 *Note that this definition coincides with the classical one when T is a densely defined operator and it allows one to define the conjugate of a non-densely defined operator.* ◇

In the following we list some basis properties of the adjoint of a linear relation.

Proposition 2.8.1 [50, Chapter 3] *Let $T \in LR(X, Y)$. Then,*

(i) *T^* is a closed linear relation in $LR(Y^*, X^*)$.*
(ii) *$G(T^* + S^*) \subset G((T + S)^*)$.*
(iii) *If S is continuous with $\mathcal{D}(T) \subset \mathcal{D}(S)$, then $(T + S)^* = T^* + S^*$.*
(iv) *$N(T^*) = R(T)^{\perp}$ and $T^*(0) = \mathcal{D}(T)^{\perp}$.*
(v) *If T is closed, then $R(T)$ is closed if and only if, $R(T^*)$ is closed if and only if, T is open if and only if, T^* is open.*
(vi) *If T is continuous, then $\|T\| = \|T^*\|$.*
(vii) *$\mathcal{D}(T^*)^{\top\perp} = \overline{\mathcal{D}(T^*)}$.* ◇

Furthermore, it follows from Proposition 2.8.1 (iv) that the adjoint of a linear operator is single valued if and only if, it is densely defined.

Lemma 2.8.1 *Let $T, S \in LR(X, Y)$, then*

$$(Q_{T+S}S)^* = S^* J^{Y^*}_{\overline{(T(0)+S(0))}^{\perp}}. \qquad ◇$$

Proof. First, we have

$$
\begin{aligned}
(Q^Y_{T+S})^* &= \left(Q^Y_{\overline{T(0)+S(0)}} \right)^* \\
&= \left(J^{Y^*}_{\overline{(T(0)+S(0))}^{\perp}} \right).
\end{aligned}
$$

Moreover, by using Proposition 2.8.1, we have

$$
\begin{aligned}
\mathcal{D}((Q^Y_{T+S})^*) &= \overline{T(0) + S(0)}^{\perp} \\
&= \left(\frac{Y}{\overline{(T(0) + S(0))}} \right)^*.
\end{aligned}
$$

Since $S \in LR(X, Y)$ and $Q_{T+S} \in LR(Y, Y/(\overline{T(0) + S(0)}))$ with $\mathcal{D}(Q_{T+S}) = Y$, then

$$(Q_{T+S}S)^* = S^* J^{Y^*}_{\overline{(T(0)+S(0))}^\perp}. \qquad \text{Q.E.D.}$$

Lemma 2.8.2 *Let H be a Hilbert space with inner product $\langle \cdot, \cdot \rangle$ and $T \in LR(H)$. If T is bounded, then for all $n \in \mathbb{N}^*$,*

$$(T^n)^* = (T^*)^n. \qquad \diamondsuit$$

Proof. Since $\mathcal{D}(T) = H$, then $\mathcal{D}(T^2) = H$. Hence $(T^2)^*$ is a single valued linear operator. Let $y \in H$, then

$$\langle (T^2)^* x, y \rangle = \langle x, t \rangle \quad \text{for all } t \in T^2(y).$$

Let $t \in T^2(y)$. Then, there exists z such that $(y, z) \in G(T)$ and $(z, t) \in G(T)$. Which implies that

$$\langle x, t \rangle = \langle T^* x, z \rangle = \langle T^*(T^* x), y \rangle.$$

This follows that $(T^2)^* = (T^*)^2$. Proceeding by induction, we can prove the result for each $n \geq 3$. \qquad Q.E.D.

Lemma 2.8.3 *Let $T \in LR(X)$ and S be a continuous linear relation such that $S(0) \subset \overline{T(0)}$ and $\mathcal{D}(S) \supset \mathcal{D}(T)$. Then, S^* is continuous, $S^*(0) \subset T^*(0)$, and $\mathcal{D}(S^*) \supset \mathcal{D}(T^*)$.* \qquad $\diamondsuit$

Proof. Since S is continuous, then by using Proposition 2.8.1 (vi), we have $\|S^*\| = \|S\|$. Hence, by using Theorem 2.5.5 (i), we have S^* is continuous. Also,

$$\mathcal{D}(S^*) = S(0)^\perp \supset \overline{T(0)}^\perp \supset \overline{T}(0)^\perp = \mathcal{D}(T^*)^{\top\perp} = \overline{\mathcal{D}(T^*)},$$

and using Proposition 2.8.1 (iv), $S^*(0) = \mathcal{D}(S)^\perp \subset \mathcal{D}(T)^\perp = T^*(0)$. Q.E.D.

2.9 Minimum Modulus of Linear Relations

Definition 2.9.1 *Let $T \in LR(X, Y)$. The minimum modulus of a linear relation T is the quantity $\gamma(T)$*

$$\gamma(T) = \sup \{\lambda \text{ such that } \|Tx\| \geq \lambda \operatorname{dist}(x, N(T)) \text{ for } x \in \mathcal{D}(T)\}. \quad \diamondsuit$$

Proposition 2.9.1   [50, Proposition 2.2.2]   *Let $T \in LR(X, Y)$. The minimum modulus of T is the quantity*

$$\gamma(T) = \begin{cases} \infty & \text{if } \mathcal{D}(T) \subset \overline{N(T)} \\ \inf \left\{ \dfrac{\|Tx\|}{\text{dist}(x, N(T))} \text{ such that } x \in \mathcal{D}(T) \backslash \overline{N(T)} \right\} & \text{otherwise.} \end{cases} \quad \diamond$$

Proposition 2.9.2   [50, Corollary 2.3.9]   *If T is open and $N(T)$ is closed, then $N(T) = N(Q_T T)$ and $\gamma(T) = \gamma(Q_T T)$.* $\diamond$

Theorem 2.9.1   [50, Theorem 2.3.11]   *Let $T \in LR(X, Y)$ and $S \in LR(Y, Z)$. Then,*

(i) $\gamma(ST) \geq \gamma(S_{|R(T)})\gamma(T)$ ($\infty.0$ excluded), with $\gamma(ST) = \infty$ whenever $\gamma(T) = \infty$ (even $\gamma(S_{|R(T)}) = 0$).
(ii) If $S^{-1}(0) \subset R(T)$, then $\gamma(ST) \geq \gamma(S)\gamma(T)$. $\diamond$

Theorem 2.9.2 *Let $S, T \in LR(X, Y)$ such that $\|T\| \neq 0$, $\mathcal{D}(S) = \mathcal{D}(T)$, and $N(S) \subset N(T)$. Then,*

$$\gamma(SG_T) \leq \frac{\gamma(T)\|S\|}{(1 + \gamma(T))\|T\|}.$$

Moreover, if $N(S) = N(T)$, then

$$\gamma(SG_T) = \frac{\gamma(T)\|S\|}{(1 + \gamma(T))\|T\|}. \quad \diamond$$

Proof.   We have

$$\begin{aligned} N(SG_T) &= \{x \in X_T \text{ such that } SG_T x = S(0)\} \\ &= \{x \in \mathcal{D}(T) \text{ such that } Sx = S(0)\} \\ &= \{x \in \mathcal{D}(S) \text{ such that } Sx = S(0)\} \text{ as } \mathcal{D}(S) = \mathcal{D}(T) \\ &= N(S). \end{aligned} \quad (2.36)$$

Since

$$\begin{aligned} \text{dist}(x, N(SG_T)) &= \inf_{z \in N(SG_T)} \|x - z\|_T \\ &= \inf_{z \in N(S)} \|x - z\| + \|T(x - z)\| \text{ see } (2.36) \\ &= \inf_{z \in N(S)} \|x - z\| + \|Tx\| \text{ as } N(S) \subset N(T), \end{aligned}$$

then

$$\text{dist}(x, N(SG_T)) = \text{dist}(x, N(S)) + \|Tx\|.$$

Hence,

$$
\begin{aligned}
\gamma(SG_T) &= \inf\left\{\frac{\|SG_Tx\|}{\operatorname{dist}(x,N(SG_T))} \text{ such that } x \in X_T\backslash N(SG_T)\right\} \\
&= \inf\left\{\frac{\|Sx\|}{\operatorname{dist}(x,N(S))+\|Tx\|} \text{ such that } x \in \mathcal{D}(T)\backslash N(T)\right\} \\
&\leq \inf\left\{\frac{\|Sx\|}{\operatorname{dist}(x,N(T))+\|Tx\|} \text{ such that } x \in \mathcal{D}(T)\backslash N(T)\right\} \\
&= \left(\sup\left\{\frac{\operatorname{dist}(x,N(T))+\|Tx\|}{\|Sx\|} \text{ such that } x \in \mathcal{D}(T)\backslash N(T)\right\}\right)^{-1} \\
&= \left(\sup\left\{\frac{\frac{\operatorname{dist}(x,N(T))}{\|Tx\|}+1}{\frac{\|Sx\|}{\|Tx\|}} \text{ such that } x \in \mathcal{D}(T)\backslash N(T)\right\}\right)^{-1}.
\end{aligned}
$$

So,

$$
\begin{aligned}
\gamma(SG_T) &\leq \left(\frac{(\gamma(T)^{-1}+1)\|T\|}{\|S\|}\right)^{-1} \\
&= \left(\frac{(\gamma(T)+1)\|T\|}{\gamma(T)\|S\|}\right)^{-1} \\
&= \frac{\gamma(T)\|S\|}{(1+\gamma(T))\|T\|}.
\end{aligned}
$$

If $N(T) = N(S)$, we have the equalities. This completes the proof.   Q.E.D.

Corollary 2.9.1 *Let $T \in LR(X,Y)$, then $\gamma(TG_T) = \dfrac{\gamma(T)}{(1+\gamma(T))}$.*   ◇

Proposition 2.9.3 [50, Proposition 2.3.2] *Let $T \in LR(X,Y)$. Then,*

(i) T is open if and only if, $\gamma(T) > 0$.
(ii) T is open if and only if, $R(T^*) = N(T)^{\perp}$.
(iii) If T is open, then $\gamma(T) = \gamma(T^*)$.   ◇

Proposition 2.9.4 [50, Proposition 2.5.17] *Let $S \in CR(X,Y)$ and $T \in LR(Y,Z)$ have a closed range and satisfy $\alpha(T) < \infty$ and $\gamma(T) > 0$, then $TS \in CR(X,Z)$.*   ◇

Corollary 2.9.2 *If $S \in LR(X,Y)$ and $T \in LR(Y,Z)$ are closed with $\alpha(T) < \infty$ and $R(T)$ is closed, then $TS \in CR(X,Z)$.*   ◇

Proof. By using Theorem 2.5.4 *(ii)* and Proposition 2.9.3 *(i)*, we have T is open. Thus, the result follows immediately from Proposition 2.9.4.   Q.E.D.

Lemma 2.9.1 [50, Theorem 2.2.5] *Let $T \in LR(X,Y)$, then $\gamma(T) = \|T^{-1}\|^{-1}$.*   ◇

Theorem 2.9.3 [50, Theorem 3.7.4] *Let* $\gamma(T) > 0$ *and* S *satisfy* $S(0) \subset \overline{T(0)}$, $\mathcal{D}(S) \supset \mathcal{D}(T)$ *and* $\|S\| < \gamma(T)$. *Then,*

(i) $\alpha(T + S) \leq \alpha(T)$.
(ii) $\overline{\beta}(T + S) \leq \overline{\beta}(T)$. $\Diamond$

Lemma 2.9.2 *Let* T, $S \in L\mathcal{R}(X)$ *such that* T *is injective, open,* $\overline{S(0)} \subset \overline{T(0)}$ *and* $\mathcal{D}(T) \subset \mathcal{D}(S)$. *Then,* $\gamma(T - S) \geq \gamma(T) - \|S\|$. *In addition, if* $\|S\| < \gamma(T)$, *then* $T - S$ *is open and injective.* $\Diamond$

Proof.   Let $x \in \mathcal{D}(T - S)$, then by using Proposition 2.5.7, we have

$$
\begin{aligned}
\|(T - S)x\| &\geq & \|Tx\| - \|Sx\| \\
&\geq & \gamma(T)\|x\| - \|S\|\|x\| \\
&\geq & (\gamma(T) - \|S\|)\|x\|.
\end{aligned}
$$

Therefore, $\gamma(T - S) \geq \gamma(T) - \|S\|$. Since $\|S\| < \gamma(T)$, then $\gamma(T - S) > 0$. Hence, by Proposition 2.9.3 (i), we have $T - S$ is open. In view of Theorem 2.9.3 (i), we have $T - S$ is injective.   Q.E.D.

Lemma 2.9.3 [50, Corollary 3.7.7] *Let* $T \in L\mathcal{R}(X)$ *be open and injective with dense range. Then, for any relation* S *such that* $S(0) \subset \overline{T(0)}$, $\mathcal{D}(S) \supset \mathcal{D}(T)$ *and* $\|S\| < \gamma(T)$, *we have* $T + S$ *is open injective with dense range.* $\Diamond$

Remark 2.9.1 *Let* $T \in L\mathcal{R}(X)$ *be open and injective with dense range. Then, for any relation* S *such that* $S(0) \subset \overline{T(0)}$, $\mathcal{D}(S) \supset \mathcal{D}(T)$ *and* $\|S\| < \gamma(T)$, *and by using Lemmas 2.9.2 and 2.9.3, we have* $T \pm S$ *is open and injective .* $\Diamond$

Proposition 2.9.5 [50, Proposition 2.6.3] *Let* X, Y *be two normed vector spaces and let* $T \in L\mathcal{R}(X,Y)$ *such that* $N(T)$ *is topologically complemented in* $\mathcal{D}(T)$ *(or in* X*). Then,*

$$0 \leq \|P\|^{-1}\gamma(T) \leq \gamma(T_{|_{R(P)}}) \leq \gamma(T),$$

where P *is any continuous projection defined on* $\mathcal{D}(T)$ *(or on* X*) with kernel* $N(T)$. $\Diamond$

Remark 2.9.2 *Let* X, Y *be two normed vector spaces and let* $T \in L\mathcal{R}(X,Y)$ *such that* $N(T)$ *is topologically complemented in* $\mathcal{D}(T)$ *(or in* X*). If* T *is open, then*

$$0 < \|P\|^{-1} \leq 1,$$

where P *is any continuous projection defined on* $\mathcal{D}(T)$ *(or on* X*) with kernel* $N(T)$. $\Diamond$

Proposition 2.9.6 *Let X be a Banach space and let T, $S \in L\mathcal{R}(X)$ such that $\gamma(T) > 0$, $S(0) \subset T(0)$ and $\alpha(T) < \infty$. Then,*

$$\gamma(T - S) \geq \|P\|^{-1}\gamma(T) - \|S\|,$$

where P is any continuous projection defined on $\mathcal{D}(T)$ (or on X) with kernel $N(T)$. ◇

Proof. Since $\alpha(T) < \infty$, then $N(T)$ is topologically complemented in $\mathcal{D}(T)$. Using the fact that $\gamma(T) > 0$ and Proposition 2.9.5, we obtain

$$0 < \|P\|^{-1}\gamma(T) \leq \gamma(T_{|R(P)}) \leq \gamma(T). \tag{2.37}$$

Let $\widehat{T} : R(P) \longrightarrow R(T)$ be the bijection associated with T. It follows from (2.37) that

$$0 < \|P\|^{-1}\gamma(T) \leq \gamma(\widehat{T}) \leq \gamma(T). \tag{2.38}$$

Consequently, $\gamma(\widehat{T}) > 0$ which implies that $\widehat{T}$ is open. Using the fact that $\widehat{T}$ is injective, $\widehat{T}(0) = T(0) \supset S(0)$ and also Lemma 2.9.2, we infer that

$$\gamma(\widehat{T} - S) \geq \gamma(\widehat{T}) - \|S\|. \tag{2.39}$$

By virtue of (2.38) and (2.39), we deduce that

$$\gamma(\widehat{T} - S) \geq \|P\|^{-1}\gamma(T) - \|S\|. \tag{2.40}$$

Now, we propose to show that $\gamma(T - S) \geq \gamma(\widehat{T} - S)$. Let $x \in N(\widehat{T} - S)$. This implies that

$$\widehat{T}x - Sx = \widehat{T}(0) - S(0) = T(0) - S(0). \tag{2.41}$$

Since $x \in N(\widehat{T} - S) \subset \mathcal{D}(\widehat{T} - S) = \mathcal{D}(\widehat{T}) \bigcap \mathcal{D}(S) \subset R(P)$, then $\widehat{T}x = Tx$. It follows from (2.41) that

$$Tx - Sx = T(0) - S(0).$$

Therefore, we deduce that $N(\widehat{T} - S) \subset N(T - S)$. This leads to

$$\text{dist}(x, N(T - S)) \leq \text{dist}(x, N(\widehat{T} - S)). \tag{2.42}$$

For $x \in \mathcal{D}(\widehat{T} - S)$, by applying (2.42), we have

$$\begin{aligned}
\gamma(\widehat{T} - S)\, \text{dist}(x, N(T - S)) &\leq \gamma(\widehat{T} - S)\, \text{dist}(x, N(\widehat{T} - S)) \\
&\leq \|(\widehat{T} - S)x\| \\
&\leq \|(T - S)x\|.
\end{aligned}$$

This leads to

$$\gamma(T - S) \geq \gamma(\widehat{T} - S). \tag{2.43}$$

Hence, by (2.40) and (2.43), we conclude that

$$\gamma(T - S) \geq \|P\|^{-1}\gamma(T) - \|S\|. \qquad \text{Q.E.D.}$$

2.10   Quantities for Linear Relations

2.10.1   A Formula for Gap Between Multivalued Linear Operators

The gap between two linear subspaces M and N of a normed space X is defined by the following formula

$$\delta(M, N) = \sup_{x \in M, \ \|x\|=1} \text{dist}(x, N), \tag{2.44}$$

in the case where $M \neq \{0\}$. Otherwise we define $\delta(\{0\}, N) = 0$ for any subspace N. Moreover, $\delta(M, \{0\}) = 1$ if $M \neq \{0\}$, as shown from (2.44). We can also define $\widehat{\delta}(M, N) = \max \{\delta(M, N), \delta(N, M)\}$. Sometimes, the latter is called the symmetric or maximal gap between M and N in order to distinguish it from the former. The gap $\delta(M, N)$ can be characterized as the smallest number δ such that $\text{dist}(x, N) \leq \delta\|x\|$, for all $x \in M$. The notion of a gap between linear subspaces and linear operators was introduced by M. G. Krein and M. A. Krasnoselski in Ref. [69] in the 1940s. Among the works in this direction we can state, for example (see Refs. [52, 67]).

Remark 2.10.1 *Let us assume that M and N be two vectorial subspaces of a Banach space X. Then,*

(i) Keeping in mind that $\text{dist}(cx, N) = |c|\text{dist}(x, N)$ *for any non-zero element c in $\mathbb{R}$. Then, we infer that*

$$\sup_{\substack{x \in M \\ \|x\| \leq |c|}} \text{dist}(x, N) = |c|\delta(M, N). \tag{2.45}$$

(ii) $\delta(M, N) = \delta(\overline{M}, \overline{N})$ and $\widehat{\delta}(M, N) = \widehat{\delta}(\overline{M}, \overline{N})$.
(iii) $\delta(M, N) = 0$ if and only if, $\overline{M} \subset \overline{N}$.
(iv) $\widehat{\delta}(M, N) = 0$ if and only if, $\overline{M} = \overline{N}$.
(v) $0 \leq \delta(M, N) \leq 1$ and $0 \leq \widehat{\delta}(M, N) \leq 1$. ◇

Definition 2.10.1 *Let T, $S \in C\mathcal{R}(X,Y)$. The gap between T and S is defined by*

$$\delta(T,S) = \begin{cases} \sup_{\substack{\varphi \in G(T) \\ \|\varphi\|=1}} \ \text{dist}(\varphi, G(S)) & \text{if } G(T) \neq \{0\} \\ 0 & \text{otherwise,} \end{cases}$$

and

$$\widehat{\delta}(T,S) = \max\left\{\delta(T,S), \delta(S,T)\right\}. \qquad \diamond$$

Theorem 2.10.1 *Let T, $S \in C\mathcal{R}(X,Y)$, then $\delta(T,S) = \delta(T^{-1}, S^{-1})$ and $\widehat{\delta}(T,S) = \widehat{\delta}(T^{-1}, S^{-1})$.* $\qquad \diamond$

Proof. Fix any $\varphi = (x,y) \in G(T)$ with $\|\varphi\| = 1$, then $\varphi^{-1} = (y,x) \in G(T^{-1})$ and $\|\varphi^{-1}\| = \|\varphi\| = 1$. Hence,

$$\begin{aligned} \delta(T,S) &= \sup_{\substack{\varphi \in G(T) \\ \|\varphi\|=1}} \ \text{dist}(\varphi, G(S)) \\ &= \sup_{\substack{\varphi^{-1} \in G(T^{-1}) \\ \|\varphi^{-1}\|=1}} \text{dist}(\varphi, G(S)). \end{aligned}$$

Let $\psi = (u,v) \in G(S)$, then $\psi^{-1} = (v,u) \in G(S^{-1})$. Therefore,

$$\begin{aligned} \|\varphi - \psi\|_2^2 &= \|(x,y) - (u,v)\|^2 \\ &= \|x - u\|^2 + \|y - v\|^2 \\ &= \|(y,x) - (v,u)\|^2 \\ &= \|\varphi^{-1} - \psi^{-1}\|_2^2. \end{aligned}$$

Which implies that $\text{dist}(\varphi, G(S)) = \text{dist}(\varphi^{-1}, G(S^{-1}))$. As a result,

$$\delta(T,S) = \delta(T^{-1}, S^{-1}).$$

Similarly, we can prove $\delta(S,T) = \delta(S^{-1}, T^{-1})$. $\qquad$ Q.E.D.

Remark 2.10.2 *In the previous theorem (Theorem 2.10.1), we notice that the gap is invariant with respect to inversion.* $\qquad \diamond$

2.10.2 Measures of Noncompactness

Definition 2.10.2 *The diameter of a set B is the number*

$$\text{diam}\,(B) := \begin{cases} \sup\left\{\text{dist}(x,y) \text{ such that } x \in B \text{ and } y \in B\right\} & \text{if } B \neq \emptyset \\ 0 & \text{otherwise.} \end{cases}$$

Remark 2.10.3 (i) $(B) = 0$ *if and only if, B is an empty set or consists of exactly one point.*

(ii) *The following properties are satisfied in complete space and are an immediate consequence of the Definition 2.10.2*

(ii_1) *If $B_1 \subset B_2$, then $(B_1) \leq (B_2)$.*

(ii_2) $(\overline{B}) = (B)$, *where $\overline{B}$ is the closure of B.*

(ii_3) *Cantor's intersection theorem: If $(B_n)_n$ is a decreasing sequence of nonempty, closed and bounded subsets of X and $\lim_{n \to +\infty}(B_n) = 0$, then the intersection B_∞ of all B_n is nonempty and consists of exactly one point. Moreover, if X is a Banach space, then*

(iv) $(\lambda B) = |\lambda|(B)$ *for any real number λ.*

(v) $(x + B) = (B)$ *for any $x \in X$.*

(vi) $(B_1 + B_2) \leq (B_1) + (B_2)$.

(vii) $(\mathrm{co}\,(B)) = (B)$, *where $\mathrm{co}\,(B)$ is the convex hull of B.*   ◇

Definition 2.10.3 *Let D be a bounded subset of X. The Kuratowski measure of noncompactness is defined by $\delta(D) = \inf\{d > 0 : D$ can be covered by a finite number of sets of diameter $\leq d\}$.*   ◇

It is obvious that $0 \leq \delta(B) \leq (B) < +\infty$ for every nonempty bounded subset B of X. In what follows, we use the following known properties of $\delta(\cdot)$.

Proposition 2.10.1 [83, Proposition 1] *For arbitrary bounded sets D, Q, $D_n \subset X$ and $\lambda \in \mathbb{C}$, we have*

(i) $\delta(D) = 0$ *if and only if, $\overline{D}$ is compact, where $\overline{D}$ is the closure of D.*

(ii) $\delta(\overline{D}) = \delta(D)$, $\delta(\lambda D) = |\lambda|\,\delta(D)$ *and if $D \subset Q$, then $\delta(D) \leq \delta(Q)$.*

(iii) *If $D_n = \overline{D_n}, D_{n+1} \subset D_n$, and $\lim_{n \to +\infty} \delta(D_n) = 0$, then*

$$D_\infty := \bigcap_{n=1}^{+\infty} D_n \neq \emptyset$$

and

$$\delta(D_\infty) = 0.$$

(iv) $\delta(D \bigcup Q) = \max\{\delta(D), \delta(Q)\}$, $\delta(D + Q) \leq \delta(D) + \delta(Q)$.

(v) $\delta(D) = \delta(\overline{\mathrm{co}}\,(D))$, *where $\overline{\mathrm{co}}\,(D)$ is the convex closure of D.*

(vi) $\delta(N_\varepsilon(D)) \leq \delta(D) + 2\varepsilon$, *where $N_\varepsilon(D) = \{x \in X : \mathrm{dist}(x, D) < \varepsilon\}$.*   ◇

Definition 2.10.4 *Let $T \in LR(X)$ and δ_1, δ_2 be two Kuratowski measure of noncompactness. Then,*

(i) T is said to be D-condensing if, for any bounded subset B of $\mathcal{D}(T)$, $Q_T T(B)$ is a bounded subset of $Y/\overline{T(0)}$ and

$$\delta_2(Q_T T B) < \delta_1(Q_D B)$$

whenever $\delta_1(Q_D B) > 0$.

(ii) If $D = \{0\}$, then T is said to be $\{0\}$-condensing linear relation or simply condensing. ◇

Definition 2.10.5 *Let X and Y be two normed spaces, and $\mathcal{I}(X)$, $\mathcal{C}(X)$, and $\mathcal{P}(X)$ denote, respectively, the infinite dimensional, finite codimensional, and closed finite codimensional subspaces of a normed linear space X. We define the quantities (also called measures of compactness) $\Gamma(T)$, $\Gamma_0(T)$, $\overline{\Gamma}_0(T)$, and $\widetilde{\Delta}(T)$ as follows:*

First case : *If* $\dim(\mathcal{D}(T)) < \infty$ *and* $\dim(Y) < \infty$, *then all quantities are zero. Thus,*

$$\Gamma(T) = \widetilde{\Delta}(T) = \Gamma_0(T) = \overline{\Gamma}_0(T) = 0.$$

Second case : *If* $\dim(\mathcal{D}(T)) = \infty$ *and* $\dim(Y) = \infty$, *then*

$$\Gamma(T) = \inf_{M \in \mathcal{I}(\mathcal{D}(T))} \|T_{|M}\|$$

$$\Gamma_0(T) = \inf_{M \in \mathcal{C}(\mathcal{D}(T))} \|T_{|M}\|$$

$$\overline{\Gamma}_0(T) = \inf_{M \in \mathcal{P}(\mathcal{D}(T))} \|T_{|M}\|$$

$$\widetilde{\Delta}(T) = \sup_{M \in \mathcal{I}(\mathcal{D}(T))} \Gamma(T_{|M}).$$

◇

The following inequalities hold

$$\Gamma(T) \leq \widetilde{\Delta}(T) \leq \Gamma_0(T) \leq \overline{\Gamma}_0(T). \tag{2.46}$$

Proposition 2.10.2 (i) [50, Exercise 4.1.5] *If T is continuous, then*

$$\overline{\Gamma}_0(T) = \Gamma_0(T).$$

(ii) [50, Proposition 4.3.4] *Let $T \in LR(X, Y)$. For each $f \in \{\widetilde{\Delta}, \Gamma, \Gamma_0\}$, we have*

$$f(T G_T) = \frac{f(T)}{1 + f(T)},$$

where $\frac{\infty}{\infty} := 1$. ◇

Proposition 2.10.3 *Let $S, T \in LR(X,Y)$ such that $\mathcal{D}(S) = \mathcal{D}(T)$. For each $f \in \{\widetilde{\Delta}, \Gamma, \Gamma_0\}$, we have*

$$\frac{f(S)}{1 + f(T)} \leq f(SG_T),$$

where $\frac{\infty}{\infty} := 1$.

Proof. We may suppose that $\dim \mathcal{D}(T) = \infty$. We have

$$
\begin{aligned}
\Gamma(SG_T) &= \inf_{M \in \mathcal{I}(\mathcal{D}(SG_T))} \|SG_T|_M\| \\
&= \inf_{M \in \mathcal{I}(X_T)} \|SG_T|_M\| \quad \text{as } \mathcal{D}(T) = \mathcal{D}(S).
\end{aligned}
$$

Hence,

$$
\begin{aligned}
\Gamma(SG_T) &= \inf_{M \in \mathcal{I}(X_T)} \sup_{m \in M} \frac{\|SG_T m\|}{\|m\|_T} \\
&= \inf_{M \in \mathcal{I}(X_T)} \sup_{m \in M} \frac{\|SG_T m\|}{\|m\| + \|Tm\|} \\
&= \inf_{M \in \mathcal{I}(X_T)} \sup_{m \in M} \frac{\|SG_T m\|}{\|G_T m\| + \|TG_T m\|} \\
&= \inf_{M \in \mathcal{I}(X_T)} \sup_{m \in M} \left(\frac{\|G_T m\|}{\|SG_T m\|} + \frac{\|TG_T m\|}{\|SG_T m\|} \right)^{-1}.
\end{aligned}
$$

Since $\mathcal{D}(T) = \mathcal{D}(S)$, then $G_T m = G_S m$ for all $m \in M$. This implies that $\|G_T m\| = \|G_S m\|$. Thus,

$$
\begin{aligned}
\Gamma(SG_T) &= \inf_{M \in \mathcal{I}(X_T)} \sup_{m \in M} \left(\frac{\|G_T m\|}{\|SG_T m\|} + \left[\frac{\|TG_T m\|}{\|SG_T m\|} \times \frac{\|G_S m\|}{\|G_T m\|} \right] \right)^{-1} \\
&= \inf_{M \in \mathcal{I}(X_T)} \sup_{m \in M} \left(\frac{\|G_T m\|}{\|SG_T m\|} + \left[\frac{\|TG_T m\|}{\|G_T m\|} \times \frac{\|G_S m\|}{\|SG_T m\|} \right] \right)^{-1} \\
&= \inf_{M \in \mathcal{I}(X_T)} \left(\inf_{m \in M} \frac{\|G_T m\|}{\|SG_T m\|} + \inf_{m \in M} \left[\frac{\|TG_T m\|}{\|G_T m\|} \times \frac{\|G_S m\|}{\|SG_T m\|} \right] \right)^{-1} \\
&= \inf_{M \in \mathcal{I}(X_T)} \left(\frac{1}{\|S_{|G_T M}\|} + \left[\inf_{m \in M} \frac{\|TG_T m\|}{\|G_T m\|} \times \frac{1}{\|S_{|G_S M}\|} \right] \right)^{-1}.
\end{aligned}
$$

Since for all $M \in \mathcal{I}(X_T)$, we have $\inf_{m \in M} \frac{\|TG_T m\|}{\|G_T m\|} \leq \|T|_{G_T M}\|$, then

$$
\left(\frac{1}{\|S_{|G_T M}\|} + \left[\inf_{m \in M} \frac{\|TG_T m\|}{\|G_T m\|} \times \frac{1}{\|S_{|G_S M}\|} \right] \right)^{-1} \geq \frac{\|S_{|G_S M}\|}{1 + \|T|_{G_T M}\|}.
$$

This implies that

$$\inf_{M \in \mathcal{I}(X_T)} \frac{\|S_{|G_S M}\|}{1 + \|T_{|G_T M}\|} \leq \Gamma(SG_T).$$

The fact that $\mathcal{D}(T) = \mathcal{D}(S)$ implies that $\Gamma(S) = \inf_{M \in \mathcal{I}(X_T)} \|S_{|G_S M}\|$. This implies that

$$\frac{\Gamma(S)}{1 + \Gamma(T)} \leq \Gamma(SG_T).$$

Similarly, for

$$\frac{\Gamma_0(S)}{1 + \Gamma_0(T)} \leq \Gamma_0(SG_T).$$

Next

$$\frac{\widetilde{\Delta}(S)}{1 + \widetilde{\Delta}(T)} \leq \widetilde{\Delta}(SG_T) = \sup_M \frac{\Gamma(S_{|G_T M})}{1 + \Gamma(S_{|G_T M})}. \qquad \text{Q.E.D.}$$

2.11 Precompact and Compact Linear Relations

Let X and Y be two normed spaces and $T \in LR(X, Y)$. We say that T is precompact (respectively compact), if the analogous property holds for $Q_T T$, i.e., T is precompact (respectively compact) if $Q_T T(B_X)$ (respectively $\overline{Q_T T(B_X)}$) is totally bounded (respectively a compact subset) in Y. A relation $S \in LR(X, Y)$ is said to be T-compact if $\mathcal{D}(T) \subset \mathcal{D}(S)$, and SG_T is compact. S is called T-precompact if $\mathcal{D}(T) \subset \mathcal{D}(S)$, and SG_T is precompact.

The families of all precompact and compact linear relations will be denoted by $\mathcal{PRK}(X, Y)$ and $\mathcal{KR}(X, Y)$, respectively. If $X = Y$, we write $\mathcal{PRK}(X) := \mathcal{PRK}(X, X)$ and $\mathcal{KR}(X) := \mathcal{KR}(X, X)$. We denote by $\mathcal{K}(X)$ the class of compact operators on X. Note that if $T \in \mathcal{PRK}(X, Y)$ and Y is complete, then $T \in \mathcal{KR}(X, Y)$.

Proposition 2.11.1 [50, Theorem 5.2.2] *T is precompact if and only if, $\overline{\Gamma}_0(T) = 0$.* ◇

Proposition 2.11.2 [50, Proposition 5.5.3] *Let T be a continuous linear relation. Then, T is precompact if and only if, T^* is compact.* ◇

Proposition 2.11.3 *Let A, B and $S \in LR(X)$ and $\lambda \in \mathbb{C}$. Then,* (*i*) $R((\lambda S - A)G_B) = R(\lambda S - A)$.

(*ii*) $N((\lambda S - A)G_B) = N(\lambda S - A)$.

(*iii*) $i((\lambda S - A)G_B) = i(\lambda S - A)$.

(*iv*) If X is complete space and B is A-precompact, then

$$i(\lambda S - A) = i(\lambda S - (A + B)). \qquad \diamond$$

Proof. (*i*) Using the fact that $G_B x = (G_B)^{-1} x = x$, $R(A) = A(\mathcal{D}(A))$ and $\mathcal{D}(AB) = B^{-1}(\mathcal{D}(A))$, we have

$$
\begin{aligned}
R((\lambda S - A)G_B) &= (\lambda S - A)G_B(\mathcal{D}((\lambda S - A)G_B)) \\
&= (\lambda S - A)(\mathcal{D}((\lambda S - A)G_B)) \\
&= (\lambda S - A)G_B(\mathcal{D}(\lambda S - A)) \\
&= (\lambda S - A)(\mathcal{D}(\lambda S - A)) \\
&= R(\lambda S - A).
\end{aligned}
$$

(*ii*) Since

$$
\begin{aligned}
N((\lambda S - A)G_B) &= \{x \in \mathcal{D}((\lambda S - A)G_B) \ : \ (\lambda S - A)G_B(x) = (\lambda S - A)G_B(0)\} \\
&= \{x \in \mathcal{D}(\lambda S - A) \ : \ (\lambda S - A)(x) = (\lambda S - A)(0)\},
\end{aligned}
$$

then $N((\lambda S - A)G_B) = N(\lambda S - A)$.

(*iii*) The assertion (*iii*) is immediately deduced from (*i*) and (*ii*).

(*iv*) Since BG_A is precompact, and X is complete, then BG_A is compact. So, by using (*iii*), we have

$$
\begin{aligned}
i(\lambda S - A) &= i((\lambda S - A)G_A), \\
&= i((\lambda S - A)G_A + BG_A) \\
&= i((\lambda S - (A + B))G_A) \\
&= i(\lambda S - (A + B)).
\end{aligned}
$$

This completes the proof. $\qquad$ Q.E.D.

2.12 Strictly Singular Linear Relations

Definition 2.12.1 *Let $T \in LR(X, Y)$. The linear relation T is called strictly singular if there is no infinite-dimensional subspace M of $\mathcal{D}(T)$ for which $T_{|M}$ is injective and open.* $\qquad \diamond$

We will denote by $SSR(X,Y)$ the set of strictly singular linear relations. If $X = Y$, then $SSR(X) := SSR(X,X)$.

The following proposition gives a characterization of the strictly singular multivalued linear operators by means of the compactness measure.

Proposition 2.12.1 [50, Theoreom 5. 2.6] *Let X and Y be two normed spaces and let $T \in LR(X,Y)$. Then, the linear relation T is strictly singular if and only if, $\widetilde{\Delta}(T) = 0$.* ◇

Lemma 2.12.1 [50, Corollary 5.2.3 and 5.2.8] *Let $T \in LR(X,Y)$ be precompact, then T is continuous strictly singular linear relation.* ◇

Lemma 2.12.2 *Let $S, T \in LR(X,Y)$. If S is T-precompact, then S is strictly singular.* ◇

Proof. Since S is T-precompact, then SG_T is precompact. By using Lemma 2.12.1, it follows that SG_T is continuous. On the one hand, by using Proposition 2.11.1, we obtain

$$\overline{\Gamma}_0(SG_T) = \Gamma_0(SG_T) = 0.$$

Then, by using (2.46), we have $\widetilde{\Delta}(SG_T) = 0$. On the other hand, by using Proposition 2.10.3, we get

$$\frac{\widetilde{\Delta}(S)}{1 + \widetilde{\Delta}(T)} = 0.$$

As a matter of fact,

$$\widetilde{\Delta}(S) = 0.$$

Therefore, the use of Proposition 2.12.1 shows that S is strictly singular. This completes the proof. Q.E.D.

Note that the sum of two strictly singular linear relation is strictly singular, but there exist strictly singular linear relations S and T such that ST is not strictly singular. We shall use the following result which gives sufficient conditions for the composition of two linear relations to be strictly singular.

Theorem 2.12.1 [50, Proposition 5.2.12] *Let $T \in L(X,Y)$ and $S \in LR(Y,Z)$ be strictly singular. Then, $ST \in SSR(X,Z)$.* ◇

Theorem 2.12.2 [50, Proposition 5.2.10] *Let $T \in LR(X,Y)$ be strictly singular (respectively, precompact) and let $S \in LR(Y,Z)$ be continuous. If $\overline{T(0)} \subset \mathcal{D}(S)$, then ST is strictly singular (respectively, precompact) linear relation.* ◇

2.13   Polynomial Multivalued Linear Operators

Our aim in this section is to show some fundamental results for a polynomial linear relation. The notion of polynomial linear operator can be naturally generalized to linear relation as follows:

Definition 2.13.1 *Let $T \in LR(X)$ and given a polynomial $P(\lambda) = \alpha_0 + \alpha_1\lambda + \cdots + \alpha_n\lambda^n$ with coefficients in $\mathbb{C}$, we define the polynomial in T by*

$$P(T) = \alpha_0 T^0 + \alpha_1 T + \cdots + \alpha_n T^n,$$

where T^0 is the identity operator defined on X. $\diamond$

Remark 2.13.1 *Let $\mu \in \mathbb{C}$, n and m_i, $1 \leq i \leq n$ be some positive integers, and let $\lambda_i \in \mathbb{K}$, $1 \leq i \leq n$ be some distinct constants. Fix $\lambda \in \mathbb{C}$ and let*

$$P(\lambda) - \mu = c \prod_{k=1}^{n} (\lambda - \lambda_k)^{m_k}.$$

Then, by using Lemma 2.5.1, the polynomial $P(T)$ in T is given by

$$P(T) - \mu = c \prod_{k=1}^{n} (T - \lambda_k)^{m_k} \tag{2.47}$$

is a linear relation in X. $\diamond$

The following properties concerning the behaviour of the domain, the range, the null space and the multivalued part of $P(T)$, were proved in [84, Theorems 3.2–3.6] by A. Sandovici.

Lemma 2.13.1 *Let $P(T)$ as in Remark 2.13.1. Then,*

(i) $\mathcal{D}(P(T)) = \mathcal{D}(T^{m_1 + m_2 + \ldots + m_n})$ *and* $P(T)(0) = T^{m_1 + m_2 + \ldots + m_n}(0)$.

(ii) $R(P(T)) = \displaystyle\bigcap_{i=1}^{n} R((T - \lambda_i)^{m_i})$.

(iii) $N(P(T)) = \displaystyle\sum_{i=1}^{n} N((T - \lambda_i)^{m_i})$ *and the sum is direct if $\mathcal{D}(T) = X$.*

(iv) For $d \in \mathbb{N}$, we have

$$N(P(T)) \cap R(P(T)^d) = \sum_{i=1}^{n} (N(T - \lambda_i)^{m_i} \cap R(T - \lambda_i)^{m_i d}). \quad \diamond$$

Lemma 2.13.2 [42, Theorem 3.3] *Let $T \in LR(X)$ and let P and Q be two polynomials in $\mathbb{C}$. Then,*

$$(QP)(T) = Q(T)P(T).$$ ◇

Lemma 2.13.3 *Let $\beta \in \mathbb{C}$. Then,*

(i) $T^n = \sum_{k=0}^{n} C_n^k \beta^k (T - \beta)^{n-k}, \quad n \in \mathbb{N}.$

(ii) Every polynomial in T, $P(T)$ of degree n, can be described as $P(T) = (T - \beta)Q(T) + \zeta$, where $Q(T)$ is a polynomial in T of degree $n - 1$ and ζ is a scalar. ◇

Proof. *(i)* We proceed by induction. For $n = 0$ it is trivial. Since $T = (T - \beta) + \beta$, we have the property is true for $n = 1$. Assume (i) holds for $n = m$. Then, $T^{m+1} = A(B + C)$, where $A = \sum_{k=0}^{m} C_m^k \beta^k (T - \beta)^{m-k}, \quad B = T - \beta$ and $C = \beta = \beta I$. Indeed,

$$T^{m+1} = T^m T = \left(\sum_{k=0}^{m} C_m^k \beta^k (T - \beta)^{m-k} \right) ((T - \beta) + \beta).$$

Moreover,

$$A(B + C) \quad = \quad AB + AC. \tag{2.48}$$

We first prove that $\mathcal{D}(A(B + C)) = \mathcal{D}(AB + AC)$. In fact, we have

$$\mathcal{D}(A(B + C)) \quad = \quad \left\{ x \in \mathcal{D}(B + C) \text{ such that } (B + C)x \bigcap \mathcal{D}(A) \neq \emptyset \right\}$$
$$= \quad \left\{ x \in \mathcal{D}(T) \text{ such that } Tx \bigcap \mathcal{D}(T^m) \neq \emptyset \right\}$$
$$= \quad \mathcal{D}(T^{m+1}).$$

On the other hand, $\mathcal{D}(AB + AC) = \mathcal{D}(AB) \bigcap \mathcal{D}(AC)$, where

$$\mathcal{D}(AB) \quad = \quad \left\{ x \in \mathcal{D}(B) \text{ such that } Bx \bigcap \mathcal{D}(A) \neq \emptyset \right\}$$
$$= \quad \left\{ x \in \mathcal{D}(T - \beta) \text{ such that } (T - \beta)x \bigcap \mathcal{D}(T^m) \neq \emptyset \right\}$$
$$= \quad \mathcal{D}(T^m(T - \beta))$$
$$= \quad \mathcal{D}(T^{m+1}),$$

and

$$\mathcal{D}(AC) \quad = \quad \left\{ x \in \mathcal{D}(C) = X \text{ such that } \beta x \bigcap \mathcal{D}(A) \neq \emptyset \right\}$$
$$= \quad \mathcal{D}(T^m).$$

Hence, $\mathcal{D}(AB + AC) = \mathcal{D}(T^{m+1}) \cap \mathcal{D}(T^m) = \mathcal{D}(T^{m+1})$. Therefore, $A(B + C)$ and $AB + AC$ have the same domain. Furthermore, we infer from Proposition 2.3.6 (iv) that $A(B + C)$ is an extension of $AB + AC$. So,

$$G(AB + AC) \subset G(A(B + C)),$$

and

$$A(B + C)(0) = AB(0) + AC(0).$$

In this situation, the property (2.48) follows immediately from Proposition 2.3.6 (iii). Now, we prove that

$$T^{m+1} = \sum_{k=0}^{m+1} C_{m+1}^k \beta^k (T - \beta)^{m+1-k}. \tag{2.49}$$

From (2.48), we have $T^{m+1} = E + F$, where

$$E = \sum_{k=0}^{m} C_m^k \beta^k (T - \beta)^{m-k}(T - \beta),$$

and

$$F = \beta \sum_{k=0}^{m} C_m^k \beta^k (T - \beta)^{m-k}.$$

If we take $S = T - \beta$,

$$Q(S) = \sum_{k=0}^{m} C_m^k \beta^k (T - \beta)^{m-k},$$

and $P(S) = T - \beta$, then in view of Lemma 2.13.2, we obtain $Q(P(S)) = Q(S)P(S)$, that is

$$E = \sum_{k=0}^{m} C_m^k \beta^k (T - \beta)^{m+1-k}.$$

Therefore, (2.49) holds. The proof of the statement (i) is completed.

(ii) Let $P(T) = \alpha_0 + \alpha_1 T + \cdots + \alpha_n T^n$. The use of (i) leads to

$$P(T) = a_0 + a_1(T - \beta) + \cdots + a_n(T - \beta)^n,$$

for some scalars $a_0, a_1, \ldots, a_n$. On the other hand, if in Lemma 2.13.2, we take $S = T - \beta$, $P(S) = a_1 S + \cdots + a_n S^{n-1}$ and $Q(S) = S$, we deduce that

$$P(T) - a_0 = (T - \beta)\left(a_1 + a_2(T - \beta) + \cdots + a_n(T - \beta)^{n-1}\right) = (T - \beta)Q(T),$$

where,

$$Q(T) = a_1 + a_2(T - \beta) + \cdots + a_n(T - \beta)^{n-1}.$$

Again, applying the part (i), we have $Q(T) = b_1 + b_2T + \cdots + b_nT^{n-1}$, for some scalars $b_1, b_2, \ldots, b_n$. So, Q is a polynomial in T of degree $n - 1$, as desired. This completes the proof.   Q.E.D.

Lemma 2.13.4 *Suppose that T is closed and there exists $\beta \in \mathbb{C}$ such that $T - \beta$ is bijective. Let $P(T) = \alpha_0 + \alpha_1T + \cdots + \alpha_nT^n$ be a polynomial in T of degree n. Then, $P(T)$ is a closed linear relation.*   ◇

Proof.   We shall proceed by induction. For $n = 1$, it is clear since $P(T) = \alpha_0 + \alpha_1T$ is closed by Proposition 2.7.2. Assume that the property is true for $n = m$. Then, if $P(T)$ is a polynomial in T of degree $m + 1$, we infer from the condition (ii) in Lemma 2.13.3 that $P(T) = (T - \beta)Q(T) + \delta$, where $Q(T)$ is a polynomial in T of degree m and δ is scalar. Then, $Q(T)$ is closed by the induction hypothesis. By using Proposition 2.7.2, we have $T - \beta$ is closed. Since $T - \beta$ is bijective, we deduce from Proposition 2.7.3 that $(T - \beta)Q(T)$ is closed. So, in view of Proposition 2.7.2, we get $P(T)$ is closed, as required. This completes the proof.   Q.E.D.

Lemma 2.13.5 *Assume that A is everywhere defined. Then,*

(i) For all nonzero scalar λ and for all $n \in \mathbb{N}$, we have

$$(A - \lambda)^n = \sum_{i=0}^{n} C_n^i(-1)^i\lambda^iA^{n-i}.$$

(ii) Suppose that $(A - \alpha)^{-1}$ is a single valued linear operator everywhere defined. Then,

(a) For all $\beta \in \mathbb{K}$ and for all $m, n \in \mathbb{N}$

$$(A - \alpha)^{-n}(A - \beta)^m \subset (A - \beta)^m(A - \alpha)^{-n}.$$

(b) For all $\beta \in \mathbb{K}$ with $\alpha \neq \beta$ and for all $n \in \mathbb{N}$, we have

$$N((A - \beta)^n) = N(((\beta - \alpha)^{-1} - (A - \alpha)^{-1})^n)$$

and

$$R((A - \beta)^n) = R(((\beta - \alpha)^{-1} - (A - \alpha)^{-1})^n).$$ ◇

Proof. (i) We proceed by induction. The case $n = 1$ is clear. Suppose the property holds for some positive integer k. Then, one deduces from Lemma 2.5.1 (i) and Proposition 2.3.6 combined with the induction hypothesis that

$$
\begin{aligned}
(A - \lambda)^{k+1} &= A(A - \lambda)^k - \lambda(A - \lambda)^k \\
&= \sum_{i=0}^{k} C_k^i (-1)^i \lambda^i A^{k+1-i} - \sum_{i=0}^{k} C_k^i (-1)^i \lambda^{i+1} A^{k-i} \\
&= A^{k+1} + \sum_{i=1}^{k} C_{k+1}^i (-1)^i \lambda^i A^{k+1-i} + (-1)^{k+1} \lambda^{k+1} \\
&= \sum_{i=0}^{k+1} C_{k+1}^i (-1)^i \lambda^i A^{k+1-i}.
\end{aligned}
$$

(ii) (a) We first note that $(A - \alpha)^n$ is bijective and everywhere defined and thus it is easy to see that $(A - \alpha)^{-n}(A - \alpha)^n = I_X \subset (A - \alpha)^n(A - \alpha)^{-n}$. These facts together with the part (c) of Lemma 2.5.1 imply that

$$
\begin{aligned}
(A - \alpha)^{-n}(A - \beta)^m &\subset (A - \alpha)^{-n}(A - \beta)^m(A - \alpha)^n(A - \alpha)^{-n} \\
&= (A - \alpha)^{-n}(A - \alpha)^n(A - \beta)^m(A - \alpha)^{-n} \\
&= (A - \beta)^m(A - \alpha)^{-n}.
\end{aligned}
$$

(b) We first verify that

$$
\begin{aligned}
A - \beta &= (\beta - \alpha)((\beta - \alpha)^{-1} - (A - \alpha)^{-1})(A - \alpha) \\
&= (\beta - \alpha)(A - \alpha)((\beta - \alpha)^{-1} - (A - \alpha)^{-1}). \quad (2.50)
\end{aligned}
$$

One deduces from Proposition 2.3.6 and the part (a) that

$$
\begin{aligned}
A - \beta &= (A - \alpha) - (\beta - \alpha) \\
&= (A - \alpha) - (\beta - \alpha)(A - \alpha)^{-1}(A - \alpha) \\
&= (I - (\beta - \alpha)(A - \alpha)^{-1})(A - \alpha) \\
&= (\beta - \alpha)((\beta - \alpha)^{-1} - (A - \alpha)^{-1})(A - \alpha) \\
&= (\beta - \alpha)(\beta - \alpha)^{-1}(A - \alpha) - (\beta - \alpha)I_X \\
&\subset (\beta - \alpha)((\beta - \alpha)^{-1}(A - \alpha) - (A - \alpha))(A - \alpha)^{-1}) \\
&= (\beta - \alpha)(A - \alpha)((\beta - \alpha)^{-1} - (A - \alpha)^{-1}).
\end{aligned}
$$

On the other hand, since $A - \alpha$ is everywhere defined and bijective it is easy to see that

$$
\mathcal{D}((\beta-\alpha)^{-1}-(A-\alpha)^{-1})(A-\alpha)) = \mathcal{D}((A-\alpha)((\beta-\alpha)((\beta-\alpha)^{-1}-(A-\alpha)^{-1})) = X
$$

and

$$((\beta - \alpha)^{-1} - (A - \alpha)^{-1})(A - \alpha)(0) \;=\; (A - \alpha)((\beta - \alpha)^{-1} - (A - \alpha)^{-1})(0)$$
$$=\; A(0).$$

Now, the use of Proposition 2.3.6, makes us to conclude (2.50).

(b) We shall prove the first equality in (b) proceeding by induction. Define $B := (\beta - \alpha)^{-1} - (A - \alpha)^{-1}$. Applying (2.50), we infer that

$$N(A - \beta) \;=\; B^{-1}(N(A - \alpha))$$
$$=\; B^{-1}(0) \quad (\text{as } N(A - \alpha) = \{0\})$$
$$=\; N(B).$$

Assume the property holds for some positive integer k. Then,

$$N(B^{k+1}) \;=\; B^{-1}(N(B^k))$$
$$=\; B^{-1}(N((A - \beta)^k))$$
$$=\; B^{-1}(N((A - \beta)^k(A - \alpha))) \quad (\text{by Lemma 2.5.1})$$
$$=\; B^{-1}(A - \alpha)^{-1}(N(A - \beta)^k)$$
$$=\; ((A - \alpha)B)^{-1}(N(A - \beta)^k)$$
$$=\; (B(A - \alpha))^{-1}(N(A - \beta)^k) \quad (\text{by (2.50)})$$
$$=\; N((A - \beta)^{k+1}).$$

The second identity in (b) is obtained by induction. The case $n = 1$ follows immediately from (2.50). Assume now that $R((A - \beta)^k) = R(B^k)$ for some positive integer k. Then,

$$R((A - \beta)^{k+1}) \;=\; (A - \beta)(R(A - \beta)^k)$$
$$=\; (A - \beta)(R(B^k))$$
$$=\; (\beta - \alpha)B(R(B^k(A - \alpha))).$$

Thus, by Lemma 2.5.1 (i) (c) and the surjectivity of $A - \alpha$, we infer that $R((A - \beta)^{k+1}) = R(B^{k+1})$. This completes the induction argument. Q.E.D.

2.14 Some Classes of Multivalued Linear Operators

2.14.1 Multivalued Fredholm and Semi-Fredholm Linear Operators

Definition 2.14.1 *A linear relation $T \in LR(X,Y)$ is said to be upper semi-Fredholm, denoted by $T \in \mathcal{F}_+(X,Y)$, if there exists a closed finite codimensional subspace M of X such that the restriction $T_{|M}$ has a single valued continuous inverse. A linear relation T is said to be lower semi-Fredholm, denoted by $T \in \mathcal{F}_-(X,Y)$, if its adjoint T^* is upper semi-Fredholm.* ◇

Lemma 2.14.1 [50, Proposition 5.5.11 and Corollary 5.7.7] *Let $T \in LR(X,Y)$ and let $M \subset Y$ such that $\dim(M) < \infty$. Then,*

(i) $T \in \mathcal{F}_+(X,Y)$ if and only if, $Q_M T \in \mathcal{F}_+(X,Y/M)$.
(ii) $T \in \mathcal{F}_-(X,Y)$ if and only if, $Q_M T \in \mathcal{F}_-(X,Y/M)$. ◇

Proposition 2.14.1 [50, Proposition 5.5.2] *The following properties are equivalent*

(i) $T \in \mathcal{F}_-(X,Y)$.
(ii) $\overline{\beta}(T) < \infty$ and $\gamma(T^*) > 0$.
(iii) $\widetilde{T} \in \Phi_-(\widetilde{X}, \widetilde{Y})$.
(iv) $Q_T T \in \mathcal{F}_-(X,Y/\overline{T(0)})$. ◇

Proposition 2.14.2 [50, Proposition 5.5.13] *Let $T \in LR(X,Y)$ and we consider the properties*

(i) $T \in \mathcal{F}_+(X,Y)$ and $\overline{\beta}(T) < \infty$.
(ii) $T^* \in \mathcal{F}_-(Y^*,X^*)$ and $\alpha(T^*) < \infty$.

Then, (i) $\Longrightarrow$ (ii) and if X is complete and $T \in CR(X,Y)$, then (ii) $\Longrightarrow$ (i). ◇

Lemma 2.14.2 [50, Proposition 5.2.16] *Let $T \in LR(X,Y)$ and $S \in LR(Y,Z)$. Then,*

(i) If T is sa single valued linear operator, $T \in \mathcal{F}_+(X,Y)$ and $S \in \mathcal{F}_+(Y,Z)$, then $ST \in \mathcal{F}_+(X,Z)$.
(ii) If $S \in BR(Y,Z)$ and $ST \in \mathcal{F}_+(X,Z)$, then $T \in \mathcal{F}_+(X,Y)$. ◇

Theorem 2.14.1 [50, Theorem 5.10.3] *Let X be a Banach space, Y be a normed vector space and $T \in LR(X,Y)$. Then, the following properties are equivalent*

(i) T is upper semi-Fredholm.

(ii) There exists a bounded linear operator A and a bounded finite rank projection operator P such that $AT = I_{\mathcal{D}(T)} - P$. ◊

Proposition 2.14.3 *(i)* [50, Proposition 5.5.27] *Let $T \in LR(X,Y)$. If T is closable, then $T \in \mathcal{F}_-(X,Y)$ if and only if, $TG_T \in \mathcal{F}_-(X_T,Y)$.*

(ii) [50, Corollary 5.2.5] *$T \in \mathcal{F}_+(X,Y)$ if and only if, $TG_T \in \mathcal{F}_+(X_T,Y)$.* ◊

Theorem 2.14.2 [50, Theorem 5.1.11] *Let X and Y be two normed spaces. Then, the following properties are equivalent*

(i) $T \notin \mathcal{F}_+(X,Y)$.

(ii) There exists a non precompact bounded subset W of $\mathcal{D}(T)$ such that $Q_T W$ is precompact.

(iii) T has a singular sequence, i.e., a sequence $(x_n)_n$ of norm one elements of $\mathcal{D}(T)$ such that $(x_n)_n$ has no Cauchy subsequence and $\lim_{n\to\infty} \|Tx_n\| = 0$. ◊

Proposition 2.14.4 [13, Theorem 2.17] *Let $S \in LR(X,Y)$, $T \in \mathcal{F}_+(X,Y)$ with $G(S) \subset G(T)$, and $\dim \mathcal{D}(S) = \infty$, then $S \in \mathcal{F}_+(X,Y)$.* ◊

Proposition 2.14.5 [92, Proposition 5.9.2] *Let $T \in \mathcal{F}_+(X,Y)$. Then, any bounded sequence $(x_n)_n$ in $\mathcal{D}(T)$ such that $(Q_T T x_n)_n$ is a Cauchy sequence has a Cauchy subsequence.* ◊

Proposition 2.14.6 [50, Theorem 5.2.4] *Let $T \in LR(X,Y)$. If $\dim \mathcal{D}(T) = \infty$, then $T \in \mathcal{F}_+(X,Y)$ if and only if, $\Gamma(T) > 0$.* ◊

2.14.2 Multivalued Fredholm and Semi-Fredholm of Closed Linear Operators in Banach Space

Let X and Y be two Banach spaces, we extend the classes of closed single valued Fredholm type operators to include closed multivalued operators, and note that the definitions of the classes $\mathcal{F}_+(X,Y)$ and $\mathcal{F}_-(X,Y)$ are consist, respectively, with

$$\Phi_+(X,Y) := \{T \in CR(X,Y) \text{ such that } \alpha(T) < \infty \text{ and } R(T) \text{ is closed in } Y\},$$

and

$$\Phi_-(X,Y) := \{T \in CR(X,Y) \text{ such that } \beta(T) < \infty \text{ and } R(T) \text{ is closed in } Y\}.$$

$\Phi_\pm(X,Y) := \Phi_+(X,Y) \bigcup \Phi_-(X,Y)$ and $\Phi(X,Y) := \Phi_+(X,Y) \bigcap \Phi_-(X,Y)$ denotes the set of Fredholm linear relations from X into Y. If $X = Y$, then $\Phi_+(X,Y)$, $\Phi_-(X,Y)$, $\Phi_\pm(X,Y)$ and $\Phi(X,Y)$ are replaced, respectively, by $\Phi_+(X)$, $\Phi_-(X)$, $\Phi_\pm(X)$ and $\Phi(X)$. For $T \in CR(X)$, a number complex λ is in Φ_{+T}, or Φ_T if $\lambda - T$ is in $\Phi_+(X)$, or $\Phi(X)$, respectively. For $T \in CR(X,Y)$ and $S \in LR(X,Y)$ such that $\lambda S - T$ is closed, a number complex λ is in $\Phi_{+T,S}$, or $\Phi_{T,S}$ if $\lambda S - T$ is in $\Phi_+(X,Y)$, or $\Phi(X,Y)$, respectively.

Definition 2.14.2 *We say that a linear relation T from X into Y is Browder, if T is a Fredholm linear relation of index zero and has finite ascent and descent.* ◇

We will denote by $\mathcal{B}(X,Y)$ the set of Browder linear relations. If $X = Y$, then $\mathcal{B}(X) := \mathcal{B}(X,X)$.

Remark 2.14.1 *Let T be a Fredholm single valued operator from X into Y of finite ascent and descent. Then, T has index zero. However, this property is not true in the context of linear relations.* ◇

Theorem 2.14.3 [11, Theorem 6.1] *Let X be a Banach space and let S, $T \in CR(X)$. Then,*

(i) If S, $T \in \Phi_+(X)$, then $ST \in \Phi_+(X)$ and $TS \in \Phi_+(X)$.
(ii) If S, $T \in \Phi_-(X)$ with TS (respectively, ST) is closed, then $TS \in \Phi_-(X)$ (respectively, $ST \in \Phi_-(X)$).
(iii) If S, $T \in \Phi(X)$, then $TS \in \Phi(X)$ and

$$i(TS) = i(T) + i(S) + \dim\left(\frac{X}{R(S) + \mathcal{D}(T)}\right) - \dim(S(0) \bigcap N(T)).$$

(iv) If S and T are everywhere defined and $TS \in \Phi_+(X)$, then $S \in \Phi_+(X)$.
(v) If S and T are everywhere defined such that $TS \in \Phi(X)$ and $ST \in \Phi(X)$, then $S \in \Phi(X)$ and $T \in \Phi(X)$. ◇

Example 2.14.1 *Let $T : l_2(\mathbb{N}) \longrightarrow l_2(\mathbb{N})$ be defined as*

$$Te_k = \begin{cases} e_{\frac{k}{2}} & \text{if } k \text{ is even,} \\ 0 & \text{if } k \text{ is odd.} \end{cases}$$

This single valued linear operator T is not Fredholm, since $\alpha(T) = +\infty$. Now, we defined

$$S : l_2(\mathbb{N}) \longrightarrow l_2(\mathbb{N})$$

as the continuous linear extension of

$$Se_k = e_{2k}, (k \in \mathbb{N}).$$

Then, $TS = I$ the identity on $l_2(\mathbb{N})$, which is clearly Fredholm single valued linear operator, but T is not.

Theorem 2.14.4 [9, Proposition 8 and Theorem 25] *Let X and Y be two Banach spaces and let $T \in \mathcal{CR}(X, Y)$. Then,*

(i) $T \in \Phi_+(X, Y)$ if and only if, $T^* \in \Phi_-(Y^*, X^*)$.
(ii) $T \in \Phi_-(X, Y)$ if and only if, $T^* \in \Phi_+(Y^*, X^*)$.
Moreover, if T is semi-Fredholm, then $\alpha(T^*) = \beta(T)$ and $\alpha(T) = \beta(T^*)$.
(iii) Let $T \in \Phi_+(X)$. Then, there exists $\varepsilon > 0$ such that for every $\lambda \in \mathbb{K}$ with $0 < |\lambda| < \varepsilon$, we have $\lambda - T \in \Phi_+(X)$, $\alpha(\lambda - T)$ is constant, $\alpha(\lambda - T) \leq \alpha(T)$ and $i(\lambda - T) = i(T)$.
(iv) Let $T \in \Phi_-(X)$. Then, there exists $\varepsilon > 0$ such that for every $\lambda \in \mathbb{K}$ with $0 < |\lambda| < \varepsilon$, we have $\lambda - T \in \Phi_-(X)$, $\beta(\lambda - T)$ is constant, $\beta(\lambda - T) \leq \beta(T)$ and $i(\lambda - T) = i(T)$. ◇

Definition 2.14.3 *(i) Let $T \in \Phi_+(X)$ and let $\varepsilon > 0$ as in Theorem 2.14.4 (iii). The jump of T is defined by*

$$j(T) := \alpha(T) - \alpha(\lambda - T), \quad 0 < |\lambda| < \varepsilon.$$

(ii) Let $T \in \Phi_-(X)$ and let $\varepsilon > 0$ as in Theorem 2.14.4 (iv). The jump of T is defined by

$$j(T) := \beta(T) - \beta(\lambda - T), \quad 0 < |\lambda| < \varepsilon.$$ ◇

Remark 2.14.2 *(i) Clearly, $j(T) = 0$ and the continuity of the index (see Theorem 2.14.4) ensures that both definitions of $j(T)$ coincide whenever $T \in \Phi_+(X) \bigcap \Phi_-(X)$, so that $j(T)$ is unambiguously defined.*
(ii) An immediate consequence of Theorem 2.14.4 (i) is that if T is semi-Fredholm, then $j(T) = j(T^*)$. ◇

Lemma 2.14.3 *Let X and Y be two Banach spaces and let $T \in \mathcal{CR}(X, Y)$. Then,*

(i) $T \in \Phi_+(X,Y)$ if and only if, $Q_T T \in \Phi_+(X, Y/T(0))$. In such case $i(T) = i(Q_T T)$.

(ii) $T \in \Phi_-(X,Y)$ if and only if, $Q_T T \in \Phi_-(X, Y/T(0))$. In such case $i(T) = i(Q_T T)$.

(iii) $T \in \Phi(X,Y)$ if and only if, $Q_T T \in \Phi(X, Y/T(0))$. In such case $i(T) = i(Q_T T)$.   ◇

Proof. First, in view of Proposition 2.7.1 (i), $Q_T T$ is closed and $T(0)$ is closed. Second, $R(T)$ is closed if and only if, $R(Q_T T)$ is closed. In such case $\beta(T) = \beta(Q_T T)$. Indeed, $T(0)$ is closed implies

$$R(Q_T T) = \left(R(T) + \overline{T(0)} \right) \big/ \overline{T(0)} = R(T)/T(0). \tag{2.51}$$

So, by using Lemma 2.1.3 (i), we have $R(T)$ is closed if and only if, $R(Q_T T)$ is closed. In such case by Lemma 2.1.3 (ii), we have $\beta(T) = \beta(Q_T T)$. Finally,

$$
\begin{aligned}
N(T) &= \{x \in \mathcal{D}(T) \text{ such that } Tx = T(0)\} \\
&= \{x \in \mathcal{D}(T) \text{ such that } Tx = \overline{T(0)}\} \text{ (since } T \text{ is closed)} \\
&= \{x \in \mathcal{D}(T) \text{ such that } Q_T Tx = 0\} \\
&= N(Q_T T).
\end{aligned}
$$

This completes the proof.   Q.E.D.

2.14.3   Atkinson Linear Relations

We begin this subsection with the notion of a Atkinson linear relation in a Banach space.

Lemma 2.14.4 *Let $S \in LR(Y, Z)$ and $T \in \Phi_-(X, Y)$ be bounded operators such that $N(S)$ and $N(T)$ are topologically complemented in Y and X, respectively. Then, $N(ST)$ is topologically complemented in X.*   ◇

Proof. Let Y_1 and X_1 be closed subspaces of Y and X, respectively, such that $Y = N(S) \oplus Y_1$ and $X = N(T) \oplus X_1$. Since $R(T)$ is a closed finite codimensional subspace we infer from Lemma 2.1.1 that there exist a finite dimensional subspaces $N \subset N(S)$ and $Y_2 \subset Y_1$ such that $N(S) = (N(S) \cap R(T)) \oplus N$ and $Y_1 = (Y_1 \cap R(T)) \oplus Y_2$. Therefore, the subspace $(N(S) \cap R(T)) \oplus (Y_1 \cap R(T))$ is a finite codimensional subspace of $R(T)$. So, there is a finite dimensional subspace $Y_3 \subset R(T)$ such that

$$R(T) = \left(\left(N(S) \cap R(T) \right) \oplus \left(Y_1 \cap R(S) \right) \right) + Y_3$$

and

$$\{0\} = \left(\left(N(S) \bigcap R(T) \right) \bigoplus \left(Y_1 \bigcap R(S) \right) \right) \bigcap Y_3.$$

We define $T_1 := T_{|X_1}$, $X_3 := T_1^{-1}(N(S) \bigcap R(T))$ and $X_4 := T_1^{-1}(Y_3 \bigoplus (R(T) \bigcap Y_1))$. Then, it is easy to see that $X = N(T) \bigoplus X_3 \bigoplus X_4$. Hence, since $T(N(ST)) = R(T) \bigcap N(S)$, we deduce that $N(ST) = N(T) \bigoplus X_3$. Thus, $X = N(ST) \bigoplus X_4$. Q.E.D.

Let X and Y be two normed spaces and $T \in C\mathcal{R}(X,Y)$. We say that T is bounded below if it is injective and open, T is left invertible if T is injective and $R(T)$ is topologically complemented in Y, and T is right invertible if T is surjective and $N(T)$ is topologically complemented in X. The sets of all left invertible linear relations, right invertible relations and invertible linear relations on X are defined, respectively, by

$$\begin{aligned}
\mathcal{GR}_l(X,Y) &= \{T \in C\mathcal{R}(X,Y) : T \text{ is a left invertible linear relation}\}, \\
\mathcal{GR}_r(X,Y) &= \{T \in C\mathcal{R}(X,Y) : T \text{ is a right invertible linear relation}\}, \\
\mathcal{GR}(X,Y) &= \mathcal{GR}_l(X,Y) \bigcap \mathcal{GR}_r(X,Y).
\end{aligned}$$

The class of α-Atkinson multivalued linear operators from X into Y is defined by

$$\mathcal{A}_\alpha(X,Y) = \{T \in \Phi_+(X,Y) : R(T) \text{ is topologically complemented in } Y\}$$

and the class of β-Atkinson multivalued linear operators from X into Y is defined by

$$\mathcal{A}_\beta(X,Y) = \{T \in \Phi_-(X,Y) : N(T) \text{ is topologically complemented in } X\}.$$

Based on the previous definitions, we defined the set of left Weyl linear relations by

$$\mathcal{W}^l(X,Y) := \{T \in \mathcal{A}_\alpha(X,Y) \text{ such that } i(T) \leq 0\},$$

and the set of right Weyl linear relations is defined by

$$\mathcal{W}^r(X,Y) := \{T \in \mathcal{A}_\beta(X,Y) \text{ such that } i(T) \geq 0\}.$$

The set of Weyl linear relations is defined by $\mathcal{W}(X,Y) := \mathcal{W}^l(X,Y) \bigcap \mathcal{W}^r(X,Y)$. Thus, we have the following inclusions:

$$\mathcal{W}(X,Y) \subset \mathcal{A}_\alpha(X,Y) \subset \Phi_+(X,Y), \text{ and}$$

$$\mathcal{W}(X,Y) \subset \mathcal{A}_\beta(X,Y) \subset \Phi_-(X,Y).$$

If $X = Y$, then the sets $\mathcal{GR}_l(X,Y)$, $\mathcal{GR}_r(X,Y)$, $\mathcal{GR}(X,Y)$, $\mathcal{A}_\alpha(X,Y)$, $\mathcal{A}_\beta(X,Y)$, $\mathcal{W}^l(X,Y)$, $\mathcal{W}^r(X,Y)$, and $\mathcal{W}(X,Y)$ are remplaced, respectively, by $\mathcal{GR}_l(X)$, $\mathcal{GR}_r(X)$, $\mathcal{GR}(X)$, $\mathcal{A}_\alpha(X)$, $\mathcal{A}_\beta(X)$, $\mathcal{W}^l(X)$, $\mathcal{W}^r(X)$, and $\mathcal{W}(X)$.

Lemma 2.14.5 [13, Proposition 3.28] *Let X and Y be two complete spaces and let $T \in C\mathcal{R}(X,Y)$. Then,*

(i) $T \in \mathcal{A}_\alpha(X,Y)$ if and only if, $T^* \in \mathcal{A}_\beta(Y^*, X^*)$.
(ii) $T \in \mathcal{A}_\beta(X,Y)$ if and only if, $T^* \in \mathcal{A}_\alpha(Y^*, X^*)$.   ◇

Remark 2.14.3 *Let X and Y be two complete spaces and let $T \in C\mathcal{R}(X,Y)$.*

(i) Using Lemma 2.14.5 and definitions of both $\mathcal{W}^l(X,Y)$ and $\mathcal{W}^r(X,Y)$, we have the following equivalences

> *(i$_1$) $T \in \mathcal{W}^l(X,Y)$ if and only if, $T^* \in \mathcal{W}^r(Y^*, X^*)$.*
> *(i$_2$) $T \in \mathcal{W}^r(X,Y)$ if and only if, $T^* \in \mathcal{W}^l(Y^*, X^*)$.*

(ii) Let $M \subset Y$ such that $dim(M) < \infty$. Then,

> *(ii$_2$) If $T(0)$ is topologically complemented in Y, then*

$$T \in \mathcal{A}_\alpha(X,Y) \text{ if and only if, } Q_M T \in \mathcal{A}_\alpha(X, Y/M).$$

> *(ii$_3$) If $T^*(0)$ is topologically complemented in Y^*, then*

$$T \in \mathcal{A}_\beta(X,Y) \text{ if and only if, } Q_M T \in \mathcal{A}_\beta(X, Y/M).$$ ◇

Theorem 2.14.5 [79, Theorem 1.1] *Let X and Y be two complete spaces and let $T \in C\mathcal{R}(X,Y)$. If T is a single valued linear operator, then we have the following equivalences*

(i) $T \in \mathcal{A}_\alpha(X,Y)$.
(ii) There exist $U \in \mathcal{L}(Y,X)$ and $E \in \mathcal{K}(X)$ such that

> *(ii$_1$) $R(U) \subset \mathcal{D}(T)$.*
> *(ii$_2$) $R(E) \subset \mathcal{D}(T)$.*
> *(ii$_3$) $UT = I - K$ on $\mathcal{D}(T)$.*   ◇

Lemma 2.14.6 [13, Corollary 3.25] *Let $T \in C\mathcal{R}(X,Y)$. The following properties are equivalent*

(i) $T \in \mathcal{A}_\beta(X,Y)$.
(ii) There exist $T_r \in \mathcal{L}(Y,X)$ and a finite rank projection $F \in \mathcal{L}(Y)$ such that $R(T_r)$ is topologically complemented in $\mathcal{D}(T)$, $T_r T$ and FT are continuous operators, $I - F \in \Phi(X,Y)$ and $I - F$ is a linear selection of TT_r.   ◇

Definition 2.14.4 *An operator $T_r \in \mathcal{L}(Y, X)$ satisfying Lemma 2.14.6 is called a right regularizer (or a right generalized inverse) of T.* ◇

Proposition 2.14.7 *Let $T \in C\mathcal{R}(X, Y)$ and let $S \in L\mathcal{R}(Y, Z)$. Then,*

(i) If $T \in \mathcal{A}_\beta(X, Y)$, $S \in \mathcal{A}_\beta(Y, Z) \cap \mathcal{L}(Y, Z)$ and $ST \in C\mathcal{R}(X, Z)$, then $ST \in \mathcal{A}_\beta(X, Z)$.
(ii) If $\mathcal{D}(S^*) = Z^*$, $R(T) \subset \mathcal{D}(S)$, $T \in \mathcal{A}_\alpha(X, Y)$ and $S \in \mathcal{A}_\alpha(Y, Z) \cap \mathcal{L}(Y, Z)$, then $ST \in \mathcal{A}_\alpha(X, Z)$. ◇

Proof. (i) Follows immediately from Lemma 3.1.5 and Theorem 2.14.4.

(ii) Let $T \in \mathcal{A}_\alpha(X, Y)$ and let $S \in \mathcal{A}_\alpha(Y, Z) \cap \mathcal{L}(Y, Z)$. Then, by using Lemma 2.14.5 (i), we have $T^* \in \mathcal{A}_\beta(Y^*, X^*)$ and $S^* \in \mathcal{A}_\beta(Z^*, Y^*) \cap \mathcal{L}(Z^*, Y^*)$. Hence, by using (i), we have $(ST)^* = T^* S^* \in \mathcal{A}_\beta(Z^*, X^*)$. By applying Lemma 2.14.5 (ii), we have $ST \in \mathcal{A}_\beta(X, Z)$. Q.E.D.

Proposition 2.14.8 *Assume that T is a single valued linear operator. Then, the following properties are equivalent*

(i) $T \in \mathcal{A}_\alpha(X, Y)$.
(ii) There exist $A \in \mathcal{L}(Y, X)$ and B a continuous linear operator in X such that $\mathcal{D}(T) \subset \mathcal{D}(B)$, $I - B \in \Phi(X)$, $R(B) \subset \mathcal{D}(T)$ and $AT = (I - B)_{|\mathcal{D}(T)}$. ◇

Proof. Let us consider two cases of T:

First case : If $\overline{\mathcal{D}(T)} = X$, then the result was proved through Theorem 2.14.5.

Second case : If $\overline{\mathcal{D}(T)} \neq X$, then for A and B satisfy the conditions in (ii), we have $AT G_T = (I - B)_{|X_T}$. Hence,

$$\|T G_T\| = \sup_{x \in X_T} \frac{\|T G_T x\|}{\|x\|_T}$$

$$= \sup_{x \in \mathcal{D}(T) \backslash \{0\}} \frac{\|Tx\|}{\|x\| + \|Tx\|}$$

$$= \sup_{x \in \mathcal{D}(T) \backslash \{0\}} \frac{\|Tx\|}{\left(1 + \frac{\|Tx\|}{\|x\|}\right) \|x\|}$$

$$= \frac{\|T\|}{1 + \|T\|}$$

$$< 1.$$

Then, $T G_T \in \mathcal{L}(X_T, Y)$ and X_T is complete, then by applying the first case to the operator $T G_T$, we have $T G_T \in \mathcal{A}_\alpha(X_T, Y)$. Therefore, $T \in \mathcal{A}_\alpha(X, Y)$, as desired. Q.E.D.

2.15   Quasi-Fredholm and Semi Regular Linear Relations

2.15.1   Quasi-Fredholm Linear Relations

In this subsection, we introduce and study the class of quasi-Fredholm linear relations in Banach spaces. Let us first recall some facts which will help to understand Definition 2.15.1 below. It is proved in Theorem 2.7.2 (i) that an everywhere defined closed linear relation in a Banach space is bounded. These remarks suggest generalizing the notion of bounded quasi-Fredholm operator due to M. Mbekhta and V. Müller [78]) to the case of multivalued linear operators as follows:

Definition 2.15.1 *Let X be a Banach space and let T be an everywhere defined closed linear relation in X with there exists $\lambda \in \mathbb{C}$ such that $(T - \lambda)^{-1}$ is a single valued linear operator everywhere defined. We say that T is quasi-Fredholm if there exists $d \in \mathbb{N}$ for which $R(T^{d+1})$ is closed and*

$$R(T) + N(T^d) = R(T) + N^\infty(T). \qquad \diamond$$

In this case T is said to be a quasi-Fredholm linear relation of degree d and we write $T \in q\phi_d(X)$.

2.15.2   Semi Regular Linear Relations

We begin with the following lemma.

Lemma 2.15.1 *[73, Lemma 3.7] Let $T \in BR(X)$. Then, the following statements are equivalent*

(i) $N(T) \subset R(T^n)$ for every $n \in \mathbb{N}$,
(ii) $N(T^m) \subset R(T)$ for every $m \in \mathbb{N}$, and
(iii) $N(A^m) \subset R(T^n)$ for every $n, m \in \mathbb{N}$. $\qquad \diamond$

Definition 2.15.2 *We say that a linear relation $T \in LR(X)$ is semi regular if $R(T)$ is closed and T verifies one of the equivalent conditions of Lemma 2.15.1.* $\qquad \diamond$

Theorem 2.15.1 *[9, Theorem 27] Let $T \in \Phi_\pm(X)$. Then, T is semi regular if and only if, $j(T) = 0$.* $\qquad \diamond$

Theorem 2.15.2 [9, Theorem 23] *Let X be a complex Banach space and let T be a semi regular linear relation in X. Then, there exists $\delta > 0$ such that $\lambda - T$ is semi regular if $|\lambda| < \delta$.* ◇

Trivial examples of regular linear relations are surjective multivalued operators as well as injective multivalued operators with closed range. For an essential version of semi regular linear relation, we use the following notations.

Definition 2.15.3 *Let $T \in LR(X)$. Assume that X_1 and X_2 are two subspaces of X such that $X = X_1 \oplus X_2$. We say that T is completely reduced by the pair (X_1, X_2) if*

$$T = T_{|X_1} \oplus T_{|X_2}.$$ ◇

In such case, we have

$$\left\{ \begin{array}{rcl}
\mathcal{D}(T) & = & \mathcal{D}(T_{|X_1}) \oplus \mathcal{D}(T_{X_2}), \\
N(T) & = & N(T_{|X_1}) \oplus N(T_{|X_2}), \\
R(T) & = & R(T_{|X_1}) \oplus R(T_{|X_2}), \\
T(0) & = & T_{|X_1}(0) \oplus T_{|X_2}(0), \text{ and} \\
T^n & = & (T_{|X_1})^n \oplus (T_{|X_2})^n \text{ for all } n \in \mathbb{N}.
\end{array} \right.$$

Definition 2.15.4 *Let $T \in LR(X)$. T is said to be a Kato linear relation of degree d, if there exists $d \in \mathbb{N}$ and a pair of closed subspaces (M, N) of E such that $E = M \oplus N$, T is completely reduced by the pair (M, N) with $T_{|M}$ is a regular linear relation and $T_{|N}$ is a nilpotent bounded operator of degree d (that is $T_{|N}^d = 0$). T is said to be a Kato linear relation if it is a Kato linear relation of degree d, for some $d \in \mathbb{N}$. The pair (M, N) is called the Kato decomposition of the linear relation T.* ◇

Proposition 2.15.1 [12, Proposition 2.5] *Let T be a Kato linear relation of degree d and let (M, N) be a decomposition of degree d associated with T. Then,*

(i) $R^\infty(T) = T(R^\infty(T)) = R(T_{|M})$. Further, $R^\infty(T)$ is closed.
(ii) For every nonnegative integer $n \geq d$, we have $N(T) \bigcap R(T^n) = N(T^n) \bigcap M = N(T) \bigcap R(T^d)$.
(iii) For every nonnegative integer $n \geq d$, we have $R(T) + N(T^n) = T(M) \oplus N$ is closed. ◇

Theorem 2.15.3 [74, Theorem 2.5] *Let T be a range space relation in a Hilbert space X (i.e., the graph of T is the range of a bounded linear operator* *from a Hilbert space X to $X \times X$), which is quasi-Fredholm linear relation of degree $d \in \mathbb{N}$. Then, T is closed and there exist two closed subspaces M and N of X such that*

(i) $X = M + N$ and $M \bigcap N = \{0\}$.
(ii) $R(T^d) \subset M$,
(iii) $N \subset N(T^d)$ and, if $d \geq 1$, then $N \nsubseteq N(A^{d-1})$. $\diamondsuit$

Proposition 2.15.2 [9, Proposition 11] *Let X be a complex Banach space. If T is a semi regular linear relation in X, then $\gamma(T^n) = \gamma(T)^n$.* $\diamondsuit$

Proposition 2.15.3 [9, Proposition 12] *Let X be a complex Banach space and let T be a regular linear relation in X. Then, for each $n \in \mathbb{N}$, we have*

(i) $N(T^n)^\perp = R((T^*)^n)$.
(ii) $N((T^*)^n)^\top = R(T^n)$. $\diamondsuit$

Proposition 2.15.4 [45, Proposition 3.2] *Let $T \in \mathcal{L}(X)$. Then, T is a quasi-Fredholm operator if and only if there exists $n \in \mathbb{N}$ such that $R(T^n)$ is closed and T^n is semi regular.* $\diamondsuit$

2.15.3   Essentially Semi Regular Linear Relations

Definition 2.15.5 *Let $T \in L\mathcal{R}(X)$. T is said essentially semi regular if $R(T)$ is closed and $N(T) \subset_e R^\infty(T)$.* $\diamondsuit$

Samuel multiplicity operators have been studied by several authors. Particularly, in Ref. [95], the authors studied the Samuel multiplicity of essentially semi regular operators. In the next we extend this study to the general case of multivalued linear operators.

Definition 2.15.6 *For any essentially semi regular linear relation A in Hilbert space X, define its shift (Samuel) multiplicity by*

$$s.mult(A) = \lim_{k \to \infty} \left(\frac{\beta(A^k)}{k} \right).$$

Similarly, define its backward shift (Samuel) multiplicity by

$$b.s.mult(A) = \lim_{k \to \infty} \left(\frac{\alpha(A^k)}{k} \right).$$ $\diamondsuit$

2.16 Perturbation Results for Multivalued Linear Operators

2.16.1 Small Perturbation Theorems of Multivalued Linear Operators

Theorem 2.16.1 [50, Theorem V.3.2] *Let* S, $T \in LR(X,Y)$ *such that* $S(0) \subset \overline{T(0)}$. *If* $\widetilde{\Delta}(S) < \Gamma(T)$, *then* $T + S \in \mathcal{F}_+(X,Y)$. $\diamond$

The following example shows that the hypothesis of $S(0) \subset \overline{T(0)}$ in Theorem 2.16.1 is crucial.

Example 2.16.1 *Let* $T = I_X$ *and let* $S \in LR(X)$ *be defined by* $G(S) = X \times X$. *Then,* S *is a compact multivalued linear operator. Now, by using Proposition 2.12.1, we have* $0 = \widetilde{\Delta}(S) < \Gamma(T) = 1$ *but* $T + S \notin \mathcal{F}_+(X)$.

Theorem 2.16.2 *Let* $T \in BR(X) \bigcap KR(X)$ *and* $\lambda \in \mathbb{K}\backslash\{0\}$. *If* $\dim(T(0)) < \infty$, *then* $\lambda - T \in \Phi(X)$ *and* $i(\lambda - T) = \dim(T(0)) < \infty$. $\diamond$

Proof. In view of Theorem 2.7.2 (iv), we have T is closed. By using (2.51), we have $R(Q_T(\lambda)) = X/T(0)$. In view of $N(Q_T(\lambda)) = T(0)$, then $Q_T(\lambda)$ is a Fredholm operator. Moreover,

$$Q_{\lambda-T}(\lambda - T) = Q_T(\lambda - T) = Q_T(\lambda) - Q_T T. \tag{2.52}$$

Since $\|Q_T T\| = \|T\| < \infty$ and $T \in KR(X)$, then $Q_T T$ is bounded compact operator. Hence, by using (2.52), we have $Q_{\lambda-T}(\lambda-T)$ is a Fredholm operator. So, by using Lemma 2.14.3 (iii), we have $\lambda - T$ is a Fredholm linear relation and $i(\lambda - T) = i(Q_{\lambda-T}(\lambda - T)) = \dim T(0)$. Q.E.D.

The following example shows that if T is a bounded compact linear relation on a Banach space X and $\lambda \in \mathbb{K} \setminus \{0\}$, then $\lambda - T$ may not be Fredholm.

Example 2.16.2 *Let* X *be an infinite-dimensional Banach space and let* T *be a linear relation whose graph is* $X \times X$. *Then,* T *is a bounded compact closed linear relation such that* $I + T$ *is not Fredholm. Indeed, we infer immediately from the definition of the sum (Definition 2.3.4) of linear relations that* $I + T = T$. *Observing that for any linear relation* U *in a vector space* E, *we have* $\mathcal{D}(U) \times E = G(U) + (\{0\} \times E)$, $\{0\} \times U(0) = G(U) \bigcap(\{0\} \times E)$ *and* $N(U) \times \{0\} = G(U) \bigcap(E \times \{0\})$. *We deduce trivially that* $\mathcal{D}(T) = N(T) =$

$T(0) = X$. Therefore, T is not an operator (since $T(0) \neq \{0\}$), $Q_T T$ is the zero operator on X (since $\mathcal{D}(T) = T(0) = X$). So, T is a bounded compact linear relation. Finally, it follows from the equalities $I + T = T$ and $N(T) = X$ that $I + T$ is not a Fredholm linear relation.

Remark 2.16.1 *We remark that in Example 2.16.2, $\dim T(0) = \infty$.*   ◇

Theorem 2.16.3 [50, Theorem V.5.12] *Let X and Y be two normed spaces and let S, $T \in LR(X, Y)$ such that $\mathcal{D}(S) \supset \mathcal{D}(T)$. Then,*

(i) If $S \in \mathcal{PRK}(X, Y)$ and $T \in \mathcal{F}_-(X, Y)$, then $T + S \in \mathcal{F}_-(X, Y)$.
(ii) If $\|S\| < \gamma(T^*)$ and $T \in \mathcal{F}_-(X, Y)$, then $T + S \in \mathcal{F}_-(X, Y)$.
(iii) If $\dim(R(S)) < \infty$ and $T \in \mathcal{F}_-(X, Y)$, then $T + S \in \mathcal{F}_-(X, Y)$.   ◇

By using Proposition 2.3.4, Theorems 2.16.1 and 2.16.3, Propositions 2.12.1 and 2.7.2, one can easily get to the following result.

Corollary 2.16.1 *Let X and Y be two complete spaces and let S, $T \in LR(X, Y)$ such that $S(0) \subset T(0)$ and $\overline{\mathcal{D}(T)} \subset \mathcal{D}(S)$. Then,*

(i) If $S \in \mathcal{PRK}(X, Y)$ and $T \in \Phi_-(X, Y)$, then $T \pm S \in \Phi_-(X, Y)$.
(ii) If $\|S\| < \gamma(T^*)$ and $T \in \Phi_-(X, Y)$, then $T \pm S \in \Phi_-(X, Y)$.
(iii) If $\dim(R(S)) < \infty$ and $T \in \Phi_-(X, Y)$, then $T + S \in \Phi_-(X, Y)$.
(iv) If $S \in \mathcal{SSR}(X, Y)$, and $T \in \Phi_+(X, Y)$, then $T \pm S \in \Phi_+(X, Y)$.
(v) If $S \in \mathcal{SSR}(X, Y)$, and $T + S \in \Phi_+(X, Y)$ or $T - S \in \Phi_+(X, Y)$, then $T \in \Phi_+(X, Y)$.
(vi) If $\|S\| < \gamma(T)$ and $T \in \Phi_+(X, Y)$, then $T \pm S \in \Phi_+(X, Y)$.
(vii) If $S \in \mathcal{SSR}(X, Y)$, then $T + S \in \Phi(X, Y)$ and $i(T + S) = i(T)$.   ◇

Corollary 2.16.2 *Let X be a Banach space and $T \in \Phi(X)$. If $0 < |\lambda| < \gamma(T)$, then $\lambda - T \in \Phi(X)$.*   ◇

Proof.   To prove this corollary, it is sufficient to replace S by λI_X in Corollary 2.16.1 (ii), (vi) and Proposition 2.9.3 (iii).   Q.E.D.

Lemma 2.16.1 *Let $A \in CR(X)$ and $K \in LR(X)$ be continuous such that $K(0) \subset A(0)$ and $\mathcal{D}(A) \subset \mathcal{D}(K)$. Then,*

(i) $A \in \Phi_+(X)$ if and only if, $A + K - K \in \Phi_+(X)$. In such case

$$i(A) = i(A + K - K).$$

(ii) $A \in \Phi_-(X)$ if and only if, $A + K - K \in \Phi_-(X)$. In such case

$$i(A) = i(A + K - K).$$ ◇

Proof. We first note that by using Proposition 2.7.2, $A + K$ and $A + K - K$ are closed. Furthermore, we have

(a) $(A + K - K)(0) = A(0)$ and $Q_{A+K-K} = Q_{A(0)/K(0)}Q_K$. Indeed,

$$(A + K - K)(0) = A(0) + K(0) = A(0),$$

then $Q_{A+K-K} = Q_A$ and by Lemma 2.1.3, we infer that

$$Q_A = Q_{A(0)/K(0)}Q_K.$$

(b) $Q_{A+K-K}(A + K - K) = Q_A(A)$. Indeed,

$$Q_K(A + K - K) = Q_K(A) + Q_K(K) - Q_K(K) = Q_K(A),$$

so, by (a) we obtain (b). Now, combining (b) and Lemma 2.14.3, we obtain the result. Q.E.D.

Lemma 2.16.2 *Let $S \in LR(X,Y)$ and $T \in \mathcal{F}_+(X,Y)$ with $\dim \mathcal{D}(T) = \infty$. If S is precompact, then S is strictly singular, with $\widetilde{\Delta}(S) < \Gamma(T)$. If, additionally, $S(0) \subset \overline{T(0)}$, then $T + S \in \mathcal{F}_+(X,Y)$.* ◇

Proof. Since $T \in \mathcal{F}_+(X,Y)$, and $\dim \mathcal{D}(T) = \infty$, then by using Proposition 2.14.6, we obtain

$$\Gamma(T) > 0.$$

Since S is precompact and using Proposition 2.11.1, we get

$$\overline{\Gamma}_0(S) = 0.$$

So, $\widetilde{\Delta}(S) \leq \overline{\Gamma}_0(S) = 0$. Hence,

$$\widetilde{\Delta}(S) = 0.$$

Then, S is strictly singular. Thus, by Proposition 2.3.4, Theorems 2.16.1 and 2.16.3, it follows that $T + S \in \mathcal{F}_+(X,Y)$. Q.E.D.

Theorem 2.16.4 *Let X be a Banach space and let $T \in CR(X)$, $S \in LR(X)$ such that $S(0) \subset T(0)$, $\overline{\mathcal{D}(T)} \subset \mathcal{D}(S)$, $\|S\| < \gamma(T)$ and $\beta(T) < \infty$. Then,*

$$\gamma(T - S) \geq \|P\|^{-1}\gamma(T) - \|S\|,$$

where P is any continuous projection defined on $\mathcal{D}(T^*)$ with kernel $N(T^*)$. ◇

Proof. Combining both T is closed and $\gamma(T) > 0$, and by using Propositions 2.8.1 (v) and 2.9.3 (i), we have $R(T)$ is closed. Accordingly, $\beta(T) < \infty$ and $T \in \Phi_-(X)$. By virtue of Theorem 2.9.3 (iii), we have

$$0 < \gamma(T^*).$$

Since

$$\|S\| < \gamma(T) = \gamma(T^*)$$

and according to $S(0) \subset T(0)$, $\overline{\mathcal{D}(T)} \subset \mathcal{D}(S)$, and Corollary 2.16.1 (ii), we obtain $T - S \in \Phi_-(X)$. This implies from Theorem 2.14.4 (i) that $(T - S)^* \in \Phi_+(X^*)$. Using the fact that $\mathcal{D}(T) \subset \mathcal{D}(S)$, and in view of Proposition 2.8.1 (iii), we can write

$$(T - S)^* = T^* - S^*. \tag{2.53}$$

Now, we have to prove that $S^*(0) \subset T^*(0)$. Indeed, since $\mathcal{D}(T) \subset \mathcal{D}(S)$, then by using Proposition 2.8.1 (iv), we have

$$S^*(0) = \mathcal{D}(S)^\perp \subset \mathcal{D}(T)^\perp = T^*(0).$$

By referring to Proposition 2.9.6 and (2.53), we obtain

$$\gamma((T - S)^*) = \gamma(T^* - S^*) \geq \|P\|^{-1}\gamma(T^*) - \|S^*\|,$$

where P is any continuous projection defined on $\mathcal{D}(T^*)$ with kernel $N(T^*)$. Finally, the use of Proposition 2.9.3 (iii) allows us to conclude that

$$\gamma(T - S) \geq \|P\|^{-1}\gamma(T) - \|S\|. \qquad \text{Q.E.D.}$$

Corollary 2.16.3 *Let X be a Banach space and let $T \in C\mathcal{R}(X)$, $S \in L\mathcal{R}(X)$ such that $S(0) \subset T(0)$, $\overline{\mathcal{D}(T)} \subset \mathcal{D}(S)$, and $\|S\| < \gamma(T)$. If $T \in \Phi_-(X)$, then*

$$\gamma(T - S) \geq \|P\|^{-1}\gamma(T) - \|S\|,$$

where P is any continuous projection defined on $\mathcal{D}(T^*)$ with kernel $N(T^*)$.$\Diamond$

Proof. The proof is a direct consequence of Theorem 2.16.4 and Proposition 2.5.3. $\qquad$ Q.E.D.

2.17 Fredholm Perturbation Classes of Linear Relations

Definition 2.17.1 *Let X and Y be two Banach spaces. Then,*

(i) The set of upper semi-Fredholm perturbations of linear relations is defined by

$$PR\left(\Phi_+(X,Y)\right) = \{S \in \mathcal{LR}(X,Y) \text{ is continuous}: \ T+S \in \Phi_+(X,Y),$$
$$\text{for all } T \in \Phi_+(X,Y), \ \mathcal{D}(T) \subset \mathcal{D}(S) \text{ and } T(0) \supset S(0)\}.$$

(ii) The set of lower semi-Fredholm perturbations of linear relations is defined by

$$PR(\Phi_-(X,Y)) = \{S \in \mathcal{LR}(X,Y) \text{ is continuous}: \ T+S \in \Phi_-(X,Y), \text{ for all}$$
$$T \in \Phi_-(X,Y), \ \mathcal{D}(T) \subset \mathcal{D}(S) \text{ and } T(0) \supset S(0)\}.$$

(iii) The set of α-Atkinson perturbations linear relations is defined by

$$PR\left(\mathcal{A}_\alpha(X,Y)\right) = \{S \in LR(X,Y): T+S \in \mathcal{A}_\alpha(X,Y) \text{ for all } T \in \mathcal{A}_\alpha(X,Y)$$
$$\text{such that } T(0) \text{ is topologically complemented in } Y,$$
$$\mathcal{D}(T) \subset \mathcal{D}(S) \text{ and } T(0) \supset S(0)\}.$$

(iv) The set of β-Atkinson perturbations linear relations is defined by

$$PR\left(\mathcal{A}_\beta(X,Y)\right) = \{S \in LR(X,Y) \text{ such that } T+S \in \mathcal{A}_\beta(X,Y) \text{ for all } T \in$$
$$\mathcal{A}_\beta(X,Y) \text{ such that } T^*(0) \text{ is topologically complemented in } Y^*,$$
$$\mathcal{D}(T) \subset \mathcal{D}(S) \text{ and } S(0) \subset T(0)\}.$$

(v) The set of Atkinson perturbations is defined by

$$PR(\mathcal{A}(X,Y)) := PR\left(\mathcal{A}_\alpha(X,Y)\right) \bigcap PR\left(\mathcal{A}_\beta(X,Y)\right). \qquad \diamondsuit$$

If $X = Y$, we denote by $PR(\Phi_+(X)) := PR(\Phi_+(X,X))$, $PR(\Phi_-(X)) := PR(\Phi_-(X,X))$, $PR(\mathcal{A}_\alpha(X)) := PR(\mathcal{A}_\alpha(X,X))$, $PR(\mathcal{A}_\beta(X)) := PR(\mathcal{A}_\beta(X,X))$ and $PR(\Phi(X)) := PR(\Phi(X,X))$.

Proposition 2.17.1 *Let X and Y be two Banach spaces, and let $F_1, F_2 \in LR(X,Y)$. Then,*

(i) If $F_1 \in PR\left(\Phi_+(X,Y)\right)$ and $F_2 \in PR\left(\Phi_+(X,Y)\right)$, then $F_1 \pm F_2 \in PR\left(\Phi_+(X,Y)\right)$.

(ii) If $F_1 \in PR\left(\Phi_-(X,Y)\right)$ and $F_2 \in PR\left(\Phi_-(X,Y)\right)$, then $F_1 \pm F_2 \in PR\left(\Phi_-(X,Y)\right)$.

(*iii*) If $F_1 \in \mathcal{PR}(\mathcal{A}_\alpha(X,Y))$ and $F_2 \in \mathcal{PR}(\mathcal{A}_\alpha(X,Y))$, then $F_1 \pm F_2 \in \mathcal{PR}(\mathcal{A}_\alpha(X,Y))$.

(*iv*) If $F_1 \in \mathcal{PR}(\mathcal{A}_\beta(X,Y))$ and $F_2 \in \mathcal{PR}(\mathcal{A}_\beta(X,Y))$, then $F_1 \pm F_2 \in \mathcal{PR}(\mathcal{A}_\beta(X,Y))$.

(*v*) If $F_1 \in \mathcal{PR}(\Phi(X,Y))$ and $F_2 \in \mathcal{PR}(\Phi(X,Y))$, then $F_1 \pm F_2 \in \mathcal{PR}(\Phi(X,Y))$. $\diamond$

Proof.   (*i*) Let $A \in \Phi_+(X,Y)$ such that

$$\mathcal{D}(A) \subset \mathcal{D}(F_1 \pm F_2) = \mathcal{D}(F_1) \bigcap \mathcal{D}(F_2), \tag{2.54}$$

and

$$A(0) \supset (F_1 \pm F_2)(0) = F_1(0) + F_2(0). \tag{2.55}$$

By using the relation (2.54), we deduce that $\mathcal{D}(A) \subset \mathcal{D}(F_1)$ and, according to the relation (2.55), we infer that $A(0) \supset F_1(0)$. However, if $F_1 \in \mathcal{PR}(\Phi_+(X,Y))$, then $A + F_1 \in \Phi_+(X,Y)$,

$$\mathcal{D}(A + F_1) = \mathcal{D}(A) \subset \mathcal{D}(F_2), \tag{2.56}$$

and

$$(F_1 + A)(0) \supset F_2(0). \tag{2.57}$$

Now, by using the relations (2.56), (2.57) and the fact that $F_2 \in \mathcal{PR}(\Phi_+(X,Y))$, we deduce that $A + F_1 \pm F_2 \in \Phi_+(X,Y)$. So, $F_1 \pm F_2 \in \mathcal{PR}(\Phi_+(X,Y))$.

The proof of the other cases is analogous to the proof of (*i*).   Q.E.D.

2.18   Spectrum and Pseudospectra of Linear Relations

2.18.1   Resolvent Set and Spectrum of Linear Relations

Definition 2.18.1 *Let $T \in LR(X)$ and $\lambda \in \mathbb{C}$. The quantites*

$$R(\lambda, T) = (\lambda I_X - T)^{-1} = (\lambda - T)^{-1}$$

is called the resolvent of T (corresponding to λ) and

$$T_\lambda := (\lambda - \widetilde{T})^{-1}$$

is called the complete resolvent of T, where $\widetilde{T}$ is given in (2.4). $\diamond$

Definition 2.18.2 *Let $T \in LR(X)$. The resolvent set and the spectrum of T are, respectively, defined as*

$$\begin{aligned} \rho(T) &= \{\lambda \in \mathbb{C} \text{ such that } T_\lambda \text{ is single valued everywhere defined}\}, \\ \sigma(T) &= \mathbb{C}\backslash\rho(T). \end{aligned}$$

The left spectrum is defined by

$$\sigma_l(T) = \{\lambda \in \mathbb{C} \text{ such that } \lambda - T \notin \mathcal{GR}_l(X)\},$$

and the right spectrum is defined by

$$\sigma_r(T) = \{\lambda \in \mathbb{C} \text{ such that } \lambda - T \notin \mathcal{GR}_r(X)\}.$$

The approximate point spectrum of T is the set defined by

$$\sigma_{ap}(T) := \{\lambda \in \mathbb{C} \text{ such that } \lambda - T \text{ is not bounded below}\}.$$

The defect spectrum of T is the set defined by

$$\sigma_\delta(T) := \{\lambda \in \mathbb{C} \text{ such that } \lambda - T \text{ is not surjective}\}. \qquad \diamondsuit$$

Remark 2.18.1 *Let $T \in LR(X)$. Then,*

(i) The resolvent set $\rho(T)$ is open, whereas the spectrum $\sigma(T)$ of a closed linear relation T is closed.
(ii) We observe that

$$\sigma(T) = \sigma(\widetilde{T}) = \sigma(\overline{T}) \quad and \quad \rho(T) = \rho(\overline{T}),$$

where $\widetilde{T}$ is given in (2.4) and $\overline{T}$ is the closure of T. $\qquad \diamondsuit$

Lemma 2.18.1 [50, Exercise 6.1.2] *Let $T \in LR(X)$. Then,*

$$\rho(T) = \{\lambda \in \mathbb{C} \text{ such that } \lambda - T \text{ is injective, open with dense range on } X\}. \diamondsuit$$

Lemma 2.18.2 *Let S be a closed linear relation in X. Then,*

(i) If S is bounded, then S^n is bounded for all $n \in \mathbb{N}$.
(ii) If $\rho(S) \neq \emptyset$, then S^n is closed for all $n \in \mathbb{N}$.
(iii) If S is everywhere defined and M is a closed subspace of X such that $S(0) \subset M$, then $S^{-1}(M)$ is a closed subspace of X. $\qquad \diamondsuit$

Remark 2.18.2 *Let $\mu \in \rho(T)$. Then, for all $\mu \neq \lambda \in \mathbb{C}$ and*

$$S = (\mu - \lambda)((\mu - \lambda)^{-1} - T_\mu),$$

we have

$$\lambda - T = S(\mu - T). \qquad \diamondsuit$$

Proposition 2.18.1 [50, Proposition 6.1.3, and 6.1.11] *Let $T \in LR(X)$. Then,*

(i) $\rho(T)$ *is an open set.*

(ii) [*Resolvent Equation*] *Let* $\lambda,\ \mu \in \rho(T)$. *Then,*

$$T_\mu - T_\lambda = (\lambda - \mu)T_\mu T_\lambda.$$

(iii) $\sigma(T) = \sigma(T^*)$.   ◇

Proposition 2.18.2 [50, Corollary 6.1.8] *The family* $\{T_\lambda$ *such that* $\lambda \in \rho(T)\}$ *of resolvent operators is holomorphic.*   ◇

Let $H(U, X)$ denote the space of all analytic functions from U into X, where U is open set of $\mathbb{C}$ and X is a complex Banach space. Let $\mathbb{D}(0, \varepsilon)$ denoted the open disc of $\mathbb{C}$ centred at 0 and with radius ε.

2.18.2   Subdivision of the Spectrum of Linear Relations

There are many different ways to subdivide the spectrum of linear relation. Some of them are motivated by applications to physics (in particular, quantum mechanics).

Definition 2.18.3 *Let $T \in LR(X)$. The point spectrum, continuous, and the residual spectrum are defined, respectively, as*

$$
\begin{aligned}
P\sigma(T) &:= \ \{\lambda \in \mathbb{C} \ such \ that \ N(T_\lambda) \neq 0\}, \\
R\sigma(T) &:= \ \left\{\lambda \in \mathbb{C} \ such \ that \ T_\lambda \ is \ injective \ and \ \overline{R(T_\lambda)} \neq X\right\}, \\
C\sigma(T) &:= \ \left\{\lambda \in \mathbb{C} : T_\lambda \ is \ injective \ and \ \overline{R(T_\lambda)} = X \ but \ is \ not \ open\right\}.
\end{aligned}
$$

                                               ◇

Example 2.18.1 *Let $X = l_p(\mathbb{N})$ $(1 \leq p \leq \infty)$, $(a_n)_n$ be a bounded sequence in $\mathbb{K}$, and $T \in \mathcal{L}(l_p(\mathbb{N}))$ be defined by*

$$T(x_1, x_2, x_3, \ldots) = (a_1 x_1, a_2 x_2, a_3 x_3, \ldots).$$

Let first $p < \infty$. Denoting by $\Theta = \{a_1, a_2, a_3, \ldots\}$ the set of all elements of the sequence $(a_n)_n$, we get $\sigma(T) = \overline{\Theta}$, $P\sigma(T) = \Theta$, $C\sigma(T) = \overline{\Theta}\backslash\Theta$ and $R\sigma(T) = \emptyset$. In particular, if $a_n \to 0$ (i.e., T is compact single valued), then $C\sigma(T) = \emptyset$. In case $p = \infty$ the continuous and residual spectrum change their role, i.e., we get $\sigma(T) = \overline{\Theta}$, $P\sigma(T) = \Theta$, $C\sigma(T) = \Theta$ and $R\sigma(T) = \overline{\Theta}\backslash\Theta$.

Remark 2.18.3 (*i*) *We remark that in Example 2.16.2, $\rho(T) = \emptyset$.*

(*ii*) *It is well known that the spectrum of a bounded operator in a Banach space is a proper subset of $\mathbb{C}$. But a non-single valued bounded linear relation may have a spectrum that coincides with the whole complex plane. But this property is not valid in the case of a linear relation. For example, let $X = \mathbb{C}^2$ and let T be defined in the terms of its graph by*

$$G(T) = G(I) + \{0\} \times E,$$

where E may be one two dimensional. Then, T is bounded, $T(0) = E$ and $\sigma(T) = \mathbb{C}$. ◇

Example 2.18.2 *Let $X = l_2$ and K be the linear single valued defined by*

$$K(x) = \left(0, x_1, \frac{1}{2}x_2, \frac{1}{3}x_3, \dots\right).$$

Then, K is an injective compact single valued linear operator with $\sigma(K) = \{0\}$. Now, we defined $T \in LR(X)$ by

$$G(T) = G(K) + \{0\} \times E.$$

Then, T is a bounded compact linear relation. Since $N(T) = K^{-1}(E)$, then T is injective as $E \bigcap R(K) = \{0\}$. Also, T is closed if and only if, E is closed. Now, suppose that E is closed and let $\lambda \in \mathbb{C}$, $\lambda \neq 0$. Since $\lambda - K$ is surjective, then for all $0 \neq e \in E$, there exists $x \neq 0$ such that $(\lambda - K)x = e$. Then, $(\lambda - T)x = E = (\lambda - T)(0)$. Hence, λ is an eigenvalue of T. We have constructed a bounded compact closed injective linear relation T such that $P\sigma(T) = \mathbb{C}\backslash\{0\}$, and $\sigma(T) = \mathbb{C}$. Moreover, in this example we can choose $\dim(T(0)) = \dim(E) < \infty$.

Lemma 2.18.3 [91, Proposition 7.3.2] *Let X, Y be two normed space and let $T \in LR(X,Y)$. If A is a linear selection of T, then $P\sigma(T) = P\sigma(A)$.* ◇

Lemma 2.18.4 [50, Theorem 6.5.4] *Let X be a normed space and $T \in LR(X)$. Then, for any complex polynomial P, we have $\sigma(P(\widetilde{T})) = \sigma(P(T))$, where $\widetilde{T}$ is the completion of a linear relation T.* ◇

Lemma 2.18.5 *Let $P(T) = \prod_{i=1}^{n}(T - \lambda_i)^{m_i}$ as in Definition 2.13.1 and let m be a positive integer. Then, $R(P(T)^m)$ is closed if and only if, $R((T-\lambda_i)^{m_i m})$ is closed for all $i, 1 \leq i \leq n$.* ◇

Proof. Note that $P(T)$ is closed and $\sigma(P(T)^m) = (\sigma(P(T))^m$ by virtue of Lemma 2.18.4. So, $P(T)^m$ is closed and bounded by Lemma 2.18.2. Assume that $R(P(T)^m)$ is closed and we write

$$U := (T - \lambda_1)^{m_1} \text{ and } V := \prod_{i=2}^{n} (T - \lambda_i)^{m_i}.$$

Then, arguing as in Lemma 2.13.1 (iv) and using Lemma 2.18.2 together with the condition $R(P(T)^m)$ is closed, we obtain that

$$V^{-m} R(P(T)^m) = R(U^m) + N(V^m) = R(U^m),$$

that is, $R((T - \lambda_1)^{m_i m})$ is closed. Now, applying the above reasoning, we obtain that $R((T - \lambda_i)^{m_i m})$ is closed, $1 \leq i \leq n$.

The converse follows immediately from Lemmas 2.5.1 and 2.13.1.   Q.E.D.

2.18.3   Essential Spectra of Linear Relations

In this book, we are concerned with the following essential spectra (see Refs. [1, 17, 26, 65, 66, 93]):

$$\sigma_{e1}(T) = \{\lambda \in \mathbb{C} \text{ such that } \lambda - T \notin \Phi_+(X)\},$$

$$\sigma_{e2}(T) = \{\lambda \in \mathbb{C} \text{ such that } \lambda - T \notin \Phi_-(X)\},$$

$$\sigma_{e3}(T) = \{\lambda \in \mathbb{C} \text{ such that } \lambda - T \notin \Phi_{\pm}(X)\},$$

$$\sigma_{e4}(T) = \{\lambda \in \mathbb{C} \text{ such that } \lambda - T \notin \Phi(X)\},$$

$$\sigma_{e5}(T) = \bigcap_{K \in \mathcal{K}_T(X)} \sigma(T + K),$$

$$\sigma_{q\phi_d}(T) = \{\lambda \in \mathbb{C} \text{ such that } \lambda - T \notin q\phi_d(X)\},$$

$$\sigma_{eb}(T) = \{\lambda \in \mathbb{C} \text{ such that } \lambda - T \notin \mathcal{B}(X)\}$$

$$\sigma_{eap}(T) = \bigcap_{K \in \mathcal{K}_T(X)} \sigma_{ap}(T + K),$$

$$\sigma_{e\delta}(T) = \bigcap_{K \in \mathcal{K}_T(X)} \sigma_\delta(T + K),$$

$$\sigma_{e\alpha}(T) = \{\lambda \in \mathbb{C} \text{ such that } \lambda - T \notin \mathcal{A}_\alpha(X)\},$$

$$\sigma_{e\beta}(T) = \{\lambda \in \mathbb{C} \text{ such that } \lambda - T \notin \mathcal{A}_\beta(X)\},$$

$$\sigma_{el}(T) = \bigcap_{K \in \mathcal{K}_T(X)} \sigma_l(T + K),$$

$$\sigma_{er}(T) = \bigcap_{K \in \mathcal{K}_T(X)} \sigma_r(T + K),$$

$$\sigma_r(T) := \bigcap_{K \in \Upsilon_T(X)} \sigma(T + K).$$

$$\sigma_l(T) = \bigcap_{K \in \Psi_T(X)} \sigma(T + K),$$

with $\mathcal{K}_T(X) = \{K \in \mathcal{KR}(X) \text{ such that } \mathcal{D}(T) \subset \mathcal{D}(K),\ K(0) \subset T(0)\}$,

$$\Lambda(X) = \{T \in L\mathcal{R}(X) : \mu T \text{ is demicompact for every } \mu \in [0,1]\},$$

$$\Upsilon_T(X) = \left\{ K \in L\mathcal{R}(X) : \begin{array}{c} \mathcal{D}(T) \subset \mathcal{D}(K), \\ K(0) \subset T(0) \text{ and } \forall \mu \in \rho(T+K), \\ -(\mu - T - K)^{-1}K \in \Lambda(X) \end{array} \right\},$$

and

$$\Psi_T(X) = \left\{ K \in L\mathcal{R}(X) : \begin{array}{c} \mathcal{D}(T) \subset \mathcal{D}(K), \\ K(0) \subset T(0) \text{ and } \forall \mu \in \rho(T+K), \\ -K(\mu - T - K)^{-1} \in \Lambda(X) \end{array} \right\}.$$

In general those spectra satisfy the following inclusions

$$\sigma_{e1}(T) \bigcap \sigma_{e2}(T) = \sigma_{e3}(T) \subset \sigma_{e4}(T) \subset \sigma_{e5}(T) \subset \sigma(T).$$

$$\sigma_{e1}(T) \subset \sigma_{eap}(T),\ \sigma_{e2}(T) \subset \sigma_{e\delta}(T),$$

$$\sigma_{e1}(T) \subset \sigma_{e\alpha}(T),\ \sigma_{e2}(T) \subset \sigma_{e\beta}(T),\ \text{and}$$

$$\sigma_{e5}(T) = \sigma_{eap}(T) \bigcup \sigma_{e\delta}(T).$$

Remark 2.18.4 (*i*) *The sets* $\sigma_{ei}(T)$ *are closed, with* $i = 1, \ldots, 5, ap,\ \delta,\ \alpha,\ \beta,\ l,\ r$.
(*ii*) *It is clear that, for all* $K \in \mathcal{K}_T(X)$, *we have* $\sigma_{ei}(T + K) = \sigma_{ei}(T)$, *with*
$i = 5,\ ap,\ \delta,\ l,\ r$.
(*iii*) *For all* $K \in \mathcal{K}_T(X)$, $\sigma_{ei}((T + K)^*) = \sigma_{ei}(T^* + K^*) = \sigma_{ei}(T^*)$, *with*
$i = 5,\ ap,\ \delta,\ l,\ r$.
(*iv*) *If* K *is a single valued compact operator, then* $\sigma_{ei}(K) = \{0\}$, *with*
$i = 1, \ldots, 5,\ \alpha,\ \beta,\ l,\ r$. *But this is not true if* K *is a compact linear*
relation. In fact, let $K \in \mathcal{KR}(X)$ *be a bounded linear relation such that*
$0 \neq \dim K(0) < \infty$. *By using Theorem 2.16.2, we get* $\lambda - K \in \Phi(X)$ *and*
$i(\lambda - K) = \dim K(0)$. *So,* $\sigma_{e5}(K) = \mathbb{C}$. $\diamondsuit$

2.19   *S*-Spectra of Linear Relations in Normed Space

Definition 2.19.1 *Let X be a normed vector space, $S, T \in LR(X)$, S be a continuous linear relation such that $S(0) \subset \overline{T(0)}$ and $\mathcal{D}(S) \supset \mathcal{D}(T)$. Then, we define the S-resolvent set of T by*

$$\rho_S(T) := \{\lambda \in \mathbb{C} \text{ such that } \lambda S - T \text{ is injective, open with dense range on } X\}.$$

We denote the S-spectrum set of T by

$$\sigma_S(T) = \mathbb{C}\backslash\rho_S(T). \qquad \diamond$$

Remark 2.19.1 *It is clear that if $S, T \in CR(X)$ and X is complete, we will return to the S-spectrum definition in a Banach space with closed linear relation. In this case*

$$\begin{aligned}\rho_S(T) &= \{\lambda \in \mathbb{C} \text{ such that } \lambda S - T \text{ is bijective}\} \\ &= \{\lambda \in \mathbb{C} : (\lambda S - T)^{-1} \text{ is a bounded linear operator on } X\}. \quad \diamond\end{aligned}$$

Definition 2.19.2 *Let $S, T \in LR(X)$ and $\lambda \in \mathbb{C}$. We recall the S-resolvent of T at λ the operator defined by*

$$R_S(\lambda, T) := (\lambda S - T)^{-1}. \qquad \diamond$$

Definition 2.19.3 *The augmented S-spectrum of T is the set*

$$\overline{\sigma}_S(T) = \begin{cases} \sigma_S(T) \bigcup \{\infty\} & \text{if } 0 \in \sigma_{S^{-1}}(T^{-1}) \\ \sigma_S(T) & \text{otherwise.} \end{cases} \qquad \diamond$$

The Mobius transformation is defined by

$$\widetilde{\eta}_\mu(\lambda) = (\mu - \lambda)^{-1},$$

where μ is a fixed point of $\mathbb{C}$, and a topological homeomorphism from $\mathbb{C}^\infty$ onto itself.

Proposition 2.19.1 *Let $\lambda \in \mathbb{C}$, $T \in CR(X)$ and $S \in \mathcal{L}(X)$. Let $\mu \in \rho_S(T)$ such that $\lambda \neq \mu$. Then,*

$$N(\lambda S - T) = N((\mu - \lambda)^{-1} - (\mu S - T)^{-1}S). \qquad \diamond$$

Proof. Let $\mu \in \rho_S(T)$ and $\lambda \in \mathbb{C}$ such that $\mu \neq \lambda$. Let $(x,y) \in G(T-T+(\mu-\lambda)S)$, then $x \in \mathcal{D}(T)$ and $y \in (T-T+(\mu-\lambda)S)x$. So, $y \in T(0)+(\mu-\lambda)Sx$. Since $T(0) \subset (\lambda S - T)x$ and S is a single valued linear operator, we obtain

$$y \in (\lambda S - T)x + (\mu - \lambda)Sx = (\mu S - T)x.$$

Hence, $(x,y) \in G(\mu S - T)$. We infer that $G(T-T+(\mu-\lambda)S) \subset G(\mu S - T)$. Furthermore, $(T-T+(\mu-\lambda)S)(0) = T(0) = (\mu S - T)(0)$ and clearly $\mathcal{D}(T-T+(\mu-\lambda)S) = \mathcal{D}(T) \bigcap \mathcal{D}(S) = \mathcal{D}(T) = \mathcal{D}(\mu S - T)$. Hence, by using Proposition 2.3.6 (iii), we have $T - T + (\mu - \lambda)S = \mu S - T$. So,

$$T(0) + (\mu - \lambda)Sx = (\lambda S - T)x + (\mu - \lambda)Sx = (\mu S - T)x. \qquad (2.58)$$

Thus, $x \in N(\lambda S - T)$ if and only if, $x \in \mathcal{D}(\lambda S - T) = \mathcal{D}(T)$ and $(\lambda S - T)x = (\lambda S - T)(0) = T(0)$ if and only if, $(\mu S - T)x = (\lambda S - T)x - \lambda Sx + \mu Sx = T(0) - \lambda Sx + \mu Sx$ if and only if, $(\mu S - T)^{-1}(\mu S - T)x = (\mu S - T)^{-1}((\mu - \lambda)Sx + T(0))$ (by (2.58)) if and only if, $x + (\mu S - T)^{-1}(0) = (\mu - \lambda)(\mu S - T)^{-1}Sx + (\mu S - T)^{-1}(\mu S - T)(0)$ if and only if, $0 = (I - (\mu - \lambda)(\mu S - T)^{-1}S)x$ (since $(I - (\mu - \lambda)(\mu S - T)^{-1}S)(0) = 0$) if and only if, $((\mu - \lambda)^{-1} - (\mu S - T)^{-1}S)x = 0$ if and only if, $x \in N((\mu - \lambda)^{-1} - (\mu S - T)^{-1}S)$. Q.E.D.

Corollary 2.19.1 *Let $\mu \in \rho(T)$ and let $\mu \neq \lambda$. Then,*

$$N(\lambda - T) = N\left((\mu - \lambda)^{-1} - (\mu - T)^{-1}\right).$$

In particular, λ is an eigenvalue of T if and only if, $(\mu - \lambda)^{-1}$ is an eigenvalue of $(\mu - T)^{-1}$. ◇

2.19.1 Some Properties of S-Resolvent of Linear Relations in Normed Space

Proposition 2.19.2 *Let $T \in LR(X)$ and S be a continuous linear relation such that $S(0) \subset \overline{T(0)}$, $\mathcal{D}(T) \subset \mathcal{D}(S)$. If $\|S\| = 0$, then $\rho_S(T) = \emptyset$ or $\mathbb{C}$.* ◇

Proof. Let $\|S\| = 0$. We suppose that $\rho_S(T) \neq \emptyset$ and $\rho_S(T) \neq \mathbb{C}$. Let $\lambda_0 \in \rho_S(T)$. Then, $\lambda_0 S - T$ is injective, open with dense range. Hence, $\gamma(T - \lambda_0 S) > 0$. It is clear to see $\|\lambda_0 S\| = |\lambda_0| \|S\| = 0$. So, $\|\lambda_0 S\| < \gamma(T - \lambda_0 S)$. Then, by using Lemma 2.9.3, we have $\lambda_0 S + T - \lambda_0 S$ is injective, open with dense range. Moreover, $\lambda_0 S(0) \subset \overline{T(0)}$ and $\mathcal{D}(T) \subset \mathcal{D}(S) = \mathcal{D}(\lambda_0 S)$. Thus, by using Proposition 2.3.4, $\lambda_0 S + T - \lambda_0 S = T$ is injective open with dense range. Now, let $\lambda \in \mathbb{C}$, then $\|\lambda S\| < \gamma(T)$. Hence, $\lambda S - T$ is injective, open with dense range. Thus, $\rho_S(T) = \mathbb{C}$, which is a contradiction. This completes the proof.   Q.E.D.

Corollary 2.19.2 *Let* $S, T \in LR(X)$ *and* S *be a continuous linear relation such that* $S(0) \subset \overline{T(0)}$ *and* $\mathcal{D}(T) \subset \mathcal{D}(S)$. *If* $S(B_{\mathcal{D}(S)}) \subset \overline{S(0)}$, *then* $\rho_S(T) = \emptyset$ *or* $\mathbb{C}$. $\diamond$

The following proposition show that the S-spectrum of linear relation may be a proper subset of $\mathbb{C}$.

Theorem 2.19.1 *Let* $T \in CR(X)$ *be injective with dense range and* S *be a continuous linear relation such that* $S(0) \subset \overline{T(0)}$, $\mathcal{D}(S) \supset \mathcal{D}(T)$ *and* $\|S\| \neq 0$. *Then,*

$$\sigma_S(T) \subset \left\{ \lambda \in \mathbb{C} \text{ such that } |\lambda| \geq \frac{\gamma(T)}{\|S\|} \right\}. $$
     $\diamond$

Proof. It suffices to show that $\left\{ \lambda \in \mathbb{C} \text{ such that } |\lambda| < \frac{\gamma(T)}{\|S\|} \right\} \subset \rho_S(T)$. Let $\lambda \in \mathbb{C}$ such that $|\lambda| < \frac{\gamma(T)}{\|S\|}$. Since $\gamma(T) > 0$, then T is open. On the other hand, $\lambda S(0) = S(0) \subset \overline{T(0)}$ and $\mathcal{D}(\lambda S) = \mathcal{D}(S) \supset \mathcal{D}(T)$. Moreover, $\|\lambda S\| = |\lambda| \|S\| < \gamma(T)$ and T is open, injective with dense range. So, by using Lemma 2.9.3, we obtain that $\lambda S - T$ is open, injective with dense range. Hence, $\lambda \in \rho_S(T)$. This completes the proof.   Q.E.D.

Theorem 2.19.2 *Let* $T \in LR(X)$ *and* S *be a continuous linear relation such that* $S(0) \subset \overline{T(0)}$ *and* $\mathcal{D}(S) \supset \mathcal{D}(T)$. *Then,* $\rho_S(T)$ *is an open set.* $\diamond$

Proof. We will discuss two cases.

First case : $\|S\| = 0$, by using Proposition 2.19.2, we have $\rho_S(T) = \emptyset$ or $\rho_S(T) = \mathbb{C}$. Thus, $\rho_S(T)$ is open.

Second case : $\|S\| \neq 0$, if $\lambda \in \rho_S(T)$, then $\gamma(\lambda S - T) > 0$. Let $\mu \in \mathbb{C}$ such that $|\mu - \lambda| < \frac{\gamma(\lambda S - T)}{\|S\|}$, then $\|(\mu - \lambda)S\| = |\mu - \lambda| \|S\| < \gamma(\lambda S - T)$. Furthermore, $(\mu - \lambda)S(0) \subset \overline{T(0)} = \overline{(\lambda S - T)(0)}$ and $\mathcal{D}((\mu - \lambda)S) = \mathcal{D}(S) \supset \mathcal{D}(T) = \mathcal{D}(\lambda S - T)$. Thus, by using Lemma 2.9.3, $\mu S - T$ is injective, open with dense range. So, $\mu \in \rho_S(T)$. Therefore, $\rho_S(T)$ is open.   Q.E.D.

Theorem 2.19.3 *Let* X *be a Banach space. Let* $T \in CR(X)$ *and* $S \in BR(X)$ *such that* $\mathcal{D}(T) = X$, $S(0) = T(0)$ *and* $\dim S(0) < \infty$. *If* $0 \in \rho(S)$, *then*

$$\rho_S(T) = \rho(S^{-1}T) \bigcap \rho(TS^{-1}). $$
     $\diamond$

Proof. First of all, it should be mentioned that $\lambda - S^{-1}T$ and $\lambda - TS^{-1}$ are closed for all $\lambda \in \mathbb{C}$. Indeed, since S is continuous and $\mathcal{D}(S)$ and $S(0)$ are closed, then by using Theorem 2.7.2 (iv), we have S is closed. Hence, the fact that $0 \in \rho(S)$ implies that $S^{-1} \in \mathcal{L}(X)$. So, S^{-1} is closed . Further, $\alpha(S^{-1}) = \dim(N(S^{-1})) = \dim(S(0) < \infty$. By using Lemma 2.9.1, we have $\gamma(S^{-1}) = \|S\|^{-1} > 0$. Hence, $R(S^{-1})$ is closed. So, we infer from Proposition 2.9.4 that $S^{-1}T$ is closed. Now, applying Proposition 2.7.2, we get $\lambda - S^{-1}T$ is closed. Moreover, $\lambda - TS^{-1}$ is closed. Indeed, since T is closed and S^{-1} is bounded operator (as $0 \in \rho(S)$), then by applying Lemma 2.7.1, we conclude that TS^{-1} is closed, which implies by using Proposition 2.7.2, that $\lambda - TS^{-1}$ is closed. Now, let $\lambda \in \rho_S(T)$. We claim that $\lambda - S^{-1}T$ and $\lambda - TS^{-1}$ are bijective. Indeed, we first verify

$$\lambda - S^{-1}T = S^{-1}(\lambda S - T), \tag{2.59}$$

and

$$\lambda S - T = S(\lambda - S^{-1}T). \tag{2.60}$$

Since S is injective, then $S^{-1}S = I_{\mathcal{D}(S)} = I_X$. Hence, by using the fact that $\mathcal{D}(S^{-1}) = X$ and Proposition 2.3.6 (v), we have

$$
\begin{aligned}
S^{-1}(\lambda S - T) &= \lambda S^{-1}S - S^{-1}T \\
&= \lambda - S^{-1}T.
\end{aligned}
$$

Hence, (2.59) holds. Now, we prove (2.60). Let $x \in X$, then $x \in \mathcal{D}(S^{-1}) = R(S) = X$. So, there exists $a \in X$ such that $(a, x) \in G(S)$ and $(x, a) \in G(S^{-1})$. Thus, $(x, x) \in G(SS^{-1})$. This implies that $G(I_X) \subset G(SS^{-1})$. Then, $\lambda S - T \subset SS^{-1}(\lambda S - T) = S(\lambda - S^{-1}T)$. Consequently,

$$\lambda S - T \subset S(\lambda - S^{-1}T). \tag{2.61}$$

On the other hand,

$$
\begin{aligned}
\mathcal{D}(S(\lambda - S^{-1}T)) &= \left\{ x \in \mathcal{D}(\lambda - S^{-1}T) : (\lambda - S^{-1}T)x \bigcap \mathcal{D}(S) \neq \emptyset \right\} \\
&= \left\{ x \in \mathcal{D}(\lambda - S^{-1}T) : (\lambda - S^{-1}T)x \bigcap X \neq \emptyset \right\} \\
&= \left\{ x \in \mathcal{D}(S^{-1}T) : S^{-1}Tx \bigcap X \neq \emptyset \right\} \\
&= \left\{ x \in \mathcal{D}(T) : Tx \bigcap \mathcal{D}(S^{-1}) \neq \emptyset \right\} \\
&= \left\{ x \in \mathcal{D}(T) : Tx \bigcap X \neq \emptyset \right\} \\
&= \left\{ x \in \mathcal{D}(\lambda S - T) : (\lambda S - T)x \bigcap X \neq \emptyset \right\} \\
&= \mathcal{D}(\lambda S - T).
\end{aligned}
$$

In addition, we have

$$
\begin{aligned}
S(\lambda - S^{-1}T)(0) &= S((\lambda - S^{-1}T(0)) \\
&= SS^{-1}T(0) \\
&= SS^{-1}S(0) \\
&= SS^{-1}(0) \\
&= S(0) \\
&= T(0) \\
&= (\lambda S - T)(0).
\end{aligned}
$$

Thus, these properties and (2.61) implies from Proposition 2.3.6 (*iii*) that (2.60) holds. Now, using (2.59), we have

$$
\begin{aligned}
R(\lambda - S^{-1}T) &= \mathcal{D}((S^{-1}(\lambda S - T))^{-1}) \\
&= \mathcal{D}((\lambda S - T)^{-1}S) \\
&= S^{-1}(\mathcal{D}((\lambda S - T)^{-1})) \\
&= S^{-1}(R(\lambda S - T)).
\end{aligned}
$$

Since $\lambda S - T$ is open with dense range, then $R(\lambda S - T) = \overline{R(\lambda S - T)} = X$. This implies that

$$
\begin{aligned}
R(\lambda - S^{-1}T) &= R(S^{-1}) \\
&= \mathcal{D}(S) \\
&= X.
\end{aligned}
$$

On the other hand,

$$
\begin{aligned}
N(\lambda - S^{-1}T) &= N(S^{-1}(\lambda S - T)), \\
&= (\lambda S - T)^{-1}S(0), \\
&= (\lambda S - T)^{-1}T(0) \\
&= (\lambda S - T)^{-1}(\lambda S - T)(0), \\
&= (\lambda S - T)^{-1}(0) \\
&= \{0\}.
\end{aligned}
$$

Then, $\lambda - S^{-1}T$ is bijective. So, $\lambda \in \rho(S^{-1}T)$. In the same ways, we can prove that $\lambda - TS^{-1}$ is bijective. Accordingly, $\lambda - TS^{-1}$ and $\lambda - S^{-1}T$ are bijective and closed, then

$$
\rho_S(T) \subset \rho(TS^{-1}) \bigcap \rho(S^{-1}T). \tag{2.62}
$$

Conversely, let $\lambda \in \rho(S^{-1}T) \bigcap \rho(TS^{-1})$, then $\lambda - S^{-1}T$ and $\lambda - TS^{-1}$ are bijective. Since S is surjective, then by using (2.60), we have $R(\lambda S - T) = SR(\lambda - S^{-1}T) = R(S) = X$ and $N(\lambda S - T) \subset N(S^{-1}(\lambda S - T)) = N(\lambda - S^{-1}T) = \{0\}$. This implies that $\lambda S - T$ is bijective. Thus,

$$\rho(TS^{-1}) \bigcap \rho(S^{-1}T) \subset \rho_S(T). \tag{2.63}$$

By using (2.62) and (2.63), we conclude that $\rho(TS^{-1}) \bigcap \rho(S^{-1}T) = \rho_S(T)$. This completes the proof. Q.E.D.

Theorem 2.19.4 [The S-resolvent identity] *Let $T \in CR(X)$ and let $S \in BR(X)$ such that $S(0) \subset T(0)$. Then, for all $\lambda, \mu \in \rho_S(T)$, we have*

$$R_S(\mu, T) - R_S(\lambda, T) = (\lambda - \mu)R_S(\mu, T)SR_S(\lambda, T). \qquad \diamond$$

Proof. Let $\lambda, \mu \in \rho_S(T)$ and $x \in X$. On the one hand, we have

$$(\lambda - \mu)R_S(\mu, T)SR_S(\lambda, T)x = R_S(\mu, T)(\lambda S - \mu S)R_S(\lambda, T)x.$$

By using Remark 2.3.3, we have $R_S(\mu, T)T(0) = (\mu S - T)^{-1}(\mu S - T)(0) = (\mu S - T)^{-1}(0) = 0$. It follows that

$$(\lambda - \mu)R_S(\mu, T)SR_S(\lambda, T)x = R_S(\mu, T)((\lambda S - \mu S)R_S(\lambda, T)x + T(0)).$$

Also, the fact that $(\lambda S - T)R_S(\lambda, T)x - (\lambda S - T)R_S(\lambda, T)x = T(0)$ leads from Proposition 2.3.6 (iv) to

$$(\lambda - \mu)R_S(\mu, T)SR_S(\lambda, T)x = $$
$$R_S(\mu, T)\Big[(\lambda S - \mu S) + (\lambda S - T) - (\lambda S - T)\Big]R_S(\lambda, T)x.$$

Since $R_S(\lambda, T)$ is a single valued linear operator, then by using Proposition 2.3.6 (iv), we have

$$(\lambda - \mu)R_S(\mu, T)SR_S(\lambda, T)x$$

$$\begin{aligned}
&= R_S(\mu, T)\Big[(\lambda S - \mu S)R_S(\lambda, T)x + (\lambda S - T)(0)\Big] \\
&= R_S(\mu, T)\Big[(\lambda S - \mu S)R_S(\lambda, T)x + T(0)\Big] \\
&= R_S(\mu, T)\Big[(\lambda S - \mu S - T + T)R_S(\lambda, T)x\Big] \\
&= R_S(\mu, T)\Big[\big[(\lambda S - T) - (\mu S - T)\big]R_S(\lambda, T)x\Big] \\
&= R_S(\mu, T)\Big[(\lambda S - T)R_S(\lambda, T)x - (\mu S - T)R_S(\lambda, T)x\Big].
\end{aligned}$$

Since $\mathcal{D}(R_S(\mu, T)) = X$, then by using Proposition 2.3.6 (v), we have

$$(\lambda - \mu)R_S(\mu, T)SR_S(\lambda, T)x =$$
$$R_S(\mu, T)(\lambda S - T)R_S(\lambda, T)x - R_S(\mu, T)(\mu S - T)R_S(\lambda, T)x.$$

Now, the use of Theorem 2.3.1 realizes that

$$R_S(\mu, T)(\lambda S - T)R_S(\lambda, T)x = R_S(\mu, T)(x + T(0)).$$

The fact that $\mu S - T$ is injective implies from Proposition 2.3.2 that

$$R_S(\mu, T)(\mu S - T)R_S(\lambda, T)x = R_S(\lambda, T)x.$$

Hence, by using Remark 2.3.3, we have $R_S(\mu, T)T(0) = R_S(\mu, T)(\mu S - T)(0) = 0$. Then, we conclude that

$$
\begin{aligned}
(\lambda - \mu)R_S(\mu, T)SR_S(\lambda, T)x &= R_S(\mu, T)x + R_S(\mu, T)T(0) - R_S(\lambda, T)x \\
&= R_S(\mu, T)x - R_S(\lambda, T)x. \qquad (2.64)
\end{aligned}
$$

On the other hand,

$$
\begin{aligned}
\mathcal{D}(R_S(\mu, T)SR_S(\lambda, T)) &= (SR_S(\lambda, T))^{-1}\mathcal{D}(R_S(\mu, T)) \\
&= (R_S(\lambda, T))^{-1}S^{-1}X \\
&= (R_S(\lambda, T))^{-1}\mathcal{D}(S) \\
&= (R_S(\lambda, T))^{-1}X \\
&= \mathcal{D}(R_S(\lambda, T)) \\
&= \mathcal{D}(R_S(\mu, T) - R_S(\lambda, T)).
\end{aligned}
$$

In addition, we have

$$
\begin{aligned}
(\lambda - \mu)R_S(\mu, T)SR_S(\lambda, T)(0) &= (\lambda - \mu)R_S(\mu, T)S(0) \\
&\subset (\lambda - \mu)R_S(\mu, T)T(0) = \{0\}.
\end{aligned}
$$

Hence,

$$(\lambda - \mu)R_S(\mu, T)SR_S(\lambda, T)(0) = (R_S(\mu, T) - R_S(\lambda, T))(0).$$

Thus, from these properties, Proposition 2.3.6 *(iii)*, and (2.64) implies that

$$R_S(\mu, T) - R_S(\lambda, T) = (\lambda - \mu)R_S(\mu, T)SR_S(\lambda, T). \qquad \text{Q.E.D.}$$

Theorem 2.19.5 *Let $T \in C\mathcal{R}(X)$ and $S \in \mathcal{L}(X)$. Assume for all $\lambda \in \rho_S(T)$, $R_S(\lambda, T)$ commutes with S. Then,*

$$\lim_{\mu \to \lambda} \left((\mu - \lambda)^{-1}(R_S(\mu, T) - R_S(\lambda, T)) \right) = -R_S(\lambda, T)SR_S(\lambda, T). \qquad \diamond$$

Proof. Let $\lambda \in \rho_S(T)$ and $\mu \neq \lambda$ in such a neighborhood, then by using Theorem 2.19.2, we have $\mu \in \rho_S(T)$. Since $R_S(\lambda, T)$ is a single valued linear operator, then it follows from Theorem 2.19.4 that

$$(\mu - \lambda)^{-1}(R_S(\mu, T) - R_S(\lambda, T)) + R_S(\lambda, T)SR_S(\lambda, T)$$
$$= -R_S(\mu, T)SR_S(\lambda, T) + R_S(\lambda, T)SR_S(\lambda, T)$$
$$= (R_S(\lambda, T)S - R_S(\mu, T)S)R_S(\lambda, T).$$

The fact that S commutes with $R_S(\lambda, T)$ for all $\lambda \in \rho_S(T)$, then

$$(\mu - \lambda)^{-1}(R_S(\mu, T) - R_S(\lambda, T)) + R_S(\lambda, T)SR_S(\lambda, T) =$$
$$(SR_S(\lambda, T) - SR_S(\mu, T))R_S(\lambda, T).$$

This implies from Proposition 2.3.6 (iv) that

$$(\mu - \lambda)^{-1}(R_S(\mu, T) - R_S(\lambda, T)) + R_S(\lambda, T)SR_S(\lambda, T) =$$
$$S(R_S(\lambda, T) - R_S(\mu, T))R_S(\lambda, T).$$

Hence,

$$\|(\mu - \lambda)^{-1}(R_S(\mu, T) - R_S(\lambda, T) + R_S(\lambda, T)SR_S(\lambda, T)\| \leq$$
$$\|S\|\|R_S(\lambda, T) - R_S(\mu, T)\|\|R_S(\lambda, T)\|.$$

Now, suppose that $|\mu - \lambda|\|S\| < \|R_S(\lambda, T)\|^{-1}$. Since $(\lambda - \mu)SR_S(\lambda, T)$ is a single valued linear operator, then

$$\sum_{n \geq 0} |\mu - \lambda|^n \|S\|^n \|R_S(\lambda, T)\|^n < \infty.$$

Hence,

$$(I - (\lambda - \mu)SR_S(\lambda, T))^{-1} = \sum_{n \geq 0} (\mu - \lambda)^n S^n R_S(\lambda, T)^n.$$

Since $R_S(\lambda, T)$ and S commute, then for $x \in \mathcal{D}(T)$, we have

$$(\lambda - \mu)R_S(\lambda, T)^{-1}SR_S(\lambda, T)x = (\lambda - \mu)R_S(\lambda, T)^{-1}R_S(\lambda, T)Sx.$$

By using Theorem 2.3.1 (iii), we get

$$(\lambda - \mu)R_S(\lambda, T)^{-1}SR_S(\lambda, T)x = (\lambda - \mu)Sx.$$

Since $\mathcal{D}(R_S(\lambda, T)^{-1}) = X$, then it follows from Proposition 2.3.6 (iv) that

$$R_S(\lambda, T)^{-1}(I - (\lambda - \mu)SR_S(\lambda, T))x$$

$$\begin{aligned} &= R_S(\lambda, T)^{-1}x + (\mu - \lambda)R_S(\lambda, T)^{-1}SR_S(\lambda, T))x \\ &= R_S(\lambda, T)^{-1}x + (\mu - \lambda)Sx \\ &= \lambda Sx - Tx + \mu Sx - \lambda Sx \\ &= (\mu S - T)x. \end{aligned}$$

Hence,

$$R_S(\lambda, T)^{-1}(I - (\lambda - \mu)SR_S(\lambda, T))x = R_S(\mu, T)^{-1}x.$$

Since $\mathcal{D}(R_S(\lambda, T)^{-1}(I - (\lambda - \mu)SR_S(\lambda, T)) = \mathcal{D}(R_S(\mu, T)^{-1}) = \mathcal{D}(T)$, then we conclude that

$$R_S(\mu, T) = (I - (\lambda - \mu)SR_S(\lambda, T))^{-1}R_S(\lambda, T).$$

So, in the first place, we have

$$R_S(\mu, T) = \sum_{n \geq 0}(\lambda - \mu)^n S^n R_S(\lambda, T)^{n+1}.$$

In the second place, we have

$$\begin{aligned} R_S(\mu, T) - R_S(\lambda, T) &= \sum_{n \geq 0}(\lambda - \mu)^n S^n R_S(\lambda, T)^{n+1} - R_S(\lambda, T) \\ &= R_S(\lambda, T)\sum_{n \geq 1}(\lambda - \mu)^n S^n R_S(\lambda, T)^n. \end{aligned}$$

Hence,

$$\|R_S(\mu, T) - R_S(\lambda, T)\| \leq \|R_S(\lambda, T)\| \sum_{n \geq 1}|\lambda - \mu|^n \|S\|^n \|R_S(\lambda, T)\|^n.$$

Arguing as above we conclude that

$$\lim_{\mu \to \lambda}(\mu - \lambda)^{-1}(R_S(\mu, T) - R_S(\lambda, T)) = -R_S(\lambda, T)SR_S(\lambda, T). \quad \text{Q.E.D.}$$

Corollary 2.19.3 *Let* $T \in CR(X)$ *and* $S \in \mathcal{L}(X)$. *If for all* $\lambda \in \rho_S(T)$, *we have* $R_S(\lambda, T)$ *commutes with* S, *then the function* $\varphi : \lambda \longrightarrow R_S(\lambda, T)$ *is holomorphic for all* $\lambda \in \rho_S(T)$. $\diamondsuit$

Proof. If $\lambda, \mu \in \rho_S(T)$, then

$$\lim_{\mu \to \lambda}\left(\frac{\varphi(\mu) - \varphi(\lambda)}{\mu - \lambda}\right) = -R_S(\lambda, T)SR_S(\lambda, T) \in \mathcal{L}(X). \quad \text{Q.E.D.}$$

Corollary 2.19.4 *Let $T \in C\mathcal{R}(X)$ and $S \in \mathcal{L}(X)$. Assume that for all $\lambda \in \rho_S(T)$, we have $R_S(\lambda, T)$ commutes with S. If $|\lambda - \mu|\|S\| < \|R_S(\lambda, T)\|^{-1}$, then*

$$R_S(\mu, T) = \sum_{n \geq 0} (\lambda - \mu)^n S^n R_S(\lambda, T)^{n+1}. \qquad \diamond$$

Proposition 2.19.3 *Let T be a continuous linear relation on X and $S \in \mathcal{L}(X)$. If for all $\lambda \in \rho_S(T)$, we have $R_S(\lambda, T)$ commutes with S and $\rho_S(T)$ is nonempty and unbounded, then*

$$\lim_{|\lambda| \to \infty} \|R_S(\lambda, T)\| = 0, \quad \text{for all } \lambda \in \rho_S(T). \qquad \diamond$$

Proof. Let $\lambda, \mu \in \rho_S(T)$. Then, in view of Theorem 2.19.4, we have

$$R_S(\lambda, T) - R_S(\mu, T) = (\mu - \lambda) R_S(\lambda, T) S R_S(\mu, T).$$

This implies that

$$R_S(\lambda, T) \Big[I - (\mu - \lambda) S R_S(\mu, T) \Big] = R_S(\mu, T). \qquad (2.65)$$

Now, suppose that $|\mu - \lambda|\|S\| < \|R_S(\mu, T)\|^{-1}$. By using the fact that $(\mu - \lambda) S R_S(\mu, T)$ is a single valued linear operator, we deduce that $I - (\mu - \lambda) S R_S(\mu, T)$ is invertible and

$$(I - (\mu - \lambda) S R_S(\mu, T))^{-1} = \sum_{n \geq 0} (\mu - \lambda)^n S^n R_S(\mu, T)^n.$$

Hence,

$$\|(I - (\mu - \lambda) S R_S(\mu, T))^{-1}\| \leq \frac{1}{1 - |\mu - \lambda|\|S\|\|R_S(\mu, T)\|}. \qquad (2.66)$$

By using (2.65), we infer that $R_S(\lambda, T) = R_S(\mu, T) (I - (\mu - \lambda) S R_S(\mu, T))^{-1}$. Thus, the fact that $\|R_S(\mu, T)\| < \infty$ implies from (2.66) that

$$0 \leq \|R_S(\lambda, T)\| \leq \frac{\|R_S(\mu, T)\|}{1 - |\mu - \lambda|\|S\|\|R_S(\mu, T)\|}.$$

As a result

$$\lim_{|\lambda| \to \infty} \|R_S(\lambda, T)\| = 0. \qquad \text{Q.E.D.}$$

Theorem 2.19.6 *Let T be a continuous linear relation on X and $S \in \mathcal{L}(X)$. If for all $\lambda \in \rho_S(T)$, we have $R_S(\lambda, T)$ commutes with S, and $\rho_S(T)$ is non-empty and unbounded, then the S-spectrum of T is non-empty.* $\diamond$

Proof. Suppose that $\rho_S(T) = \mathbb{C}$. For $x \in X$ and $x' \in X^*$, we have

$$\lim_{\mu \to \lambda} \left(\frac{x' R_S(\lambda, T)x - x' R_S(\mu, T)x}{\lambda - \mu} \right) = x' R_S(\lambda, T) S R_S(\lambda, T)x.$$

Thus, the single valued function $f(\lambda) = x' R_S(\lambda, T) S R_S(\lambda, T)x$ is an analytic function. Moreover,

$$|f(\lambda)| \leq \|x'\| \|R_S(\lambda, T)\| \|S\| \|R_S(\lambda, T)\| \|x\|.$$

By using Proposition 2.19.3, it is clear that $\lim_{|\lambda| \to \infty} \|R_S(\lambda, T)\| = 0$, then $f(\lambda) = 0$ for all λ. Since x' is arbitrary, we have $R_S(\lambda, T)x = 0$ for all $x \in X$. Thus,

$$X = N(R_S(\lambda, T)) = (\lambda S - T)(0) = T(0) \neq \{0\}.$$

But T is not bounded below, then $0 \in \sigma_S(T)$, which contradicts the assumption $\rho_S(T) = \mathbb{C}$.   Q.E.D.

Theorem 2.19.7 *Let $T \in CR(X)$ and S be a continuous linear relation such that $S(0) \subset T(0)$ and $\mathcal{D}(S) \supset \overline{\mathcal{D}(T)}$. Then, $\sigma_S(T) = \sigma_{S^*}(T^*)$.*   ◇

Proof. Let $\lambda \in \rho_S(T)$, then $\lambda S - T$ is injective, open with dense range. Using Proposition 2.8.1 (iv), we have

$$N((\lambda S - T)^*) = (R(\lambda S - T))^{\perp} = (\overline{R(\lambda S - T)})^{\perp} = X^{\perp} = \{0\}.$$

By referring to Proposition 2.9.3 (i), (iii), it follows that $\gamma((\lambda S - T)^*) = \gamma(\lambda S - T) > 0$ and Proposition 2.8.1 (iv),

$$R((\lambda S - T)^*) = N(\lambda S - T)^{\perp} = \{0\}^{\perp} = X^*.$$

Therefore, $(\lambda S - T)^*$ is injective, open with dense range. It remains to show that $(\lambda S - T)^* = \lambda S^* - T^*$. Using Proposition 2.8.1 (vi), λS is continuous, and $\mathcal{D}(\lambda S) = \mathcal{D}(S) \supset \mathcal{D}(T)$. This allows us to conclude that

$$(\lambda S - T)^* = (\lambda S)^* - T^* = \lambda S^* - T^*. \tag{2.67}$$

So, by using (2.67), we have $\lambda \in \rho_{S^*}(T^*)$. Then, $\sigma_{S^*}(T^*) \subset \sigma_S(T)$. Conversely, let $\lambda \in \rho_{S^*}(T^*)$, then $\lambda S^* - T^* = (\lambda S - T)^*$ is injective, open with dense range. In view of Proposition 2.8.1 (i), $(\lambda S - T)^*$ is closed. Then, by applying (2.67), we infer that $\lambda S^* - T^*$ is a closed linear relation. Using the fact that $(\lambda S - T)^*$ is open and Theorem 2.5.4 (ii), $R((\lambda S - T)^*)$ is closed. This implies that $R((\lambda S - T)^*) = X^*$. Moreover, by using Proposition 2.8.1,

$$\overline{R(\lambda S - T)} = (R(\lambda S - T)^{\perp})^{\top} = N((\lambda S - T)^*)^{\top} = \{0\}^{\top} = X.$$

On the other hand, by referring to Proposition 2.8.1,

$$N(\overline{\lambda S - T}) = R((\lambda S - T)^*)^{\top} = X^{*\top} = \{0\}.$$

Since $\lambda S - T$ is closed (see Proposition 2.7.2), then $N(\overline{\lambda S - T}) = N(\lambda S - T) = \{0\}$. So,

$$R((\lambda S - T)^*) = \overline{R((\lambda S - T)^*)} = X^*.$$

In addition, based on the assumption that $\lambda S - T$ is closed, we can conclude that $(\lambda S - T)^{**} = \lambda S - T$. Therefore, $\lambda S - T$ is injective, open with dense range. Thus, $\lambda \in \rho_S(T)$. As a result $\sigma_S(T) \subset \sigma_{S^*}(T^*)$. $\hspace{2cm}$ Q.E.D.

2.19.2 The Augmented S-Spectrum

Theorem 2.19.8 *Let $T \in \mathcal{CR}(X)$ and $S \in \mathcal{L}(X)$. Let $\mu \in \rho_S(T)$, then*

$$\widetilde{\eta}_{\mu}(\overline{\sigma}_S(T)) = \sigma(SR_S(\mu, T)). \hspace{2cm} \Diamond$$

Proof. Without loss of generality, we assume that X is complete, T is closed and S is continuous. Let $\lambda \in \mathbb{C}$, and $\mu \in \rho_S(T)$ with $\lambda \neq \mu$. Then,

$$\begin{aligned}
\lambda S - T &= (\mu S - T) - (\mu - \lambda)S \text{ since } S \text{ is single valued} \\
&= (I - (\mu - \lambda)SR_S(\mu, T))(\mu S - T) \text{ since } \mu \in \rho_S(T) \\
&= (\mu - \lambda)((\mu - \lambda)^{-1} - S(\mu S - T)^{-1})(\mu S - T) \\
&= A(\mu S - T), \hspace{4cm} (2.68)
\end{aligned}$$

where $A := (\mu - \lambda)((\mu - \lambda)^{-1} - SR_S(\mu, T))$. We shall verify that A is injective. Suppose that $\lambda \in \rho_S(T)$. Let $x \in N(A)$, then $Ax = 0$ (since $A(0) = 0$). This implies $(\mu - \lambda)^{-1}x - SR_S(\mu, T)x = 0$. Hence, $(\mu - \lambda)^{-1}x = SR_S(\mu, T)x$. So, $(\mu - \lambda)^{-1}(\mu S - T)x = Sx + (\mu S - T)(0)$. Thus, $(\mu S - T)x = (\mu - \lambda)Sx + (\mu S - T)(0)$. This gives that $\mu Sx - Tx = \mu Sx - \lambda Sx + T(0)$. Therefore, $(\lambda S - T)x = (\lambda S - T)(0)$. Then, $x \in N(\lambda S - T)$. Thus, $x = 0$. So, A is injective. Next, the use of (2.68) allows us to conclude that

$$X = R(\lambda S - T) = R(A(\mu S - T)) \subset R(A).$$

Hence, A is surjective. As required, since A is both bijective and open, then it follows that $(\mu - \lambda)^{-1} \in \rho(SR_S(\mu, T))$. Conversely, let $(\mu - \lambda)^{-1} \in$

$\rho(SR_S(\mu, T))$. For $x \in \mathcal{D}(T)$, by using (2.68), we have

$$\|(\lambda S - T)x\| = \|A(\mu S - T)x\| \geq \gamma(A(\mu S - T))\text{dist}(x, R_S(\mu, T)A^{-1}(0)).$$

Since A is injective, then $R_S(\mu, T)A^{-1}(0) = R_S(\mu, T)(0) = \{0\}$. Therefore,

$$\|(\lambda S - T)x\| \geq \gamma(A(\mu S - T))\|x\|. \tag{2.69}$$

As A is injective, then $A^{-1}(0) = \{0\} \subset R(\mu S - T)$. In view of Theorem 2.9.1 (ii), we have $\gamma(A(\mu S - T)) \geq \gamma(A)\gamma(\mu S - T)$. Hence, $\gamma(A) > 0$ from the hypothesis, and $\gamma(\mu S - T) > 0$ since $\mu \in \rho_S(T)$. Hence, it is easy to verify from (2.68) and (2.69) that $\lambda S - T$ is injective, surjective and open. We deduce that $\lambda \in \rho_S(T)$.   Q.E.D.

Proposition 2.19.4 *Let S and $T \in LR(X)$ be commute such that $\mathcal{D}(T) = \mathcal{D}(S)$. Then, for all λ and $\mu \in \mathbb{C}$, we have $\lambda S - T$ and $\mu S - T$ commutes.* $\Diamond$

Proof. On the one hand, we have

$$
\begin{aligned}
\mathcal{D}((\mu S - T)(\lambda S - T)) &= (\lambda S - T)^{-1}\mathcal{D}((\mu S - T)) \\
&= (\lambda S - T)^{-1}\mathcal{D}(T) \\
&= \mathcal{D}(T(\lambda S - T)) \\
&= \mathcal{D}((\lambda TS - T^2)) \\
&= \mathcal{D}(TS) \bigcap \mathcal{D}(T^2).
\end{aligned}
$$

Since S commute with T, we infer that

$$\mathcal{D}(TS) = \mathcal{D}(ST) = T^{-1}\mathcal{D}(S) = T^{-1}\mathcal{D}(T) = \mathcal{D}(T^2).$$

Hence, the domains of $(\mu S - T)(\lambda S - T)$ and $(\lambda S - T)(\mu S - T)$ are each equal to $\mathcal{D}(T^2)$. On the other hand, for $x \in \mathcal{D}(T^2)$, we have

$$
\begin{aligned}
(\lambda S - T)(\mu S - T)x &= (\lambda S - T)(\mu Sx - Tx) \\
&= (\lambda S - T)\{\mu y_1 - y_2 : y_1 \in Sx \text{ and } y_2 \in Tx\} \\
&= \{\lambda\mu Sy_1 - \lambda Sy_2 - \mu Ty_1 + Ty_2 : y_1 \in Sx \text{ and } y_2 \in Tx\} \\
&= \lambda\mu S^2x - \lambda STx - \mu TSx + T^2x.
\end{aligned}
$$

Since T and S commute, then

$$
\begin{aligned}
(\lambda S - T)(\mu S - T)x &= \lambda\mu S^2x - \lambda TSx - \mu STx + T^2x \\
&= (\mu S - T)(\lambda S - T)x.
\end{aligned}
$$

This completes the proof.   Q.E.D.

2.19.3 *S*-Essential Resolvent Sets of Multivalued Linear Operators

Definition 2.19.4 *Let X be a normed vector space, $T \in LR(X)$, S be a continuous linear relation such that $S(0) \subset \overline{T(0)}$ and $\mathcal{D}(S) \supset \mathcal{D}(T)$. The S-essential resolvent sets of T, where we define by $\rho_{ei,S}(T)$, $i = 1, 2, 3, 4, 5, 6$ as follows*

$$
\begin{aligned}
\rho_{e1,S}(T) &= \{\lambda \in \mathbb{C} : \lambda S - T \in \Phi_+(X)\}, \\
\rho_{e2,S}(T) &= \{\lambda \in \mathbb{C} : \lambda S - T \in \Phi_-(X)\}, \\
\rho_{e3,S}(T) &= \{\lambda \in \mathbb{C} : \lambda S - T \in \Phi_+(X) \bigcup \Phi_-(X)\}, \\
\rho_{e4,S}(T) &= \{\lambda \in \mathbb{C} : \lambda S - T \in \Phi_+(X) \bigcap \Phi_-(X)\}, \\
\rho_{e5,S}(T) &= \{\lambda \in \mathbb{C} : \lambda S - T \in \Phi_+(X) \bigcap \Phi_-(X) \text{ and } i(\lambda S - T) = 0\}, \text{ and} \\
\rho_{e6,S}(T) &= \{\lambda \in \mathbb{C} : \lambda \in \rho_{e5,S}(T) \text{ such that all scalars near } \lambda \text{ are in } \rho_S(T)\}.
\end{aligned}
$$

The S-essential spectra $\sigma_{ei,S}(T)$, $i = 1, 2, 3, 4, 5, 6$ are the complement of respective $\rho_{ei,S}(T)$. ◇

Definition 2.19.5 *Let $T \in CR(X)$ and $S \in BR(X)$ such that $S(0) \subset T(0)$ and $S \neq 0$. We define*

$$
\sigma_{eap,S}(T) = \bigcap_{K \in \mathcal{K}_T(X)} \sigma_{ap,S}(T + K)
$$

and

$$
\sigma_{e\delta,S}(T) = \bigcap_{K \in \mathcal{K}_T(X)} \sigma_\delta(T + K)
$$

with

$$
\sigma_{ap,S}(T) = \{\lambda \in \mathbb{C} \text{ such that } \lambda S - T \text{ not bounded below}\},
$$

where bounded below is injective and open and

$$
\sigma_{\delta,S}(T) := \{\lambda \in \mathbb{C} \text{ such that } \lambda S - T \text{ is not surjective}\}.
$$ ◇

2.20 Pseudospectra and Essential Pseudospectra of Linear Relations

2.20.1 Pseudospectra of Linear Relations

Definition 2.20.1 *Let $T \in LR(X)$, where X is a normed vector space and $\varepsilon > 0$. We define the pseudospectra of T by*

$$\sigma_\varepsilon(T) = \sigma(T) \bigcup \left\{ \lambda \in \mathbb{C} \ such \ that \ \|T_\lambda\| > \frac{1}{\varepsilon} \right\}.$$

We denote the pseudoresolvent set of T by

$$\rho_\varepsilon(T) = \mathbb{C} \backslash \sigma_\varepsilon(T) = \rho(T) \bigcap \left\{ \lambda \in \mathbb{C} \ such \ that \ \|T_\lambda\| \leq \frac{1}{\varepsilon} \right\}. \qquad \diamond$$

Remark 2.20.1 *Let X be a normed vector space, $T \in LR(X)$, and $\varepsilon > 0$. Then,*

(i) If $\varepsilon_1 < \varepsilon_2$, then $\sigma_{\varepsilon_1}(T) \subset \sigma_{\varepsilon_2}(T)$.
(ii) If X is complete and $T \in CR(X)$, then

$$\sigma_\varepsilon(T) = \sigma(T) \bigcup \left\{ \lambda \in \mathbb{C} \ such \ that \ \|(\lambda - T)^{-1}\| > \frac{1}{\varepsilon} \right\}. \qquad \diamond$$

Definition 2.20.2 *Let X be a normed vector space, $T \in LR(X)$ and $\varepsilon > 0$. The ε-pseudospectrum of T is defined by*

$$\Sigma_\varepsilon(T) := \sigma(T) \bigcup \left\{ \lambda \in \mathbb{C} \ such \ that \ \|(\lambda - \widetilde{T})^{-1}\| \geq \frac{1}{\varepsilon} \right\}. \qquad \diamond$$

Remark 2.20.2 *(i) Let X be a normed vector space, $T \in LR(X)$ and $\varepsilon > 0$. Then,*

(i_1) $\Sigma_\varepsilon(T) \backslash \sigma_\varepsilon(T) = \left\{ \lambda \in \mathbb{C} \ such \ that \ \|(\lambda - \widetilde{T})^{-1}\| = \frac{1}{\varepsilon} \right\}$.
(i_2) $\Sigma_\varepsilon(T)$ is a closed subset of the complex plane.
(i_3) If $0 < \varepsilon_1 < \varepsilon_2$, then $\sigma(T) \subset \Sigma_{\varepsilon_1}(T) \subset \Sigma_{\varepsilon_2}(T)$.
(ii) Let X be a non zero normed space, $\varepsilon > 0$ and let T be a continuous and densely defined linear relation in $LR(X)$. Then, $\sigma(T) \neq \emptyset$. Hence,

$$\sigma_\varepsilon(T) \neq \emptyset \ and \ \Sigma_\varepsilon(T) \neq \emptyset.$$

(iii) If X is a Banach space and $T \in CR(X)$, then

$$\Sigma_\varepsilon(T) = \sigma(T) \bigcup \left\{ \lambda \in \mathbb{C} \ such \ that \ \|(\lambda - T)^{-1}\| \geq \frac{1}{\varepsilon} \right\}. \qquad \diamond$$

Example 2.20.1 *Let us notice that the boundary value problem for an elliptic-parabolic equation can be formulated, in a Hilbert space $L^2(\Omega)$ with Ω is a bounded domain in $\mathbb{R}^2$ of class C^2, as the following multivalued Cauchy problem on $L^2(\Omega)$*

$$\begin{cases} \dfrac{\partial v}{\partial t} \in LM^{-1}v + \displaystyle\sum_{i=1}^{n} a_i(x)\dfrac{\partial v}{\partial x_i} + f(t), & t \in (0, \tau] \\ v(0) = v_0, \end{cases}$$

where $f(t) := f(t, \cdot)$, $v_0 = v_0(\cdot)$, M^{-1} is the inverse of the operator M defined by $Mu = m(\cdot)u$, for all $u \in X$,

$$Lu = \nabla \bullet (a(\cdot)\nabla u) + c_0(\cdot), \quad \text{for all} \ \ u \in H^2(\Omega) \bigcap H_0^1(\Omega),$$

∇ denotes the gradient vector with respect to x variable, $m(x) \geq 0$ on $\overline{\Omega}$ and $m \in C^2(\overline{\Omega})$, $a(x)$, $a_i(x)$, $c_0(x)$ are real-valued smooth functions on $\overline{\Omega}$. Putting $Sv = \sum_{i=1}^n a_i(x)\dfrac{\partial v}{\partial x_i}$, for all $v \in H^1(\Omega)$ and $T = LM^{-1}$. Then,

$$\Sigma_\varepsilon(T) \subset (]-\infty, -c_1[+i\mathbb{R}) \bigcup \overline{\mathbb{D}}(0, \varepsilon^2 c_2^2), \tag{2.70}$$

where $c_1, c_2 > 0$, $i^2 = -1$, and $\overline{\mathbb{D}}(0, \varepsilon^2 c_2^2) = \{\lambda \in \mathbb{C} : |\lambda| \leq \varepsilon^2 c_2^2\}$. Indeed, it follows from Ref. [51] that, for any given $\lambda \in \mathbb{C}$ satisfying

$$-c_1 (1 + |Im\lambda|) \leq Re\lambda, \quad \text{for some} \ \ c_1 > 0,$$

we have $(\lambda - T)^{-1}$ is an operator. Hence, there exists $c_2 > 0$ such that

$$\|(\lambda - T)^{-1}\| \leq c_2(1 + |\lambda|)^{-\frac{1}{2-r}}, \quad \text{for some} \ \ r \in (0, 1). \tag{2.71}$$

Keeping in mind that $-\dfrac{1}{2-r} \ln(1 + |\lambda|) \leq 0$. Then, we infer from (2.71) that

$$\|(\lambda - T)^{-1}\| \leq c_2.$$

Therefore,

$$\{\lambda \in \mathbb{C} : -c_1 (1 + |Im\lambda|) \leq Re\lambda\} \subset \rho(T).$$

This implies that

$$\Sigma_\varepsilon(T) \subset \{\lambda \in \mathbb{C} : -c_1 (1 + |Im\lambda|) > Re\lambda\} \bigcup \left\{\lambda \in \mathbb{C} : \|(\lambda - T)^{-1}\| \geq \frac{1}{\varepsilon}\right\}. \tag{2.72}$$

We divide this part of the proof into two steps.

Step 1. We show that

$$\{\lambda \in \mathbb{C} : -c_1 (1 + |Im\lambda|) > Re\lambda\} \bigcup \left\{\lambda \in \mathbb{C} : \|(\lambda - T)^{-1}\| \geq \frac{1}{\varepsilon}\right\} \subset \Gamma_\varepsilon, \tag{2.73}$$

where $\Gamma_\varepsilon = \{\lambda \in \mathbb{C} : -c_1 (1 + |Im\lambda|) > Re\lambda\} \bigcup \{\lambda \in \mathbb{C} : |\lambda| \leq \varepsilon^2 c_2^2 - 1\}$. Let $\lambda \in \{\lambda \in \mathbb{C} : -c_1 (1 + |Im\lambda|) > Re\lambda\} \bigcup \left\{\lambda \in \mathbb{C} : \|(\lambda - T)^{-1}\| \geq \frac{1}{\varepsilon}\right\}$. This is equivalent to saying that $-c_1 (1 + |Im\lambda|) > Re\lambda$ or $\|(\lambda - T)^{-1}\| \geq \frac{1}{\varepsilon}$. If

$-c_1 (1 + |Im\lambda|) > Re\lambda$, then it is clear that $\lambda \in \Gamma_\varepsilon$. At this level, let us assume that $-c_1 (1 + |Im\lambda|) \leq Re\lambda$ and $\|(\lambda - T)^{-1}\| \geq \dfrac{1}{\varepsilon}$. Then, it follows from (2.71) that

$$\frac{1}{\varepsilon} \leq c_2 (1 + |\lambda|)^{-\frac{1}{2-r}}.$$

This leads to

$$(1 + |\lambda|)^{\frac{1}{2-r}} \leq \varepsilon c_2. \tag{2.74}$$

Since $(1 + |\lambda|)^{\frac{1}{2}} \leq (1 + |\lambda|)^{\frac{1}{2-r}}$, *then by using (2.74), we have* $(1 + |\lambda|)^{\frac{1}{2}} \leq \varepsilon c_2$. *Hence,*

$$|\lambda| \leq \varepsilon^2 c_2^2 - 1.$$

This implies that (2.73) holds.

Step 2. *We show that*

$$\Gamma_\varepsilon \subset (] - \infty, -c_1[+ i\mathbb{R}) \bigcup \overline{\mathbb{D}}(0, \varepsilon^2 c_2^2),$$

where $i^2 = -1$. *Let us assume that* $\lambda \in \Gamma_\varepsilon$. *We discuss two cases.*

First case : *If* $Re\lambda < -c_1 (1 + |Im\lambda|)$, *then* $Re\lambda < -c_1$. *This implies that* $\lambda \in (] - \infty, -c_1[+ i\mathbb{R})$.

Second case : *If* $|\lambda| \leq \varepsilon^2 c_2^2 - 1$, *then* $|\lambda| \leq \varepsilon^2 c_2^2$. *Hence,* $\lambda \in \overline{\mathbb{D}}(0, \varepsilon^2 c_2^2)$. *Thus, the use of (2.72) and (2.73) makes us conclude that*

$$\Sigma_\varepsilon(T) \subset \Gamma_\varepsilon \subset (] - \infty, -c_1[+ i\mathbb{R}) \bigcup \overline{\mathbb{D}}(0, \varepsilon^2 c_2^2).$$

Therefore, (2.70) holds.

2.20.2   Essential Pseudospectra of Linear Relations

Definition 2.20.3 *Let* $\varepsilon > 0$ *and* $T \in CR(X)$. *We define the essential pseudospectra of* T *by*

$$
\begin{aligned}
\sigma_{e1,\varepsilon}(T) &= \mathbb{C} \backslash \{\lambda \in \mathbb{C} \text{ such that } \lambda - T + S \in \Phi_+(X), \text{ for all } S \in \mathcal{U}_T(X)\}, \\
\sigma_{e2,\varepsilon}(T) &= \mathbb{C} \backslash \{\lambda \in \mathbb{C} \text{ such that } \lambda - T + S \in \Phi_-(X), \text{ for all } S \in \mathcal{U}_T(X)\}, \\
\sigma_{e3,\varepsilon}(T) &= \mathbb{C} \backslash \{\lambda \in \mathbb{C} \text{ such that } \lambda - T + S \in \Phi_\pm(X), \text{ for all } S \in \mathcal{U}_T(X)\}, \\
\sigma_{e4,\varepsilon}(T) &= \mathbb{C} \backslash \{\lambda \in \mathbb{C} \text{ such that } \lambda - T + S \in \Phi(X), \text{ for all } S \in \mathcal{U}_T(X)\}, \\
\sigma_{e5,\varepsilon}(T) &= \bigcap_{K \in \mathcal{K}_T(X)} \sigma_\varepsilon(T + K),
\end{aligned}
$$

where $\mathcal{U}_T(X) = \{S \in LR(X) \text{ is continuous such that } \|S\| < \varepsilon,\ S(0) \subset T(0) \text{ and } \mathcal{D}(S) \supset \mathcal{D}(T)\}$. $\Diamond$

Remark 2.20.3 *Let* $\varepsilon > 0$ *and* $T \in CR(X)$, *then one may order pseudospectra as*

$$\sigma_{e3,\varepsilon}(T) = \sigma_{e1,\varepsilon}(T) \bigcap \sigma_{e2,\varepsilon}(T) \subset \sigma_{e4,\varepsilon}(T) \subset \sigma_{e5,\varepsilon}(T). \qquad \diamondsuit$$

2.20.3 The Essential ε-Pseudospectra of Linear Relations

The aim of this subsection is to introduce and study the essential ε-pseudospectra of linear relations in a Banach space.

Definition 2.20.4 *Let* X *be a Banach space,* $\varepsilon > 0$ *and let* $T \in CR(X)$. *The essential ε-pseudospectra sets of* T *are defined as follows*

(i) $\Sigma_{e1,\varepsilon}(T) = \{\lambda \in \mathbb{C} : \alpha(\lambda - T) = \infty\} \bigcup \left\{\lambda \in \mathbb{C} : \|(\lambda - T)^{-1}\| \geq \dfrac{1}{\varepsilon}\right\}.$

(ii) $\Sigma_{e2,\varepsilon}(T) = \{\lambda \in \mathbb{C} : \beta(\lambda - T) = \infty\} \bigcup \left\{\lambda \in \mathbb{C} : \|(\lambda - T)^{-1}\| \geq \dfrac{1}{\varepsilon}\right\}.$

(iii) $\Sigma_{e3,\varepsilon}(T) = (\{\lambda \in \mathbb{C} : \alpha(\lambda - T) = \infty\} \bigcap \{\lambda \in \mathbb{C} : \beta(\lambda - T) = \infty\}) \bigcup$
$$\left\{\lambda \in \mathbb{C} : \|(\lambda - T)^{-1}\| \geq \dfrac{1}{\varepsilon}\right\}.$$

(iv) $\Sigma_{e4,\varepsilon}(T) = \{\lambda \in \mathbb{C} : \alpha(\lambda - T) = \infty\} \bigcup \{\lambda \in \mathbb{C} : \beta(\lambda - T) = \infty\} \bigcup$
$$\left\{\lambda \in \mathbb{C} : \|(\lambda - T)^{-1}\| \geq \dfrac{1}{\varepsilon}\right\}.$$

(v) $\Sigma_{e5,\varepsilon}(T) = \{\lambda \in \mathbb{C} : \alpha(\lambda - T) = \infty\} \bigcup \{\lambda \in \mathbb{C} : \beta(\lambda - T) = \infty\} \bigcup$
$$\{\lambda \in \mathbb{C} : \alpha(\lambda - T) \neq \beta(\lambda - T)\} \bigcup \left\{\lambda \in \mathbb{C} : \|(\lambda - T)^{-1}\| \geq \dfrac{1}{\varepsilon}\right\}. \qquad \diamondsuit$$

Remark 2.20.4 *Let* X *be a Banach space and let* $T \in CR(X)$. *Then,*

(i) *If* $0 < \varepsilon_1 < \varepsilon_2$, *then* $\Sigma_{ei,\varepsilon_1}(T) \subset \Sigma_{ei,\varepsilon_2}(T)$ *for* $i = 1, \ldots, 5$.
(ii) *If* $\varepsilon > 0$, *then* $\Sigma_{ei,\varepsilon}(T) \subset \Sigma_\varepsilon(T)$ *for* $i = 1, \ldots, 5$.
(iii) *If* $\varepsilon > 0$, *then* $\Sigma_{e3,\varepsilon}(T) = \Sigma_{e1,\varepsilon}(T) \bigcap \Sigma_{e2,\varepsilon}(T)$.
(iv) *If* $\varepsilon > 0$, *then* $\Sigma_{e4,\varepsilon}(T) = \Sigma_{e1,\varepsilon}(T) \bigcup \Sigma_{e2,\varepsilon}(T)$.
(v) *If* $\varepsilon > 0$, *then* $\Sigma_{e5,\varepsilon}(T) = \Sigma_{e4,\varepsilon}(T) \bigcup \{\lambda \in \mathbb{C} : \alpha(\lambda - T) \neq \beta(\lambda - T)\}$. $\diamondsuit$

Example 2.20.2 *Let* T *be the operator defined on* $L^2(\mathbb{R})$ *by*

$$T = \dfrac{d^2}{dx^2} + ix,$$

where $i^2 = -1$. *Then,* $\left\{\lambda \in \mathbb{C} : \max\left\{c_1; \left(\log \dfrac{1}{\varepsilon}\right)^{\frac{2}{3}} c_2\right\} \leq \mathrm{Re}\lambda\right\} \subset \Sigma_{ei,\varepsilon}(T)$, *for* $i = 1, \ldots, 5$, *where* $c_1, c_2 > 0$. *Indeed, the operator* T *is closed and non self-adjoint when considered on its maximal domain*

$$\mathcal{D}(T) = \{\psi \in L^2(\mathbb{R}) : -\psi'' + i\psi \in L^2(\mathbb{R})\}.$$

Moreover, $\rho(T) = \mathbb{C}$ and

$$\left\{ \lambda \in \mathbb{C} : \max\left\{ c_1; \left(\log\frac{1}{\varepsilon}\right)^{\frac{2}{3}} c_2 \right\} \leq Re\lambda \right\} \subset \left\{ \lambda \in \mathbb{C} : \|(\lambda - T)^{-1}\| \geq \frac{1}{\varepsilon} \right\},$$

where $c_1, c_2 > 0$. Hence,

$$\left\{ \lambda \in \mathbb{C} : \max\left\{ c_1; \left(\log\frac{1}{\varepsilon}\right)^{\frac{2}{3}} c_2 \right\} \leq Re\lambda \right\} \subset \Sigma_{ei,\varepsilon}(T), \quad for \ i = 1, \ldots, 5.$$

2.20.4   *S*-Pseudospectra and *S*-Essential Pseudospectra of Linear Relations

Definition 2.20.5 *We define the S-pseudospectra of T by*

$$\sigma_{\varepsilon,S}(T) \quad = \quad \sigma_S(T) \bigcup \left\{ \lambda \in \mathbb{C} \ such\ that\ \|(\lambda S - T)^{-1}\| > \frac{1}{\varepsilon} \right\}.$$

We denote the S-pseudoresolvent set of T by

$$\rho_{\varepsilon,S}(T) = \mathbb{C}\backslash\sigma_{\varepsilon,S}(T) = \rho_S(T) \bigcap \left\{ \lambda \in \mathbb{C} \ such\ that\ \|(\lambda S - T)^{-1}\| \leq \frac{1}{\varepsilon} \right\}. \ \Diamond$$

Remark 2.20.5 *(i) Note that if $S = I$, then we find the usual definition of pseudospectra of a linear relation.*
(ii) If $0 < \varepsilon_1 < \varepsilon_2$, then it is clear that $\sigma_{\varepsilon_1,S}(T) \subset \sigma_{\varepsilon_2,S}(T)$. $\Diamond$

Definition 2.20.6 *Let T be a linear relation in $CR(X)$ and $\varepsilon > 0$. The S-essential pseudospectrum of T is the set*

$$\sigma_{e5,\varepsilon,S}(T) = \bigcap_{K \in \mathcal{K}_T(X)} \sigma_{\varepsilon,S}(T + K).$$

We define the S-essential pseudoresolvent set by

$$\rho_{e5,\varepsilon,S}(T) = \mathbb{C}\backslash\sigma_{e5,\varepsilon,S}(T). \qquad \Diamond$$

Chapter 3

The Stability Theorems of Multivalued Linear Operators

This chapter studies the stability theorems of multivalued linear operators. In first time, we investigate the closedness of linear relations. Secondly, enlightened by the notion of gap between two linear subspaces found in Ref. [67], we generalize Kato's results, and apply it to generalized convergence between closed linear relations, which represents the convergence between their graphs in a certain distance. Hence, we investigate the stability of the index, the nullity and the deficiency of a linear relation T in Banach spaces spaces under perturbations by strictly singular and T-strictly singular linear relations. Therefore, we extend the concept of demicompactness and k-set-contactive linear operators on multivalued linear operators and we develop some properties. Finally, we study the essentially semi regular linear relation operators everywhere defined in Hilbert space. The obtained results is then applied to study some properties of the Samuel-multiplicity.

3.1 Stability of Closeness of Multivalued Linear Operators

3.1.1 Sum of Two Closed Linear Relations

The sum (respectively, the product) of two closed linear relations is not necessarily closed. We shall use the following result which gives sufficient conditions for the sum (respectively, the product) of two closed linear relations to be closed.

Theorem 3.1.1 *Let X and Y be two Banach spaces, S, $T \in LR(X,Y)$ such that $\mathcal{D}(S) \supset \mathcal{D}(T)$, $S(0) \subset \overline{T(0)}$ and*

$$\|Sx\| \le a\|x\| + b\|Tx\|, \quad x \in \mathcal{D}(T), \tag{3.1}$$

for some constants a, b with b < 1. Then, $T \in C\mathcal{R}(X,Y)$ if and only if, $T + S \in C\mathcal{R}(X,Y)$. ◇

Proof. Suppose $T \in C\mathcal{R}(X,Y)$. Let us consider two cases for S and T.

First case : If S and T are single valued. Then,

$$\|Tx\| = \|(T + S - S)x\| \le \|(T + S)x\| + \|Sx\|, \quad x \in \mathcal{D}(T),$$

which in turn yields by using (3.1)

$$(1 - b)\|Tx\| \le \|(T + S)x\| + a\|x\|, \quad x \in \mathcal{D}(T).$$

Thus, if $(x_n)_n$ is sequence in $\mathcal{D}(T + S) = \mathcal{D}(T)$ such that $x_n \to x$ and $(T + S)x_n \to \psi$, then there exist $N_0 \in \mathbb{N}$ such that for all $n, m \ge N_0$, we have

$$(1 - b)\|Tx_n - Tx_m\| \le a\|x_n - x_m\| + \|(T + S)x_n - x_n\|.$$

Therefore, $(Tx_n)_n$ is a Cauchy sequence in the Banach space Y. Thus, $Tx_n \to x_0$. Since T is closed, then $x \in \mathcal{D}(T)$ and $Tx = x_0$. Moreover, $Sx_n = (T + S)x_n - Tx_n \to \psi - x_0$ as $n \to \infty$. But

$$\|S(x_n - x)\| \le (a\|x_n - x\| + b\|Tx_n - x_0\|) \to 0 \quad \text{as} \quad n \to \infty,$$

which shows that $Sx = \psi - x_0$. Hence, $(T + S)x = \psi$ and $x \in \mathcal{D}(T + S)$.

Second case : S and T are linear relations. Since $T \in C\mathcal{R}(X,Y)$ and by using Proposition 2.7.1 (i), we have $T(0) = \overline{T(0)}$. In view of $S(0) \subset \overline{T(0)}$, then $S(0) \subset T(0)$. So, $Q_T = Q_{T+S}$ and $Q_{T+S}(T + S) = Q_T T + Q_T S$. By using Proposition 2.5.4 (iii), we deduce that $Q_T S$ is single valued and

$$\|Q_T Sx\| \le \|Q_S Sx\| = \|Sx\| \le a\|x\| + b\|Tx\|, \quad x \in \mathcal{D}(T).$$

Hence,

$$\|Q_T Sx\| \le a\|x\| + b\|Q_T Tx\|, \quad x \in \mathcal{D}(T).$$

On the other hand, since $Q_T T$ is closed, then in view of the first case, $Q_T T + Q_T S$ is closed single valued. This means, since $(T + S)(0) = T(0)$ is closed, then $T + S$ is closed. Conversely, assume that $T + S$ is closed. It follows from (2.31) that

$$\| - Sx\| = \|Sx\| \le \left(\frac{a}{1 - b}\right)\|x\| + \left(\frac{b}{1 - b}\right)\|(T + S)x\|, \quad x \in \mathcal{D}(T).$$

In view of the above, we have $T + S - S$ is closed. Now, by Proposition 2.3.4, we have T is closed. Q.E.D.

Theorem 3.1.2 *Let X and Y be two Banach spaces, $T \in LR(X,Y)$ and $S \in LR(X,Y)$ such that $S(0) \subset T(0)$. If S is T-bounded with T-bound δ, then*

(*i*) *If T is closed and $\delta < 1$, then $T + S$ is closed.*
(*ii*) *If $T + S$ is closed, and $\delta < \frac{1}{2}$, then T is closed.* ◇

Proof. (*i*) Since T is closed, then by Proposition 2.7.1 (*i*), $T(0)$ is closed, and $Q_T T$ is closed. In view of $S(0) \subset T(0)$, then $(T + S)(0) = T(0)$. Hence, $(T+S)(0)$ is closed. Since S is T-bounded with T-bound $\delta < 1$, then by using Remark 2.6.1 (*v*), we obtain $Q_T S$ is $Q_T T$-bounded with $Q_T T$-bound < 1. Therefore, since $S(0) \subset T(0)$, then $Q_T = Q_{T+S}$. Thus,

$$Q_{S+T}(T + S) = Q_T(T + S) = Q_T T + Q_T S$$

is closed. Applying Theorem 3.1.1, we have $Q_{S+T}(T+S)$ is closed. Therefore, by using Proposition 2.7.1 (*i*), it follows that $T + S$ is closed.

(*ii*) Since $T + S$ is closed, S is T-bounded with T-bound $\delta < \frac{1}{2}$, and by using Proposition 2.3.4, we get
$$T = T + S - S.$$

Then, by Proposition 2.6.1, we obtain S is $(T + S)$-bounded with $(T + S)$-bound < 1. Applying (*i*), it follows that T is closed. Q.E.D.

3.1.2 Sum of Three Closed Linear Relations

The sum (respectively, the product) of three closed linear relations is not necessarily closed. We shall use the following result which gives sufficient conditions for the sum (respectively, the product) of three closed linear relations to be closed.

Theorem 3.1.3 *Let X and Y be two Banach spaces, $T \in CR(X,Y)$, and S, $K \in LR(X,Y)$ such that $K(0) \subset S(0) \subset T(0)$. If S is T-bounded with T-bound $\delta_1 < 1$, and K is S-bounded with S-bound δ_2 such that $\frac{\delta_1 \delta_2}{1-\delta_1} < 1$. Then, $T + S + K$ is a closed linear relation.* ◇

Proof. Since T is closed and S is T-bounded with T-bound $\delta_1 < 1$, then by using Theorem 3.1.1, we have $T + S$ is closed. On the one hand, since S is T-bounded with T-bound $\delta_1 < 1$, then in view of Proposition 2.6.2, we obtain S is $(T + S)$-bounded with $(T + S)$-bound $\leq \frac{\delta_1}{1-\delta_1}$. On the other hand, since K is S-bounded with S-bound δ_2, then by using Proposition 2.6.2, we get K is $(T + S)$-bounded with $(T + S)$-bound $\leq \frac{\delta_1 \delta_2}{1-\delta_1}$. Using the fact that K is $(T + S)$-bounded with $(T + S)$-bound < 1, and $T + S$ is closed, we infer from Theorem 3.1.1 that $T + S + K$ is closed which completes the proof. Q.E.D.

Lemma 3.1.1 *Let X and Y be two Banach spaces, A, B and $C \in LR(X, Y)$ satisfy $B(0) \bigcup C(0) \subset A(0)$. Suppose that B is A-bounded with A-bound δ_1, C is A-bounded with A-bound δ_2, and Y is complete. Then,*

(i) If $\delta_1 + \delta_2 < 1$ and A is closed, then $A + B + C$ is closed.
(ii) If $\delta_1 + \delta_2 < \frac{1}{2}$ and $A + B + C$ is closed, then A is closed. $\diamondsuit$

Proof. (i) Since B is A-bounded with A-bound δ_1, and C is A-bounded with A-bound δ_2, then by Proposition 2.6.2 (ii), $B + C$ is A-bounded with A-bound $\delta_1 + \delta_2 < 1$. Applying Theorem 3.1.2 (i), we obtain $A + B + C$ is closed.

(ii) Since B is A-bounded with A-bound δ_1, and C is A-bounded with A-bound δ_2, then for all $x \in \mathcal{D}(A)$, we have

$$
\begin{aligned}
\|(B + C)x\| &\leq \|Bx\| + \|Cx\| \\
&\leq (a_1 + a_2)\|x\| + (b_1 + b_2)\|Ax\| \\
&\leq (a_1 + a_2)\|x\| + (b_1 + b_2)\|(A + B + C - B - C)x\| \\
&\leq (a_1 + a_2)\|x\| + (b_1 + b_2)\|(A + B + C)x\| + (b_1 + b_2)\|(B + C)x\| \\
&\leq \frac{(a_1 + a_2)}{1 - (b_1 + b_2)}\|x\| + \frac{(b_1 + b_2)}{1 - (b_1 + b_2)}\|(A + B + C)x\|.
\end{aligned}
$$

Therefore, $B + C$ is $(A + B + C)$-bounded with $(A + B + C)$-bound < 1. On the other side, by Proposition 2.3.4, we get $A = A + B + C - B - C$. Thus, $A + B + C$ is closed, and $B + C$ is $(A + B + C)$-bounded with $(A + B + C)$-bound < 1. Finally applying (i), it follows that A is closed. Q.E.D.

Theorem 3.1.4 *Let X and Y be two Banach spaces, S, T, $K \in LR(X, Y)$ such that $\mathcal{D}(T) \subset \mathcal{D}(S) \subset \mathcal{D}(K)$, $K(0) \subset S(0) \subset \overline{T(0)}$ and*

(i) there exist two constants a_1, $b_1 > 0$ such that

$$\|Sx\| \leq a_1\|x\| + b_1\|Tx\|, \quad x \in \mathcal{D}(T),$$

(ii) there exist two constants a_2, $b_2 > 0$ such that $b_1(1 + b_2) < 1$ and

$$\|Kx\| \leq a_2\|x\| + b_2\|Sx\|, \quad x \in \mathcal{D}(S).$$

Then, $T \in C\mathcal{R}(X, Y)$ if and only if, $T + S + K \in C\mathcal{R}(X, Y)$. ◇

Proof. Let $T \in C\mathcal{R}(X, Y)$, we will show that $T + S + K \in C\mathcal{R}(X, Y)$. To do this, we will consider two cases for S, T and K.

<u>First case</u> : If S, T and K are operators, then for all $x \in \mathcal{D}(T)$, we have

$$
\begin{aligned}
\|(T + S + K)x\| &\leq \|Tx\| + \|Sx\| + \|Kx\| \\
&\leq \|Tx\| + a_1\|x\| + b_1\|Tx\| + a_2\|x\| + b_2\|Sx\| \\
&= (1 + b_1)\|Tx\| + (a_1 + a_2)\|x\| + b_2(a_1\|x\| + b_1\|Tx\|).
\end{aligned}
$$

Hence,

$$\|(T + S + K)x\| \leq (a_1 + a_2 + a_1 b_2)\|x\| + (1 + b_1 + b_1 b_2)\|Tx\|. \tag{3.2}$$

Similarly, for all $x \in \mathcal{D}(T)$, we have

$$
\begin{aligned}
\|(S + K)x\| &\leq \|Sx\| + \|Kx\| \\
&\leq a_1\|x\| + b_1\|Tx\| + a_2\|x\| + b_2\|Sx\| \\
&= b_1\|Tx\| + (a_1 + a_2)\|x\| + b_2(a_1\|x\| + b_1\|Tx\|).
\end{aligned}
$$

So,

$$\|(S + K)x\| \leq (a_1 + a_2 + a_1 b_2)\|x\| + b_1(1 + b_2)\|Tx\|. \tag{3.3}$$

It follows from (3.3) that for all $x \in \mathcal{D}(T)$, we have

$$
\begin{aligned}
\|(T + S + K)x\| &\geq \|Tx\| - \|Sx + Kx\| \\
&\geq \|Tx\| - (a_1 + a_2 + a_1 b_2)\|x\| - b_1(1 + b_2)\|Tx\| \\
&= -(a_1 + b_2 + a_1 b_2)\|x\| + (1 - b_1(1 + a_2))\|Tx\|
\end{aligned}
$$

which yields

$$(1 - b_1(1 + a_2))\|Tx\| \leq \|(T + S + K)x\| + (a_1 + b_2 + a_1 b_2)\|x\|. \tag{3.4}$$

Setting $\Theta = 1 - b_1(1 + a_2)$ $(0 < \Theta < 1)$ and $\Psi = a_1 + b_2 + a_1 b_2$ $(\Psi > 0)$, then (3.4) gives

$$\|Tx\| \leq \Theta^{-1}\left(\|(T + S + K)x\| + \Psi\|x\|\right). \tag{3.5}$$

Let $(x_n)_n$ be sequence in $\mathcal{D}(T + S + K) = \mathcal{D}(T)$ such that $x_n \to x$ and $(T + S + K)x_n \to \psi$. Then, by (3.5), there exist $N_1 \in \mathbb{N}$ such that for all n, $m \geq N_1$, we have

$$\|Tx_n - Tx_m\| \leq \Theta^{-1}\left(\|(T + S + K)(x_n - x_m)\| + \Psi\|x_n - x_m\|\right).$$

So, $(Tx_n)_n$ is a Cauchy sequence in the Banach space Y. Therefore, there exist $\psi_1 \in Y$ such that $Tx_n \to \psi_1$. Since T is closed, then $x \in \mathcal{D}(T)$ and $Tx = \psi_1$. Now, from (3.2), we have

$$\|(T+S+K)(x_n-x)\| \le ((a_1 + a_2 + a_1 b_2)\|x_n - x\| + (1 + b_1 + b_1 b_2)\|Tx_n - x\|) \to 0,$$

which, by letting $n \to +\infty$, implies that $(T+S+K)x_n \to \psi = (T+S+K)x$. Therefore, $T + S + K$ is a closed linear operator.

<u>Second case</u> : T, S and K are linear relations. Since $K(0) \subset S(0) \subset T(0)$, it is clear that $Q_{T+S+K} = Q_T$. Then, $Q_{T+S+K}(T + S + K) = Q_T(T + S + K) = Q_T T + Q_T S + Q_T K$. The fact that $Q_T T$ is a closed operator and from Proposition 2.5.4 (*iii*), lead to the following $Q_T S$ and $Q_T K$ are single valued. Hence, $\|Q_T S\| \le \|Q_S S\|$ and $\|Q_T K\| \le \|Q_K K\|$. So, $\|Q_T S\| \le \|Q_S S\| = \|Sx\| \le a_1\|x\| + b_1\|Tx\|$, $x \in \mathcal{D}(T)$. Thus,

$$\|Q_T S\| \le \|Q_S Sx\| \le a_1\|x\| + b_1\|Q_T Tx\|, \quad x \in \mathcal{D}(T).$$

Similarly, we have $\|Q_T K\| \le a_2\|x\| + b_2\|Q_S Sx\|$, $x \in \mathcal{D}(S)$, where $b_1(1+a_2) < 1$. Consequently, $Q_T T + Q_T S + Q_T K$ is a closed operator. Since $T(0) + S(0) + K(0) = T(0)$ is closed, then by using the first case, $T + S + K$ is a closed linear relation. Conversely, assume that $T + K + S$ is closed. Then,

$$\| - Sx\| \le (a_1 + b_1 \Psi)\|x\| + b_1 \Theta^{-1}\|(T + K + S)x\|, \quad x \in \mathcal{D}(T),$$

and

$$\| - Kx\| \le a_2\|x\| + b_2\| - Sx\|, \quad x \in \mathcal{D}(S) \text{ and } b_1 \Theta^{-1}(1 + a_2) < 1.$$

Now, applying from what precedes to the relations $T + S + K$, $-K$ and $-S$, that $T + S + K - K - S$ is a closed operator. Since $(K + S)(0) \subset T(0)$, then $T = T + (K + S) - (K + S)$. Therefore, by Proposition 2.3.4, T is closed. This completes the proof.　　　　　　　　　　　　　　Q.E.D.

3.1.3　Sum of Two Closable Linear Relations

Proposition 3.1.1 *Let S, $T \in LR(X, Y)$ such that $S(0) \subset T(0)$ and $\overline{\mathcal{D}(T)} \subset \mathcal{D}(S)$. If T is closable and S is continuous, then*

(i) $\overline{T} + S$ is closed and $\overline{T + S} \subset \overline{T} + S$.
(ii) $T + S$ is closable and $\overline{S + T} = S + \overline{T}$.　　　　　　　　　　　◇

Proof. (*i*) Since T is closable, then in view of Lemma 2.7.3, $T(0) = \overline{T}(0)$. Hence, by using Lemma 2.7.2, $\overline{T(0)} \subset \overline{T}(0)$. So, $T(0) = \overline{T}(0) = \overline{T(0)}$. Thus, $S(0) \subset \overline{T}(0)$. Moreover, since $\mathcal{D}(T) \times Y = G(T) + (\{0\} \times Y)$, then $\mathcal{D}(\overline{T}) \subset \overline{\mathcal{D}(T)}$. Thus,

$$\overline{G(T)} + (\{0\} \times Y) \subset \overline{G(T) + (\{0\} \times Y)} = \overline{\mathcal{D}(T)} \times Y \qquad (3.6)$$

and clearly,

$$\overline{G(T)} + (\{0\} \times Y) = G(\overline{T}) + (\{0\} \times Y) = \mathcal{D}(\overline{T}) \times Y. \qquad (3.7)$$

Using (3.6), (3.7) and Proposition 2.7.2, we have $\overline{T} + S$ is closed, as desired.

Now, we prove $\overline{T + S} \subset \overline{T} + S$. Clearly, $T \subset \overline{T}$ and consequently, $T + S \subset \overline{T} + S$. Thus, since $\overline{T} + S$ is closed, then

$$\overline{T + S} \subset \overline{\overline{T} + S} \subset \overline{T} + S. \qquad (3.8)$$

(*ii*) Since

$$(T + S)(0) = T(0) + S(0) = T(0) = \overline{T}(0) = \overline{T(0)} = \overline{(T + S)(0)} \qquad (3.9)$$

and, by using (3.8), we have

$$\overline{(T + S)(0)} \subset \overline{(T + S)}(0) \subset \overline{T}(0) + S(0) = \overline{T(0)} + S(0) = (T + S)(0). \quad (3.10)$$

So, in view of (3.9) and (3.10), we have $(T + S)(0) = (\overline{T + S})(0)$. Thus, it follows from Lemma 2.7.3 that $T + S$ is closable.

Now, we prove $\overline{T + S} = \overline{T} + S$. By (3.8), it only remains to see that

$$\overline{T} + S \subset \overline{T + S}.$$

Using Proposition 2.3.4, we have $T = T + S - S \subset \overline{T + S} - S$. Moreover, by the hypotheses we can apply the part (*i*) to ensure that $\overline{T + S} - S$ is closed. So, the fact that $\overline{T} \subset \overline{\overline{T + S} - S} = \overline{T + S} - S$ and by Proposition 2.3.4, lead to $\overline{T} + S \subset \overline{T + S}$. Q.E.D.

Corollary 3.1.1 *Let $S, T \in LR(X, Y)$ such that $S(0) \subset T(0)$, $\overline{\mathcal{D}(T)} \subset \mathcal{D}(S)$, and $\lambda \in \mathbb{C}$. Then,*

(*i*) *If $T \in CR(X, Y)$ and S is continuous, then $\lambda S - T \in CR(X, Y)$.*
(*ii*) *If T is closable and S is continuous, then $\lambda S - T$ is closable and $\overline{\lambda S - T} = \lambda S - \overline{T}$.* ◇

Lemma 3.1.2 *Let S, T, A and $B \in LR(X,Y)$. Then,*

(i) *If S and T are closable, and $\mathcal{D}(T) \subset \mathcal{D}(S)$, then S is T-bounded with T-bound α if and only if, $\overline{S}$ is $\overline{T}$-bounded with $\overline{T}$-bound α.*

(ii) *If $S = A + B$ is T-bounded with T-bound β, B is A-bounded with A-bound $\delta < 1$, and $B(0) \subset A(0)$, then A is T-bounded with T-bound $\gamma \leq \frac{\beta}{1-\delta}$.*   ◇

Proof. (i) <u>First case</u> : If S and T are two operators, then since S is T-bounded with T-bound α, we have for any $b > \alpha$,

$$\|Sx\| \leq a\|x\| + b\|Tx\|, \tag{3.11}$$

for some $a \geq 0$ and all $x \in \mathcal{D}(T) \subset \mathcal{D}(S)$. From (3.11) it readily follows that $\mathcal{D}(\overline{T}) \subset \mathcal{D}(\overline{S})$ and

$$\|\overline{S}x\| \leq a\|x\| + b\|\overline{T}x\|, \quad \text{for } x \in \mathcal{D}(\overline{T}).$$

Thus, $\overline{S}$ is $\overline{T}$-bounded and, since $G(S) \subset G(\overline{S})$, it follows that $\overline{S}$ has $\overline{T}$-bound α. Conversely, if $\overline{S}$ has $\overline{T}$-bound α, then S has T-bound $\leq \alpha$. The first part of the proof now implies that S has T-bound α.

<u>Second case</u> : If S and T are two relations, then $Q_S S$ is $Q_T T$-bounded with $Q_T T$-bound α. The latter is equivalent to $\overline{Q_S S}$ is $\overline{Q_T T}$-bounded with $\overline{Q_T T}$-bound α. Since $\mathcal{D}(T) = \mathcal{D}(Q_T T)$, and $\overline{Q_T T} = Q_{\overline{T}} \overline{T}$, the latter holds if, and only if, $\overline{S}$ is $\overline{T}$-bounded with $\overline{T}$-bound α.

(ii) Since B is A-bounded with A-bound $\delta < 1$, and $B(0) \subset A(0)$, then if we apply Proposition 2.6.1, we obtain B is S-bounded with S-bound $\alpha \leq \frac{\delta}{1-\delta}$. Also, S is T-bounded with T-bound β. Hence, by using Proposition 2.6.2 (i), it follows that B is T-bounded with T-bound $\alpha\beta \leq \frac{\delta\beta}{1-\delta}$. So, B is T-bounded with T-bound $\alpha\beta \leq \frac{\delta\beta}{1-\delta}$, and knowing that S is T-bounded with T-bound β and $A = S - B$, the just given argument implies by Proposition 2.6.2 (ii) that A is T-bounded with T-bound $\gamma \leq \frac{\beta}{1-\delta}$.   Q.E.D.

Lemma 3.1.3 *Let S, $T \in LR(X,Y)$ such that $S(0) \subset T(0)$. Suppose that S is T-bounded with T-bound δ. Then,*

(i) *If T is closable, and $\delta < 1$, then $T + S$ is closable.*
(ii) *If $T + S$ is closable, and $\delta < \frac{1}{2}$, then T is closable.*

If, in addition, S is closable, then $\overline{S + T} = \overline{S} + \overline{T}$ and $\mathcal{D}(\overline{T + S}) = \mathcal{D}(\overline{T})$. ◇

Proof. (i) Since S is T-bounded with T-bound $\delta < 1$, then $Q_T S$ is $Q_T T$-bounded with $Q_T T$-bound < 1. Moreover, T is closable, then by Lemma 2.7.3,

$Q_T T$ is closable. In view of $(T+S)(0) = T(0)$ and $Q_{S+T}(T+S) = Q_T(T+S) = Q_T T + Q_T S$, we have $Q_{S+T}(T+S)$ is closable and $(T+S)(0)$ is closed. Now, pplying Lemma 2.7.3, it follows that $T+S$ is closable.

(ii) Since $T+S$ is closable, S is T-bounded with T-bound $\delta < \frac{1}{2}$, and by appliying Proposition 2.3.4, we get $T = T+S-S$. Also, S is $(T+S)$-bounded with $(T+S)$-bound < 1. So, using (i), it follows that T is closable. Since T and S are closable, and S is T-bounded with T-bound < 1, then by Lemma 3.1.2 (i), we obtain $\overline{S}$ is $\overline{T}$-bounded with $\overline{T}$-bound < 1. So, $\overline{S} + \overline{T}$ is closed. Since $T \subset \overline{T}$, and $S+T \subset \overline{S} + \overline{T}$, then $\overline{T+S} \subset \overline{\overline{T} + \overline{S}} = \overline{T} + \overline{S}$ ($\overline{S} + \overline{T}$ is closed). On the one hand, we obtain by Proposition 2.3.4 and $T = T+S-S$ that $\overline{T} = \overline{T+S-S} \subset \overline{T+S} - \overline{S}$. On the other hand, S is T-bounded with T-bound $\delta < \frac{1}{2}$, then S is $(T+S)$-bounded with $(T+S)$-bound < 1. Using Lemma 3.1.2 (i), we obtain $\overline{S}$ is $(\overline{T+S})$-bounded with $(\overline{T+S})$-bound < 1. So, $\overline{T+S} - \overline{S}$ is closed. Then, $\overline{T} \subset \overline{T+S} - \overline{S} = \overline{T+S} - \overline{S}$, and we obtain $\overline{T} + \overline{S} \subset \overline{T+S}$. Wherefrom, $\overline{T} + \overline{S} = \overline{T+S}$, and $\mathcal{D}(\overline{T+S}) = \mathcal{D}(\overline{T} + \overline{S}) = \mathcal{D}(\overline{T})$.   Q.E.D.

3.1.4   Product of Closable Linear Relations

Theorem 3.1.5 *Let $S \in LR(Y, Z)$ be closable and $T \in \mathcal{L}(X, Y)$. Then,*

(i) *$\overline{S}T$ is closed.*
(i) *ST is closable and $\overline{ST} \subset \overline{S}T$.*
(ii) *If T is bijective, then $\overline{ST} = \overline{S}T$.*   ◇

Proof. (i) Since $\overline{S}$ is closed and $T \in \mathcal{L}(X, Y)$, then by using Lemma 2.7.1, we have $\overline{S}T$ is closed.

(ii) Since $S \subset \overline{S}$, then $ST \subset \overline{S}T$ which implies that

$$\overline{ST} \subset \overline{\overline{S}T} = \overline{S}T. \tag{3.12}$$

Now, we prove that ST is closable. Indeed, ST is closable if and only if, $\overline{ST}(0) = ST(0)$. But, using Lemma 2.7.2, we have

$$\overline{ST(0)} \subset \overline{S}T(0). \tag{3.13}$$

We infer from (3.12) and (3.13) that

$$\overline{ST(0)} \subset \overline{S}T(0). \tag{3.14}$$

On the other hand,

$$\overline{ST}(0) \; = \; \overline{S}(0) \text{ (as } T \text{ is single valued)}$$
$$= \; S(0) \text{ (as } S \text{ is closable)} \qquad\qquad (3.15)$$
$$= \; ST(0).$$

Applying (3.14) and (3.15), we obtain that $\overline{ST}(0) = \overline{ST}(0) = S(0) = ST(0)$. So, we conclude that ST is closable.

(*iii*) By (3.12), it only remains to see that $\overline{ST} \subset \overline{ST}$. But $T^{-1}(0) = N(T) = \{0\}$, then T^{-1} is single valued, $R(T) = Y$ and $\mathcal{D}(T^{-1}) = R(T)$. Thus, $\mathcal{D}(T^{-1}) = Y$. Since T is closed, then by using Lemma 2.7.3, we have T^{-1} is closed. Thus, by the closed graph theorem for linear relations (see Theorem 2.7.2 (*i*)), T^{-1} is bounded single valued. Since ST is closable by (*i*), we can apply (*i*) to ST instead S and T^{-1} instead of T to obtain STT^{-1} is closable. Hence, $\overline{STT^{-1}} \subset \overline{ST}T^{-1}$. So, $T^{-1}T = I_{\mathcal{D}(T)}$ and $TT^{-1} = I_{R(T)} = I_Y$. Thus, $S \subset STT^{-1}$ which implies that $\overline{S} \subset \overline{STT^{-1}} \subset \overline{ST}T^{-1}$. Therefore, $\overline{S}T \subset \overline{ST}T^{-1}T = \overline{ST}$.   Q.E.D.

Corollary 3.1.2 *Let $T \in \mathcal{L}(X,Y)$ be bijective and let $S \in \mathcal{LR}(Z,Y)$ be closable. Then, $T^{-1}S$ is closable and $\overline{T^{-1}S} = T^{-1}\overline{S}$.*   ◇

Proof. Since T is bounded single valued, then T is closed. Hence, by using Proposition 2.7.1, we have T^{-1} is closed. So, T^{-1} is bounded single valued. Indeed, $T^{-1}(0) = T(0) = \{0\}$ and $R(T^{-1}) = \mathcal{D}(T) = X$. Then, by using Corollary 2.9.2, we have $T^{-1}\overline{S}$ is closed. Therefore,

$$\overline{T^{-1}\overline{S}}(0) \; = \; T^{-1}\overline{S}(0) \qquad\qquad (3.16)$$
$$= \; T^{-1}S(0) \text{ (as } S \text{ is closable).}$$

Moreover, $T^{-1}S \subset T^{-1}\overline{S}$ and

$$\overline{T^{-1}S} \subset \overline{T^{-1}\overline{S}} = T^{-1}\overline{S}.$$

Hence, $\overline{T^{-1}S} \subset T^{-1}\overline{S}$. A combination of (3.16) and Lemma 2.7.2 leads to

$$\overline{T^{-1}S(0)} \subset \overline{T^{-1}S}(0) \subset \overline{T^{-1}\overline{S}}(0) = T^{-1}S(0).$$

Hence,

$$T^{-1}S(0) \subset \overline{T^{-1}S(0)} \subset \overline{T^{-1}S}(0) \subset T^{-1}S(0),$$

Consequently,

$$\overline{T^{-1}S(0)} = \overline{T^{-1}S}(0) = T^{-1}S(0).$$

Therefore, $T^{-1}S$ is closable. The proof of $\overline{T^{-1}S} \subset T^{-1}\overline{S}$ may be checked in the same way as the proof of Theorem 3.1.5 (ii).   Q.E.D.

Theorem 3.1.6 *Let* $T \in BR(X,Y)$ *be a single valued linear operator bijective and assume that* $H \in BR(Z,W)$ *is bijective with* $H(0)$ *closed. Then,*

(i) If $S \in LR(Y,Z)$ *is closable, then* HST *is closable and* $\overline{HST} = \overline{R}\overline{H}T$.

(ii) If H *is bounded single valued, then* $S \in LR(Y,Z)$ *is closable if and only if,* HST *is closable and* $\overline{HST} = H\overline{S}T$.   ◇

Proof. Let us consider two cases of H.

First case : If H is single valued, then by virtue of Theorem 3.1.5 (ii), we have ST is closable and $\overline{ST} = \overline{S}T$. On the other hand, H^{-1} is bounded single valued bijective. Indeed, $H^{-1}(0) = N(H) = \{0\}$ (as H is injective) implies H^{-1} is a closed single valued and $\mathcal{D}(H^{-1}) = R(H) = W$ (as H is surjective). Then, H is open. Since H^{-1} is continuous, then H^{-1} is bounded single valued. Clearly, $N(H^{-1}) = H(0) = \{0\}$ (since H is single valued) and $R(H^{-1}) = \mathcal{D}(H) = Z$. So, $H = (H^{-1})^{-1}$. This leads to $HST = (H^{-1})^{-1}ST$ is closable with $\overline{HST} = H\overline{ST} = H\overline{S}T$.

Second case : If H is linear relation, then

(a) $Q_H HST$ is closable and $\overline{Q_H HST} = Q_H H\overline{S}T$.

(b) $H\overline{S}$ is closed by virtue of Proposition 2.9.4.

(c) $H\overline{S}T$ is closed equivalently $(H\overline{S}T)^{-1}$ is closed. Indeed, $(H\overline{S}T)^{-1} = T^{-1}(H\overline{S})^{-1}$ and, we have $H\overline{S}$ is closed. Therefore, $(H\overline{S})^{-1}$ is closed. Since T is bounded single valued bijective, we deduce from Corollary 3.1.2 that $T^{-1}(H\overline{S})^{-1}$ is closable and $\overline{T^{-1}(H\overline{S})^{-1}} = T^{-1}\overline{(H\overline{S})^{-1}}$. But

$$\overline{T^{-1}(H\overline{S})^{-1}} = \overline{(H\overline{S}T)^{-1}}$$

and

$$\begin{aligned} T^{-1}\overline{(H\overline{S})^{-1}} &= T^{-1}(H\overline{S})^{-1} \\ &= (H\overline{S}T)^{-1}, \end{aligned}$$

then by applying Proposition 2.7.5, we have $\overline{(H\overline{S})^{-1}} = (H\overline{S})^{-1}$. So, we obtain (c).

(d) $HST(0) = HS(0) = H\overline{S}(0) = H\overline{S}T(0)$ is closed. Indeed,

$$\begin{aligned} HST(0) &= HS(0) \text{ (again } T \text{ is single valued so that } T(0) = \{0\}) \\ &= H\overline{S}(0) \text{ (as } S \text{ is closable)} \\ &= H\overline{S}T(0) \text{ (again } T \text{ is single valued).} \end{aligned}$$

Therefore, $HST(0)$ is closed by virtue of (c) and Lemma 2.7.3.

(e) $Q_{HST}HST$ is a a closable operator. Indeed, using Lemma 2.1.3, we have

$$W \big/ HS(0) \equiv (W/H(0)) \big/ (HS(0)/H(0)),$$

and $Q_{HS} = Q_A Q_H$ with $A = HS(0)/H(0)$, $HS(0) = HST(0)$ is closed (by (d)), and $H(0)$ is closed by hypothesis. Moreover, by using (d), we have $Q_{HS} = Q_{HST} = Q_{\overline{HS}} = Q_{H\overline{ST}}$. Clearly,

$$\begin{aligned} \mathcal{D}(Q_{HST}HST) &= \mathcal{D}(Q_A Q_H HST) \\ &= \mathcal{D}(Q_H HST) \\ &= \mathcal{D}(HST). \end{aligned}$$

On the other hand, by (a), we have

$$Q_H HST = Q_H \overline{HST}\big|_{\mathcal{D}(Q_H HST)} = Q_H \overline{HST}\big|_{\mathcal{D}(HST)}.$$

So, $Q_{H\overline{ST}}\overline{HST} = Q_{HST}\overline{HST}$ is an extension of $Q_{HST}HST$. Since $\overline{HST}$ is closed (by (c)), then by applying Lemma 2.7.3, we have $Q_{HST}\overline{HST}$ is closed. Therefore, $Q_{HST}\overline{HST}$ is closed extension of $Q_{HST}HST$. Hence, (e) is true.

(f) HST is closable. Follows immediately form the properties (d), (e) and Lemma 2.7.3 (iii). It only remains to see that $\overline{HST} = H\overline{ST}$.

(g) $\overline{HST} \subset H\overline{ST}$. Indeed, since $HST \subset H\overline{ST}$, then $\overline{HST} \subset \overline{RST} = H\overline{ST}$.

(h) $H\overline{ST} \subset \overline{HST}$. Indeed, the equality $\overline{Q_H HST} = Q_H H\overline{ST}$ implies $Q_A \overline{Q_H HST} = Q_A Q_H H\overline{ST}$, where $Q_A Q_H = Q_{HST}$. But

$$Q_A \overline{Q_H HST} \subset \overline{Q_A Q_H HST}$$

implies that $Q_{HST}H\overline{ST} \subset \overline{Q_{HST}HST}$ with $\overline{Q_{HST}HST} = Q_{HST}\overline{HST}$. Hence, $Q_{HST}H\overline{ST} \subset Q_{HST}\overline{HST}$. By using (g), we have the other inclusion. So,

$$Q_{HST}H\overline{ST} = Q_{HST}\overline{HST}.$$

Thus, in particular

$$\begin{aligned} \mathcal{D}(H\overline{ST}) &= \mathcal{D}(Q_{HST}H\overline{ST}) \\ &= \mathcal{D}(Q_{HST}\overline{HST}) \\ &= \mathcal{D}(\overline{HST}) \end{aligned}$$

and $\mathcal{D}(\overline{HST}) = \mathcal{D}(H\overline{ST})$. Since $H\overline{ST}(0) = HST(0)$ by (d), $\overline{HST(0)} = H\overline{ST}(0) \subset \overline{HST}(0)$, and $\overline{HST}(0) \subset H\overline{ST}(0)$ by (g), we deduce $\mathcal{D}(\overline{HST}) =$

$\mathcal{D}(H\overline{ST})$, $\overline{HST(0)} = H\overline{ST}(0)$ and $\overline{HST} \subset H\overline{ST}$. Thus, by virtue of Proposition 2.3.6, we have $\overline{HST} = H\overline{ST}$. Conversely, if H is single valued bijective and HST is closable, then from Corollary 3.1.2, we have $H^{-1}HST = ST$ is closable and $\overline{ST} = \overline{H^{-1}HST} = H^{-1}\overline{HST} = H^{-1}H\overline{ST} = \overline{ST}$. By the same token $\overline{S}(0) = \overline{ST}(0) = \overline{ST}(0) = ST(0) = S(0)$, we conclude that S is closable. This completes the proof.   Q.E.D.

3.2   Sequence of Multivalued Linear Operators Converging in the Generalized Sense

3.2.1   The Gap Between Two Linear Relations

Theorem 3.2.1 *Let $T \in BC\mathcal{R}(X,Y)$. Then,*

$$\delta(T, O_X) \leq \frac{\|T\|}{(1 + \|T\|^p)^{\frac{1}{p}}}, \ p \geq 1. \qquad \diamond$$

Proof.   Since $T \in BC\mathcal{R}(X,Y)$, then we can conclude from Theorem 2.5.5 (*i*) that $\|Q_T T\| < \infty$, which yields that $Q_T T \in \mathcal{L}(X, Y/T(0))$. Hence,

$$\delta(Q_T T, O_X) = \frac{\|Q_T T\|}{(1 + \|Q_T T\|^p)^{\frac{1}{p}}}.$$

Let $\varphi = (x, y) \in G(T)$ such that

$$\|\varphi\| = \|(x,y)\|_p^p = \|x\|^p + \|y\|^p = 1. \qquad (3.17)$$

It follows that

$$
\begin{aligned}
\delta(T, O_X) &= \sup_{\substack{\varphi \in G(T) \\ \|\varphi\|=1}} \operatorname{dist}(\varphi, G(O_X)) \\
&= \sup_{\substack{x \in X \\ \|x\|^p + \|y\|^p = 1}} \left[\inf_{z \in X} \|(x,y) - (z, 0)\| \right] \\
&= \sup_{\substack{x \in X \\ \|x\|^p + \|y\|^p = 1}} \left[\inf_{z \in X} (\|x - z\|^p + \|y\|^p)^{\frac{1}{p}} \right] \\
&= \sup_{\substack{x \in X \\ \|x\|^p + \|y\|^p = 1}} \|y\|. \qquad (3.18)
\end{aligned}
$$

By using Lemma 2.5.7 (i), we infer that

$$\|Tx\| = \inf_{t \in Tx} \|t\|.$$

Since $x \in \mathcal{D}(T) = X$ and $y \in Tx$, then in view of definition of the lower bound, for any given $\varepsilon > 0$, we obtain that

$$\|y\| < \|Tx\| + \varepsilon. \tag{3.19}$$

This implies that

$$
\begin{aligned}
\|x\|^p + \|y\|^p \ &< \ \|x\|^p + (\|Tx\| + \varepsilon)^p \\
&= \ \|x\|^p + \|Tx\|^p + \sum_{k=0}^{p-1} C_p^k \|Tx\|^k \varepsilon^{p-k}.
\end{aligned}
$$

Since ε is arbitrary, then

$$
\begin{aligned}
\|x\|^p + \|y\|^p \ &\leq \ \|x\|^p + \|Tx\|^p \\
&= \ \|x\|^p + \|Q_T Tx\|^p.
\end{aligned}
$$

Hence, according to the condition (3.17), we deduce that

$$\|x\|^p + \|Q_T Tx\|^p = 1. \tag{3.20}$$

Putting $x = cw$, where $w \in X$ and $c = \dfrac{1}{(1 + \|Q_T Tw\|^p)^{\frac{1}{p}}}$, we get

$$
\begin{aligned}
\sup_{\substack{x \in X \\ \|x\|^p + \|Q_T Tx\|^p = 1}} \|Q_T Tx\| \ &= \ \sup_{\substack{w \in X \\ \|w\| = 1}} \frac{\|Q_T Tw\|}{(1 + \|Q_T Tw\|^p)^{\frac{1}{p}}} \\
&= \ \frac{\|Q_T T\|}{(1 + \|Q_T T\|^p)^{\frac{1}{p}}} \\
&= \ \delta(Q_T T, O_X).
\end{aligned}
$$

This implies from (3.18) and (3.19) that, for any given $\varepsilon > 0$,

$$
\begin{aligned}
\delta(T, O_X) \ &= \ \sup_{\substack{x \in X \\ \|x\|^p + \|y\|^p = 1}} \|y\| \\
&\leq \ \sup_{\substack{x \in X \\ \|x\|^p + \|Q_T Tx\|^p = 1}} \|Q_T Tx\| + \varepsilon \quad \text{see (3.20)} \\
&\leq \ \delta(Q_T T, O_X) + \varepsilon.
\end{aligned}
$$

By the arbitrariness of ε, we conclude that

$$\delta(T, O_X) \leq \delta(Q_T T, O_X) = \frac{\|Q_T T\|}{(1 + \|Q_T T\|^p)^{\frac{1}{p}}} = \frac{\|T\|}{(1 + \|T\|^p)^{\frac{1}{p}}}. \qquad \text{Q.E.D.}$$

Remark 3.2.1 *If* $T \in BC\mathcal{R}(X, Y)$, *then*

$$\delta(T, O_X) \leq \delta(Q_T T, O_X). \qquad \diamond$$

Theorem 3.2.2 *Let* $T \in BC\mathcal{R}(X, Y)$. *Then,*

$$\widehat{\delta}(T, O_X) \leq \frac{\|T\|}{\sqrt{1 + \|T\|^2}}. \qquad \diamond$$

Proof. If we take $p = 2$ in Theorem 3.2.1, we have

$$\delta(T, O_X) \leq \frac{\|T\|}{(1 + \|T\|^2)^{\frac{1}{2}}}.$$

Then, it is sufficient to prove that

$$\delta(O_X, T) \leq \frac{\|T\|}{\sqrt{1 + \|T\|^2}}.$$

Let us observe that

$$\begin{aligned}
\delta(O_X, T) &= \sup_{\substack{x \in X \\ \|x\| = 1}} \ \text{dist}((x, 0), G(T)) \\
&= \sup_{\substack{x \in X \\ \|x\| = 1}} \left[\inf_{(y,z) \in G(T)} \|(x, 0) - (y, z)\| \right] \\
&= \sup_{\substack{x \in X \\ \|x\| = 1}} \left[\inf_{(y,z) \in G(T)} (\|x - y\|^2 + \|z\|^2)^{\frac{1}{2}} \right].
\end{aligned}$$

Moreover, in view of Lemma 2.5.7 (i), we have

$$\|Ty\| = \inf_{t \in Ty} \|t\|.$$

Since $y \in \mathcal{D}(T) = X$ and $z \in Ty$, then for any given $\varepsilon > 0$, we have $\|z\| \leq \|Ty\| + \varepsilon$. So, for any $x, y \in X$, we obtain that

$$\begin{aligned}
\|x - y\|^2 + \|z\|^2 &\leq \|x - y\|^2 + (\|Ty\| + \varepsilon)^2 \\
&= \|x - y\|^2 + \|Ty\|^2 + 2\varepsilon\|Ty\| + \varepsilon^2.
\end{aligned}$$

Since ε is arbitrary, we get

$$(\|x - y\|^2 + \|z\|^2)^{\frac{1}{2}} \leq (\|x - y\|^2 + \|Ty\|^2)^{\frac{1}{2}}. \qquad (3.21)$$

Consider the function $f(x, y) = \|x - y\|^2 + \|Ty\|^2$. Let $x \in X$ such that $\|x\| = 1$ and $y = \dfrac{x}{1 + \|Tx\|^2}$. Then,

$$
\begin{aligned}
f(x, y) = f\left(x, \frac{x}{1 + \|Tx\|^2}\right) &= \left\|x - \frac{x}{1 + \|Tx\|^2}\right\|^2 + \frac{\|Tx\|^2}{(1 + \|Tx\|^2)^2} \\
&= \frac{\|x\|^2 \|Tx\|^4 + \|Tx\|^2}{(1 + \|Tx\|^2)^2} \\
&= \frac{\|Tx\|^2(\|x\|^2 \|Tx\|^2 + 1)}{(1 + \|Tx\|^2)^2} \\
&= \frac{\|Tx\|^2(\|Tx\|^2 + 1)}{(1 + \|Tx\|^2)^2} \quad \text{since } \|x\| = 1 \\
&= \frac{\|Tx\|^2}{1 + \|Tx\|^2}.
\end{aligned}
$$

Hence, by using (3.21), we deduce that

$$
\begin{aligned}
\operatorname{dist}((x, 0), G(T)) &\leq (\|x - y\|^2 + \|z\|^2)^{\frac{1}{2}} \\
&\leq \frac{\|Tx\|}{\sqrt{1 + \|Tx\|^2}}.
\end{aligned}
$$

Since the function $g(t) = \frac{t}{1+t}$ is increasing, it follows that

$$
\begin{aligned}
\delta(O_X, T) &= \sup_{\substack{x \in X \\ \|x\| = 1}} \operatorname{dist}((x, 0), G(T)) \\
&\leq \sup_{\substack{x \in X \\ \|x\| = 1}} \frac{\|Tx\|}{\sqrt{1 + \|Tx\|^2}} \\
&\leq \frac{\|T\|}{\sqrt{1 + \|T\|^2}}.
\end{aligned}
$$

Finally, we obtain the following result

$$
\begin{aligned}
\widehat{\delta}(T, O_X) &= \max\{\delta(T, O_X), \delta(O_X, T)\} \\
&\leq \frac{\|T\|}{\sqrt{1 + \|T\|^2}}.
\end{aligned}
$$

This completes the proof. $\hfill$ Q.E.D.

Now, we give a relationship between $\delta(S + A, T + A)$ and $\delta(S, T)$.

Theorem 3.2.3 *Let T, $S \in C\mathcal{R}(X, Y)$ and let $A : X \longrightarrow Y$ be a continuous linear relation satisfies $\mathcal{D}(A) \supset \mathcal{D}(T) \cup \mathcal{D}(S)$ and $A(0) \subset T(0) \cap S(0)$. Then,*

$$
\delta(S + A, T + A) \leq (2 + \|A\|^2)\delta(S, T) \tag{3.22}
$$

and

$$\widehat{\delta}(S + A, T + A) \leq (2 + \|A\|^2)\widehat{\delta}(S, T). \tag{3.23}$$

$$\diamondsuit$$

Proof. Since $\mathcal{D}(T + A) = \mathcal{D}(T)$ and $\mathcal{D}(S + A) = \mathcal{D}(S)$, then by referring to Proposition 2.7.2, we infer that $T + A$ and $S + A$ are closed. Let $\varphi = (x, y) \in G(S + A)$ with $\|\varphi\| = 1$. Then, there exists $(x, y_1) \in G(S)$ and $(x, y_2) \in G(A)$ such that $y = y_1 + y_2$. Hence,

$$\|\varphi\|_2^2 = \|x\|^2 + \|y_1 + y_2\|^2 = 1. \tag{3.24}$$

Since $y_2 \in Ax$, then by using Lemma 2.5.7 (i), we infer, for any given $\varepsilon > 0$, that

$$\begin{aligned} \|y_2\| &< \|Ax\| + \varepsilon \\ &< \|A\|\|x\| + \varepsilon. \end{aligned} \tag{3.25}$$

Hence, by using (3.25), we have

$$\begin{aligned} \|y_1\| &= \|y_1 + y_2 - y_2\| \\ &\leq \|y_1 + y_2\| + \|y_2\| \\ &\leq \|y_1 + y_2\| + \|A\|\|x\| + \varepsilon. \end{aligned}$$

Thus,

$$\begin{aligned} \|y_1\|^2 &\leq ((\|y_1 + y_2\| + \|A\|\|x\|) + \varepsilon)^2 \\ &\leq (\|y_1 + y_2\| + \|A\|\|x\|)^2 + 2\varepsilon(\|y_1 + y_2\| + \|A\|\|x\|) + \varepsilon^2. \end{aligned}$$

Therefore, for any given $\varepsilon > 0$ and by Cauchy-Schwarz inequality, we obtain

$$\|y_1\|^2 \leq (1 + \|A\|^2)(\|x\|^2 + \|y_1 + y_2\|^2) + 2\varepsilon(\|y_1 + y_2\| + \|A\|\|x\|) + \varepsilon^2.$$

It follows from (3.24) that

$$\|y_1\|^2 \leq (1 + \|A\|^2) + 2\varepsilon(\|y_1 + y_2\| + \|A\|\|x\|) + \varepsilon^2,$$

and by arbitrariness of ε, we deduce that

$$(\|x\|^2 + \|y_1\|^2)^{\frac{1}{2}} \leq (2 + \|A\|^2)^{\frac{1}{2}}. \tag{3.26}$$

In addition, for any $u \in \mathcal{D}(T + A)$, there exists $(u, v_1) \in G(T)$ and $(u, v_2) \in G(A)$ such that $v = v_1 + v_2 \in (T + A)u$. On the one hand, since $y_2 \in Ax$ and $v_2 \in Au$, then $y_2 - v_2 \in A(x - u)$. Hence, for any given $\varepsilon_0 > 0$, we have

$$\|y_2 - v_2\| \leq \|A\|\|x - u\| + \varepsilon_0. \tag{3.27}$$

On the other hand, by using (3.27), we get

$$\|y_1 + y_2 - v_1 - v_2\| \leq \|y_1 - v_1\| + \|y_2 - v_2\|$$
$$\leq \|y_1 - v_1\| + \|A\|\|x - u\| + \varepsilon_0.$$

Similarly, for all $u \in \mathcal{D}(T + A) = \mathcal{D}(T)$ and by arbitrariness of ε, we conclude that

$$\left(\|x - u\|^2 + \|y_1 + y_2 - v_1 - v_2\|^2\right)^{\frac{1}{2}} \leq \left(2 + \|A\|^2\right)^{\frac{1}{2}} \left(\|x - u\|^2 + \|y_1 - v_1\|^2\right)^{\frac{1}{2}}.$$

Noting that $\psi = (u, v) \in G(T + A)$, $\varphi_1 = (x, y_1) \in G(S)$ and $\psi_1 = (u, v_1) \in G(T)$, we get

$$\mathrm{dist}(\varphi, G(T + A)) \leq \left(2 + \|A\|^2\right)^{\frac{1}{2}} \mathrm{dist}(\varphi_1, G(T)). \qquad (3.28)$$

Therefore, it follows from (2.45), (3.26) and (3.28) that

$$
\begin{aligned}
\delta(S + A, T + A) &= \sup_{\substack{\varphi \in G(S+A) \\ \|x\|^2 + \|y_1 + y_2\|^2 = 1}} \mathrm{dist}\left(\varphi, G(T + A)\right) \\
&\leq \left(2 + \|A\|^2\right)^{\frac{1}{2}} \sup_{\substack{\varphi_1 \in G(S) \\ (\|x\|^2 + \|y_1\|^2)^{\frac{1}{2}} \leq (2 + \|A\|^2)^{\frac{1}{2}}}} \mathrm{dist}\left(\varphi_1, G(T)\right) \\
&\leq \left(2 + \|A\|^2\right) \delta(S, T).
\end{aligned}
$$

Thus, (3.22) holds and it also implies (3.23) as a direct consequence. Q.E.D.

The following result can be easily derived from Theorems 3.2.2 and 3.2.3.

Theorem 3.2.4 *Let T, $S \in LR(X, Y)$. If T and S are two closed linear relations everywhere defined satisfying $T(0) = S(0)$, then*

$$\widehat{\delta}(T, S) \leq (2 + \min\{\|T\|, \|S\|^2\}) \frac{\|T - S\|}{\sqrt{1 + \|T - S\|^2}}. \qquad \diamondsuit$$

Proof. It follows from Proposition 2.7.2 that $T - S$ and $S - T$ are closed, and in view of $T(0) = S(0)$, we have $T = T - S + S$ and $S = S - T + T$. Then, $T \in BR(X, Y)$. Hence, by using Proposition 2.3.4 and Theorem 3.2.3, we deduce that

$$\widehat{\delta}(T, S) = \widehat{\delta}((T - S) + S, O_X + S) \leq (2 + \|S\|^2)\widehat{\delta}(T - S, O_X)$$

and

$$\widehat{\delta}(T, S) = \widehat{\delta}(O_X + T, (S - T) + T) \leq (2 + \|T\|^2)\widehat{\delta}(S - T, O_X)$$

which, together with Theorem 3.2.2, yields that

$$\widehat{\delta}(T, S) \leq (2 + \min\{\|T\|, \|S\|\}^2) \frac{\|T - S\|}{\sqrt{1 + \|T - S\|^2}}. \qquad \text{Q.E.D.}$$

Theorem 3.2.5 *Let* $S, T \in CR(X, Y)$ *satisfy that* $\|T\| < \infty$, $T(0) \subset S(0)$ *and* $\mathcal{D}(T) \supset \mathcal{D}(S)$. *If* $\widehat{\delta}(S, T) < (1 + \|T\|^2)^{-\frac{1}{2}}$, *then* $\|S\| < \infty$, $\mathcal{D}(S) = \mathcal{D}(T)$ *and*

$$\|S - T\| \leq \frac{1 + \|T\|^2}{1 - \sqrt{1 + \|T\|^2}\delta(S, T)} \delta(S, T). \qquad \diamondsuit$$

Proof. We divide this proof into two steps.

Step 1. We show that $S - T$ is continuous.

For any given $x \in \mathcal{D}(S - T)$ with $x \neq 0$, there exist $y \in (S - T)x$ and $z \in Tx$. Then, by using Lemma 2.5.7 (i), we infer, for any given $\varepsilon > 0$, that

$$\|y\| < \|(S - T)x\| + \frac{\varepsilon}{2} \quad \text{and} \quad \|z\| < \|T\|\|x\| + \frac{\varepsilon}{2}. \qquad (3.29)$$

Based on the assumptions that $T(0) \subset S(0)$ and $\mathcal{D}(S) \subset \mathcal{D}(T)$, and by using Proposition 2.3.4, we can write $S = S - T + T$. Then, we deduce that $(x, y + z) \in G(S)$. Moreover, for any $(u, v) \in G(T)$, we get $(x - u, z - v) \in G(T)$ and for any $\varepsilon_0 > 0$, there is $w \in T(x - u)$ such that

$$\|w\| < \|T\|\|x - u\| + \varepsilon_0. \qquad (3.30)$$

Hence, we deduce that $(0, z - v - w) \in G(T)$. The fact that $T(0) \subset (S - T)(0)$ implies that $(0, z - v - w) \in G(S - T)$. Thus, $(x, y + z - v - w) \in G(S - T)$. Consequently, for any $(u, v) \in G(T)$, we infer from (3.30) that

$$\begin{aligned}
\|y + z - v - w\| &\leq \|y + z - v\| + \|w\| \\
&< \|y + z - v\| + \|T\|\|x - u\| + \varepsilon_0.
\end{aligned}$$

By Cauchy-Schwarz inequality, we obtain

$$\|y + z - v - w\| < \left(1 + \|T\|^2\right)^{\frac{1}{2}} \left(\|x - u\|^2 + \|y + z - v\|^2\right)^{\frac{1}{2}} + \varepsilon_0.$$

Hence,

$$\|y + z - v - w\| < \left(1 + \|T\|^2\right)^{\frac{1}{2}} \inf_{(u,v) \in G(T)} \left(\|x - u\|^2 + \|y + z - v\|^2\right)^{\frac{1}{2}} + \varepsilon_0.$$

Thus, by arbitrariness of ε_0, we conclude that

$$\|(S - T)x\| \leq \delta(S, T)(1 + \|T\|^2)^{\frac{1}{2}}. \qquad (3.31)$$

Now, let us assume that

$$\|x\|^2 + \|y + z\|^2 = 1. \tag{3.32}$$

Then, by using (3.29), we have

$$
\begin{aligned}
1 &\leq \|x\|^2 + (\|y\| + \|z\|)^2 \\
&< \|x\|^2 + (\|(S - T)x\| + \|T\|\|x\| + \varepsilon)^2 \\
&< (1 + \|T\|^2)\|x\|^2 + 2\|T\|\|x\|\|(S - T)x\| + \|(S - T)x\|^2 + \\
&\quad 2\varepsilon(\|(S - T)x\| + \|T\|\|x\|) + \varepsilon^2.
\end{aligned}
$$

Then, by arbitrariness of ε, we get

$$
\begin{aligned}
1 &\leq \left((1 + \|T\|^2)\|x\|^2 + 2\|T\|\|x\|\|(S - T)x\| + \|(S - T)x\|^2\right)^{\frac{1}{2}} \\
&\leq \left((1 + \|T\|^2)\|x\|^2 + 2(1 + \|T\|^2)^{\frac{1}{2}}\|x\|\|(S - T)x\| + \|(S - T)x\|^2\right)^{\frac{1}{2}} \\
&\leq \left(\left((1 + \|T\|^2)^{\frac{1}{2}}\|x\| + \|(S - T)x\|\right)^2\right)^{\frac{1}{2}} \\
&\leq (1 + \|T\|^2)^{\frac{1}{2}}\|x\| + \|(S - T)x\|. \tag{3.33}
\end{aligned}
$$

Therefore, it follows from (3.31) and (3.33) that

$$\|(S - T)x\| \leq (1 + \|T\|^2)\delta(S, T)\|x\| + (1 + \|T\|^2)^{\frac{1}{2}}\delta(S, T)\|(S - T)x\|.$$

Hence, for all $x \in \mathcal{D}(S - T)$ satisfying (3.32), we conclude that

$$\|(S - T)x\| \leq \frac{1 + \|T\|^2}{1 - (1 + \|T\|^2)^{\frac{1}{2}}\delta(S, T)}\delta(S, T)\|x\|. \tag{3.34}$$

Since (3.34) is homogeneous in x, then it is true for any $x \in \mathcal{D}(S - T) = \mathcal{D}(S)$ without normalization (3.32). So, $S - T$ is continuous.

Step 2. We prove that $\mathcal{D}(S)$ is dense in X.

Let u be any given non-zero vector. Then, for any $\varepsilon_1 > 0$, there is $v \in Tu$ such that

$$
\begin{aligned}
\|v\| &< \|Tu\| + \varepsilon_1 \\
&< \|T\|\|u\| + \varepsilon_1. \tag{3.35}
\end{aligned}
$$

Let us assume that $u \in \mathcal{D}(T)$ so normalized that (u, v) is in unit sphere of $G(T)$;

$$\|u\|^2 + \|v\|^2 = 1. \tag{3.36}$$

Let δ be such that $\delta(S,T) < \delta < (1 + \|T\|^2)^{-\frac{1}{2}}$. Then, there exists $(x,y) \in G(S)$ satisfying

$$\|x - u\|^2 + \|y - v\|^2 < \delta^2.$$

This implies that

$$\|x - u\| < \delta \quad \text{for all } x \in \mathcal{D}(S). \tag{3.37}$$

Since $\text{dist}\,(u, \mathcal{D}(S)) = \text{dist}\left(u, \overline{\mathcal{D}(S)}\right)$, then from (3.37), we have

$$\text{dist}(u, \overline{\mathcal{D}(S)}) < \delta.$$

In addition, it follows from (3.35) and (3.36) that

$$
\begin{aligned}
1 \quad &< \quad \|u\|^2 + (\|T\|\|u\| + \varepsilon_1)^2 \\
&< \quad (1 + \|T\|^2)\|u\|^2 + 2\varepsilon_1\|T\|\|u\| + \varepsilon_1^2.
\end{aligned}
$$

Then, by arbitrariness of ε_1, we obtain $1 \leq (1 + \|T\|^2)^{\frac{1}{2}}\|u\|$. Hence,

$$\text{dist}(u, \overline{\mathcal{D}(S)}) < (1 + \|T\|^2)^{\frac{1}{2}}\delta\|u\|. \tag{3.38}$$

Since (3.38) is homogeneous in u, then it is true for any $u \in \mathcal{D}(T)$. The fact that $(1 + \|T\|^2)^{\frac{1}{2}}\delta < 1$ implies from Lemma 2.1.6 that $\overline{\mathcal{D}(S)} = \mathcal{D}(T)$.

Step 3. We prove that S is continuous and $\mathcal{D}(S)$ is closed. By referring to Proposition 2.5.4 (ii), we conclude that

$$
\begin{aligned}
\|S\| \quad &= \quad \|S - T + T\| \\
&\leq \quad \|S - T\| + \|T\| < \infty.
\end{aligned}
$$

As a result, the use of Theorem 2.5.4 (i) makes us conclude that $\mathcal{D}(S)$ is closed. This implies that $\mathcal{D}(S) = \mathcal{D}(T)$, as desired.   Q.E.D.

The following result can be derived from Theorem 3.2.5.

Theorem 3.2.6 *Let $T, S \in CR(X,Y)$ and $T^{-1} \in \mathcal{L}(Y,X)$. If*

$$\widehat{\delta}(S,T) < (1 + \|T^{-1}\|^2)^{-\frac{1}{2}},$$

then $S^{-1} \in \mathcal{L}(Y,X)$. $\diamond$

Proof. In view of Theorem 2.10.1, we have $\delta(S^{-1}, T^{-1}) = \delta(S,T)$. Then, from Theorem 3.2.5, S^{-1} is continuous and $\mathcal{D}(S^{-1}) = \mathcal{D}(T^{-1})$. Thus, it remains to prove that S is injective, which yields that S^{-1} is an operator. Let S be such that $\delta(S,T) < (1 + \|T^{-1}\|^2)^{-\frac{1}{2}}$, and suppose, by contradiction, that $(x, 0) \in G(T)$ such that $\|x\| = 1$ can be found. Then, $(x, 0)$ is in the unit sphere of $G(S)$. Hence, there is $(y, z) \in G(T)$ such that

$$\|x - y\|^2 + \|z\|^2 < \delta'^2, \tag{3.39}$$

for some number δ' satisfy $\delta(S, T) < \delta' < (1 + \|T^{-1}\|^2)^{-\frac{1}{2}}$. Hence,

$$
\begin{aligned}
1 = \|x\|^2 &\leq (\|x - y\| + \|y\|)^2 \\
&\leq (\|x - y\| + \|T^{-1}\|\|z\|)^2.
\end{aligned}
$$

By Cauchy-Schwarz inequality and (3.39), we get

$$
\begin{aligned}
1 &\leq (1 + \|T^{-1}\|^2)(\|x - y\|^2 + \|z\|^2) \\
&< \delta'^2(1 + \|T^{-1}\|^2) \\
&< 1,
\end{aligned}
$$

which is a contradiction. Q.E.D.

Corollary 3.2.1 *Let $T, S \in CR(X, Y)$ and $0 \in \rho(T)$. If*

$$\widehat{\delta}(S, T) < (1 + \|T^{-1}\|^2)^{-\frac{1}{2}},$$

then $0 \in \rho(S)$. ◇

3.2.2 Relationship Between the Gap of Linear Relation and their Selection

Now, we give a relationship between $\delta(T, S)$ and $\delta(T_1, S_1)$, where T_1 and S_1 are selections of T and S, respectively.

Theorem 3.2.7 *Let $T, S \in CR(X, Y)$. If T_1 and S_1 are closed selections of T and S respectively, then*

$$\delta(T, S) \leq 2\delta(T_1, S_1).$$ ◇

Proof. Fix any δ' such that

$$\delta(T_1, S_1) = \delta(G(T_1), G(S_1)) < \delta'. \tag{3.40}$$

We shall show that

$$\delta(\varphi, S) \leq 2\delta', \text{ for all } \varphi \in G(T) \text{ with } \|\varphi\| = 1.$$

For this purpose, let $\varphi = (x, y) \in G(T)$ with $\|\varphi\| = 1$. Then, there exist $(x, T_1 x) \in G(T_1)$ and $(0, \alpha_1) \in \{0\} \times T(0)$ such that $y = T_1 x + \alpha_1$. Hence,

$$\|\varphi\|^2 = \|(x, T_1 x + \alpha_1)\|_2^2 = \|x\|^2 + \|T_1 x + \alpha_1\|^2 = 1. \tag{3.41}$$

Put $\gamma^2 = \|(x, T_1 x)\|_2^2 = \|x\|^2 + \|T_1 x\|^2$. It is clear that $\gamma > 0$, then we can suppose that

$$\varphi_1 = \gamma^{-1}(x, T_1 x) \in G(T_1) \text{ with } \|\varphi_1\| = 1.$$

Therefore, by (3.40), we get $\text{dist}(\varphi_1, G(S_1)) < \delta'$. Hence, there is a $z \in \mathcal{D}(S_1)$ such that

$$\|x - z\|^2 + \|T_1 x - S_1 z\|^2 < \gamma^2 \delta'^2. \tag{3.42}$$

Assume that $\psi = (z, t) \in G(S)$ such that $(z, t) = (z, S_1 z) + (0, \alpha_2)$, where $\alpha_2 \in S(0)$. On the one hand, by (3.42), we obtain that

$$
\begin{aligned}
\|\varphi - \psi\|_2^2 &= \| ((x, T_1 x) + (0, \alpha_1)) - ((z, S_1 z) + (0, \alpha_2)) \|_2^2 \\
&= \|(x - z, T_1 x - S_1 z) + (0, \alpha_1 - \alpha_2)\|_2^2 \\
&= \|x - z\|^2 + \|(T_1 x - S_1 z) + (\alpha_1 - \alpha_2)\|^2 \\
&\leq \|x - z\|^2 + 2\|T_1 x - S_1 z\|^2 + 2\|\alpha_1 - \alpha_2\|^2 \\
&< 2\gamma^2 \delta'^2 + 2\|\alpha_1 - \alpha_2\|^2.
\end{aligned}
$$

On the other hand, from (3.41), we get

$$
\begin{aligned}
\gamma^2 &= \|x\|^2 + \|T_1 x\|^2 \\
&= \|x\|^2 + \|T_1 x + \alpha_1 - \alpha_1\|^2 \\
&\leq \|x\|^2 + 2\|T_1 x + \alpha_1\|^2 + 2\|\alpha_1\|^2 \\
&\leq 2(1 + \|\alpha_1\|^2).
\end{aligned}
$$

Hence, $\|\varphi - \psi\|^2 \leq 4(1 + \|\alpha_1\|^2)\delta'^2 + 2\|\alpha_1 - \alpha_2\|^2$. Since $\alpha_1 \in T(0)$, then for any given $\varepsilon > 0$, we infer that $\|\alpha_1\| < \|T(0)\| + \varepsilon$. This yields from Proposition 2.5.3 that $\|\alpha_1\| < \varepsilon$. Consequently, for any given $\varepsilon_0 > 0$, we obtain $\|\alpha_1 - \alpha_2\| \leq \varepsilon_0$. Thus,

$$\|\varphi - \psi\|_2^2 \leq 4(1 + \varepsilon^2)\delta'^2 + 2\varepsilon_0^2.$$

Using the fact that $\psi \in G(S)$ and the arbitrariness of ε and ε_0, we can get that $\text{dist}(\varphi, G(S)) \leq 2\delta'$. Hence, by the hypothesis, $\varphi \in G(T)$ and $\|\varphi\| = 1$, we deduce that $\delta(T, S) \leq 2\delta'$. Since δ' may be arbitrarily close to $\delta(T_1, S_1)$, we conclude that $\delta(T, S) \leq 2 \delta(T_1, S_1)$. Q.E.D.

The following result can be easily derived from Theorem 3.2.7.

Corollary 3.2.2 *Let T, $S \in CR(X,Y)$. If T_1 and S_1 are two closed selections of T and S, respectively, then*

$$\widehat{\delta}(T,S) \leq 2\widehat{\delta}(T_1,S_1). \qquad \diamond$$

3.2.3   Generalized Convergence of Closed Linear Relations

Definition 3.2.1 *Let $(T_n)_n$ be a sequence of closed linear relations from X into Y and $T \in CR(X,Y)$. The sequence $(T_n)_n$ is said to be converge in the generalized sense to T, denoted by $T_n \xrightarrow{g} T$, if $\widehat{\delta}(T_n,T)$ converges to 0 when $n \to \infty$.* $\qquad \diamond$

Theorem 3.2.8 *Let $(T_n)_n$ be a sequence of closed linear relations from X into Y and $T \in CR(X,Y)$. Then,*

(i) If S is a continuous relation from X into Y satisfy $\mathcal{D}(S) \supset \mathcal{D}(T_n) \bigcup \mathcal{D}(T)$ and $S(0) \subset T_n(0) \bigcap T(0)$, for all $n \geq 1$, then $T_n \xrightarrow{g} T$ if and only if, $T_n + S \xrightarrow{g} T + S$.

(ii) If $\mathcal{D}(T_n) \subset \mathcal{D}(T)$ and $T(0) \subset T_n(0)$, for all $n \geq 1$, then $T \in BR(X,Y)$ and $T_n \xrightarrow{g} T$ if and only if, $T_n \in BR(X,Y)$ for sufficiently large n and T_n converges to T.

(iii) Let $T_n \xrightarrow{g} T$. Then, $T^{-1} \in \mathcal{L}(Y,X)$ if and only if, $T_n^{-1} \in \mathcal{L}(Y,X)$ for sufficiently large n and T_n^{-1} converges to T^{-1}. $\qquad \diamond$

Proof. (i) Since $\mathcal{D}(S) \supset \mathcal{D}(T_n) \bigcup \mathcal{D}(T)$ and $S(0) \subset T_n(0) \bigcap T(0)$, for all $n \geq 1$, then from Proposition 2.7.2, we infer that $T_n + S$ and $T + S$ are closed. So, by applying Theorem 3.2.3, we conclude that

$$\widehat{\delta}(T_n + S, T + S) \leq (2 + \|S\|^2)\widehat{\delta}(T_n, T). \qquad (3.43)$$

Let us assume that $T_n \xrightarrow{g} T$, then $\widehat{\delta}(T_n,T) \to 0$. So, by using (3.43), we conclude that, $\widehat{\delta}(T_n + S, T + S) \to 0$ when $n \to \infty$. This is equivalent to say that $T_n + S \xrightarrow{g} T + S$. Conversely, we suppose that $T_n + S \xrightarrow{g} T + S$. By referring to Proposition 2.3.4, we have $T_n = T_n - S + S$ and $T = T - S + S$. Hence, we can write $\widehat{\delta}(T_n,T) = \widehat{\delta}((T_n + S) - S, (T + S) - S)$. Thus, again by Theorem 3.2.3, we deduce that

$$\widehat{\delta}((T_n + S) - S, (T + S) - S) \leq 2(1 + \|S\|^2)\widehat{\delta}((T_n+S)-S+S, (T+S)-S+S).$$

This implies that

$$\widehat{\delta}((T_n,T) \leq 2(1 + \|S\|^2)\widehat{\delta}(T_n + S, T + S). \qquad (3.44)$$

According to (3.44), we infer that $\widehat{\delta}(T_n, T) \to 0$ as $n \to \infty$. Hence, T_n converges in the generalized sense to T.

(ii) Assume that $T_n \xrightarrow{g} T$ and $T \in BCR(X, Y)$. Let $n_0 \in \mathbb{N}$ such that $\widehat{\delta}(T_n, T) < (1 + \|T\|^2)^{\frac{1}{2}}$ holds for all $n \geq n_0$. Then, by using Theorem 3.2.5, we deduce that $\|T_n\| < \infty$, $\mathcal{D}(T_n) = \mathcal{D}(T) = X$, and

$$\|T_n - T\| \leq \left(\frac{1 + \|T\|^2}{1 - \sqrt{1 + \|T\|^2}\delta(T_n, T)} \right) \delta(T_n, T), \ \forall n \geq n_0. \tag{3.45}$$

Using the fact that $\widehat{\delta}(T_n, T) \to 0$ as $n \to \infty$, and from (3.45), we deduce that T_n converges to T. Conversely, we suppose that $T_n \in BR(X, Y)$ and T_n converges to T. By hypothesis, $X = \mathcal{D}(T_n) \subset \mathcal{D}(T)$, then by using Theorem 2.7.2 (i), T is bounded. Now, from Proposition 2.3.4, we can write

$$\widehat{\delta}(T_n, T) = \widehat{\delta}((T_n - T) + T, T + O_X).$$

Then, by applying Theorems 3.2.2 and 3.2.3, we get the following inequality

$$\widehat{\delta}(T_n, T) \leq \left(2 + \|T\|^2\right) \frac{\|T_n - T\|}{\sqrt{1 + \|T\|^2}}.$$

As a result, $T_n \xrightarrow{g} T$, as desired.

(iii) Since $T_n \xrightarrow{g} T$, then there exists $N \in \mathbb{N}$ such that

$$\widehat{\delta}(T_n, T) < (1 + \|T^{-1}\|^2)^{-\frac{1}{2}}, \text{ for all } n \geq N.$$

By referring to Theorem 3.2.6, we obtain $T_n^{-1} \in \mathcal{L}(Y, X)$ for sufficiently large n. Since the gap is invariant under inversion, then by applying Theorem 3.2.5 to the pair T_n^{-1}, T^{-1}, we deduce that

$$\|T_n^{-1} - T^{-1}\| \leq \left(\frac{1 + \|T^{-1}\|^2}{1 - \sqrt{1 + \|T^{-1}\|^2}\delta(T_n^{-1}, T^{-1})} \right) \delta(T_n^{-1}, T^{-1}), \ \forall n \geq N. \tag{3.46}$$

Hence, the use of Theorem 2.10.1, and also the estimation (3.46), makes us conclude that

$$\|T_n^{-1} - T^{-1}\| \leq \left(\frac{1 + \|T^{-1}\|^2}{1 - \sqrt{1 + \|T^{-1}\|^2}\delta(T_n, T)} \right) \delta(T_n, T). \tag{3.47}$$

Consequently, T_n^{-1} converges to T^{-1}. Conversely, let us assume that $0 \in \rho(T_n)$ for sufficiently large n and $\|T_n^{-1} - T^{-1}\| \to 0$ as $n \to \infty$. Since

$$\|T^{-1}\| = \|T^{-1} + T_n^{-1} - T_n^{-1}\| \leq \|T^{-1} - T_n^{-1}\| + \|T_n^{-1}\|,$$

then $T^{-1} \in \mathcal{L}(Y, X)$. Now, we demonstrate that $T_n \xrightarrow{g} T$. Keeping in mind that $\widehat{\delta}(T_n, T) = \widehat{\delta}(T_n^{-1}, T^{-1})$, then $\widehat{\delta}(T_n, T)$ can be expressed in the form

$$\widehat{\delta}(T_n, T) = \widehat{\delta}((T_n^{-1} - T^{-1}) + T^{-1}, 0 + T^{-1}).$$

Hence, the use of Theorems 3.2.2 and 3.2.3 allows us to conclude that

$$\widehat{\delta}(T_n, T) \leq \left(2 + \|T^{-1}\|^2\right) \frac{\|T_n^{-1} - T^{-1}\|}{\sqrt{1 + \|T^{-1}\|^2}}.$$

As a result, $(T_n)_n$ converges in the generalized sense to T, as desired. Q.E.D.

Corollary 3.2.3 *Let T_n, $S_n \in C\mathcal{R}(X, Y)$ and T, $S \in BC\mathcal{R}(X, Y)$ such that $T(0) \subset T_n(0)$ and $S(0) \subset S_n(0)$. If $T_n \xrightarrow{g} T$ and $S_n \xrightarrow{g} S$, then $T_n + S_n \xrightarrow{g} T + S$.* ◇

Proof. Let us assume that $T_n \xrightarrow{g} T$ and $S_n \xrightarrow{g} S$. Then, by using (ii) of Theorem 3.2.8, we deduce that T_n, $S_n \in B\mathcal{R}(X, Y)$, $\|T_n - T\| \to 0$ and $\|S_n - S\| \to 0$ when $n \to \infty$. Now, by applying Proposition 2.5.4 (ii), we infer that

$$\|(T_n + S_n) - (T + S)\| \leq \|T_n - T\| + \|S_n - S\| \to 0 \text{ as } n \to \infty.$$

Using the fact that $(T + S)(0) = T(0) + S(0) \subset T_n(0) + S_n(0) = (T_n + S_n)(0)$, and $\mathcal{D}(T_n + S_n) \subset X = \mathcal{D}(T + S)$, we conclude from Theorem 3.2.8 (ii) that $T_n + S_n$ converges in the generalized sense to $T + S$. Q.E.D.

3.3 Perturbation Classes of Multivalued Linear Operators

The main aim of this section, is to carry on the generalizations of some perturbation results of linear operators in Banach spaces to the multivalued case.

3.3.1 Fredholm and Semi-Fredholm Perturbation Classes of Multivalued Linear Operators

Theorem 3.3.1 *Let X be a Banach space and let $T \in C\mathcal{R}(X)$ and S be continuous such that $S(0) \subset T(0)$, $\mathcal{D}(T) \subset \mathcal{D}(S)$, and $\|S\| \neq 0$. If $0 < |\lambda| < \frac{\gamma(T)}{\|S\|}$, then*

(i) $\alpha(T - \lambda S) \leq \alpha(T)$.

(ii) $\overline{\beta}(T - \lambda S) \leq \overline{\beta}(T)$.

(iii) *If $R(T)$ is closed, then $R(T - \lambda S)$ is closed.* $\diamond$

Proof. Since $T \in CR(X)$ and S is continuous with $S(0) \subset T(0)$ and $\mathcal{D}(T) \subset \mathcal{D}(S)$, then by using Proposition 2.7.2, we have $T - \lambda S$ is closed. Hence, by Proposition 2.7.1, we infer that $\overline{(T - \lambda S)(0)} = (T - \lambda S)(0)$. Moreover,

$$
\begin{aligned}
N(T - \lambda S) &= \{x \in \mathcal{D}(T - \lambda S) \text{ such that } (T - \lambda S)x = (T - \lambda S)(0)\} \\
&= \{x \in \mathcal{D}(T - \lambda S) \text{ such that } (T - \lambda S)x = \overline{(T - \lambda S)(0)}\} \\
&= N(Q_{T-\lambda S}(T - \lambda S)) \\
&= N(Q_T(T - \lambda S)). \qquad (3.48)
\end{aligned}
$$

Let $x \in N(T - \lambda S)$ such that $x \neq 0$. Then, by using (3.48), we have

$$
\begin{aligned}
\gamma(T)\mathrm{dist}(x, N(T)) &\leq \|Tx\| = \|Q_T(Tx)\| \\
&\leq \|\lambda Q_T(Sx)\| \quad (\text{since } Q_T(Tx - \lambda Sx) = 0) \\
&\leq |\lambda|\|S\|\|x\| \\
&\leq \gamma(T)\|x\|.
\end{aligned}
$$

Thus, $\mathrm{dist}(x, N(T)) < \|x\|$ for all $x \in N(T - \lambda S)$. Hence, by using Lemma 2.1.2, we have $\alpha(T - \lambda S) \leq \alpha(T)$.

(ii) By using Proposition 2.8.1 (iii) and (vi), we have

$$
\|\lambda S\| = |\lambda|\|S^*\| < \gamma(T) = \gamma(T^*).
$$

Since λS is continuous and $\lambda \neq 0$, then $\mathcal{D}(\lambda S) = \mathcal{D}(S) \supset \mathcal{D}(T)$ and

$$
(\lambda S - T)^* = (\lambda S)^* - T^* = \lambda S^* - T^*.
$$

Replacing X by $\mathcal{D}(S)$ if necessary, we may assume S to everywhere defined, thus making S^* single valued. Applying Proposition 2.8.1 (iv), we have

$$
\overline{\beta}(T - \lambda S) = \alpha((T - \lambda S)^*) = \alpha(T^* - \lambda S^*) \leq \alpha(T^*) = \overline{\beta}(T).
$$

(iii) Using Proposition 2.7.1 (ii), it suffices to show that $R(Q_{T-\lambda S}(T - \lambda S))$ is closed. Since $Q_{T-\lambda S} = Q_T$, then

$$
R(Q_{T-\lambda S}(T - \lambda S)) = R(Q_T(T - \lambda S)) = R(Q_T T - \lambda Q_T S).
$$

So, $R(Q_T T - \lambda Q_T S)$ is closed. Thus, $R(Q_{T-\lambda S}(T - \lambda S))$ is closed.   Q.E.D.

Corollary 3.3.1 *If the hypothesis of Theorem 3.3.1 is satisfied, then*

(i) *If* $T \in \Phi_+(X,Y)$, *then* $\lambda S - T \in \Phi_+(X,Y)$ *and* $i(\lambda S - T) = i(T)$.
(ii) *If* $T \in \Phi_-(X,Y)$, *then* $\lambda S - T \in \Phi_-(X,Y)$ *and* $i(\lambda S - T) = i(T)$.
(iii) *If* $T \in \Phi(X,Y)$, *then* $\lambda S - T \in \Phi(X,Y)$ *and* $i(\lambda S - T) = i(T)$. ◇

Theorem 3.3.2 *Let* $T \in C\mathcal{R}(X,Y)$ *and* $S \in L\mathcal{R}(X,Y)$. *Then,*

(i) *If* $T \in \Phi_+(X,Y)$ *and* $S \in \mathcal{PR}(\Phi_+(X,Y))$, *then* $T + S \in \Phi_+(X,Y)$ *and* $i(T + S) = i(T)$.
(ii) *If* $T \in \Phi_-(X,Y)$ *and* $S \in \mathcal{PR}(\Phi_-(X,Y))$, *then* $T + S \in \Phi_-(X,Y)$ *and* $i(T + S) = i(T)$.
(iii) *If* $T \in \Phi(X,Y)$ *and* $S \in \mathcal{PR}(\Phi(X,Y))$, *then* $T + S \in \Phi(X,Y)$ *and* $i(T + S) = i(T)$. ◇

Proof. (i) Assume that $T \in \Phi_+(X,Y)$ and $S \in \mathcal{PR}(\Phi_+(X,Y))$. Since T is closed, then by applying Proposition 2.7.1, we have $T(0)$ is closed. Hence, $(T + S)(0) = T(0)$ is closed. Moreover, $Q_{T+S} = Q_T$. Hence, $Q_{T+S}(T+S) = Q_T(T + S) = Q_T T + Q_T S$. Since $S(0) \subset T(0)$, then $\overline{S(0)} \subset \overline{T(0)} = T(0)$. Furthermore, applying Lemma 2.1.3 (ii), we have $Y/\overline{S(0)} \Big/ T(0)/\overline{S(0)} \equiv Y/T(0)$, and $Q_T = Q_R Q_S$, where $R = T(0)/\overline{S(0)}$. This enables us to conclude that

$$Q_{T+S}(T + S) = Q_T T + Q_R Q_S S. \tag{3.49}$$

By using Proposition 2.7.1, we have $Q_T T$ is closed single valued. Clearly, $Q_R Q_S S$ is single valued continuous, so applying Proposition 2.7.1, we infer that $Q_T T + Q_R Q_S S$ is closed. In view of (3.49), we have $Q_{T+S}(T + S)$ is a closed single valued. So, by using Proposition 2.7.1, we obtain that $T + S$ is closed. Let us consider two cases for S.

<u>First case</u> : If $\|S\| < \gamma(T)$, then in view of Lemma 2.14.3 (i), we have $T \in \Phi_+(X,Y)$ if and only if, $Q_T T$ is single valued upper semi-Fredholm. Since $\mathcal{D}(Q_T T) = \mathcal{D}(T) \subset \mathcal{D}(S) = \mathcal{D}(Q_R Q_S S)$ and

$$\|Q_R Q_S S\| \leq \|Q_S S\| = \|S\| < \gamma(T) = \gamma(Q_T T),$$

then $Q_T T + Q_R Q_S S$ is upper semi-Fredholm and $i(Q_T T + Q_R Q_S S) = i(Q_T T)$. Now, applying (3.49), we obtain that $Q_{T+S}(T + S) \in \Phi_+(X, Y/(T + S)(0))$ with $i(Q_T T) = i(Q_{T+S}(T + S))$. By applying Lemma 2.14.3 (i), to conclude that $T + S \in \Phi_+(X,Y)$ and $i(T) = i(T + S)$.

<u>Second case</u> : If $\|S\| \geq \gamma(T)$, then since $S \in \mathcal{PR}\left(\Phi_+(X,Y)\right)$ we have $\lambda S \in \mathcal{PR}\left(\Phi_+(X,Y)\right)$. So, by using Definition 2.17.1, we have

$$T + \lambda S \in \Phi_+(X,Y) \tag{3.50}$$

for all $\lambda \in \mathbb{C}$. Then, $T + S \in \Phi_+(X,Y)$. Now, we prove $i(T+S) = i(T)$. For this, we shall prove that the map

$$\varphi : \mathbb{I} := [0,1] \longrightarrow \mathbb{Z}\bigcup\{-\infty\}$$
$$\lambda \longrightarrow \varphi(\lambda) = i(T + \lambda S)$$

is continuous. Let $\lambda_0 \in \mathbb{I}$ arbitrary but fixed. Then, for $\lambda \in \mathbb{I}$ such that $|\lambda - \lambda_0| < \frac{\gamma(T+\lambda_0 S)}{\|S\|}$, and by using (3.50) and Corollary 2.16.1 (iv), we have $A := T + \lambda_0 S - (\lambda_0 S - \lambda S) \in \Phi_+(X,Y)$ and

$$i(A) = i(T + \lambda_0 S). \tag{3.51}$$

Since $S(0) \subset T(0)$, then $Q_T = Q_{T+\lambda S} = Q_{T+\lambda_0 S} = Q_{T+\lambda S+\lambda_0 S - \lambda_0 S}$ and $Q_T = Q_R Q_S$ (as in (3.49)). So,

$$\begin{aligned} Q_S(T + \lambda S + \lambda_0 S - \lambda_0 S) &= Q_S(T + \lambda S) + \lambda_0 Q_S S - \lambda_0 Q_S S \\ &= Q_S(T + \lambda S) \end{aligned}$$

and $Q_T A = Q_A A = Q_R Q_S A = Q_R Q_S(T + \lambda S) = Q_T(T + \lambda S) = Q_T T + \lambda S(T + \lambda S)$. Then, we obtain that

$$Q_A A = Q_{T+\lambda S}(T + \lambda S). \tag{3.52}$$

The use of Lemma 2.14.3 (i) and (3.51) and (3.52), implies $i(Q_A A) = i(A) = i(Q_{T+\lambda S}(T + \lambda S)) = i(T + \lambda S) = i(T + \lambda_0 S)$. Thus, $i(T + \lambda_0 S) = i(T + \lambda S)$. Now, let $\varepsilon > 0$, then there exists $\frac{\gamma(T+\lambda_0 S)}{\|S\|} = \delta$ such that for all $\lambda \in \mathbb{I}$ with $|\lambda - \lambda_0| < \delta$, we have $0 = |i(T+\lambda S) - i(T+\lambda_0 S)| < \varepsilon$. So, φ is continuous. So, $\varphi(\mathbb{I})$ is a connected set which therefore, consists of only one point. It follows that $i(T) = \varphi(0) = \varphi(1) = i(T + S)$. This completes the proof of (i).

(ii) Let $T \in \Phi_-(X,Y)$ and $S \in \mathcal{PR}\left(\Phi_-(X,Y)\right)$, then for all $\lambda \in \mathbb{C}$, $T + \lambda S \in \Phi_-(X,Y)$ equivalently $(T + \lambda S)^* \in \Phi_+(Y^*,X^*)$. Since $(T + \lambda S)^* = T^* + (\lambda S)^* = T^* + \lambda S^*$ (obvious that $(\lambda S)^* = \lambda S^*$, $\lambda \neq 0$), then $T^* + \lambda S^* \in \Phi_+(Y^*,X^*)$. Furthermore, since S^* is continuous, then λS^* is continuous, $\lambda S^*(0) = S^*(0) \subset T^*(0)$ and $\mathcal{D}(T^*) \subset \mathcal{D}(S^*) = \mathcal{D}(\lambda S^*)$. So, we can apply (i) obtaining that $i(T^*) = i(T^* + S^*)$ with T^* and $T^* + S^* \in \Phi_+(Y^*,X^*)$.

Thus, $-i(T) = i(T^*) = i(T^* + S^*) = i((T + S)^*) = -i(T + S)$. Therefore, $i(T) = i(T + S)$, as required.

(*iii*) The proof of (*iii*) follows from the same reasoning as previously. Q.E.D.

Theorem 3.3.3 *Let* S, T, $K \in LR(X,Y)$ *such that* $\mathcal{D}(T) \subset \mathcal{D}(S) \subset \mathcal{D}(K)$, $K(0) \subset S(0) \subset \overline{T(0)}$ *and let* $\widehat{T}$ *be the bijection associated with* T. *Suppose*

(*a*) *there exists a constant* α_1 *such that*

$$\|Sx\| \leq \alpha_1 (\|x\| + \|Tx\|), \quad x \in \mathcal{D}(T),$$

(*b*) *there exists a constant* β_1 *such that* $\alpha_1(1 + \beta_1) < 1$, $(1 + \beta_1) < \gamma(\widehat{T})$ *and*

$$\|Kx\| \leq \beta_1 (\|x\| + \|Sx\|), \quad x \in \mathcal{D}(T),$$

If $T \in \Phi(X,Y)$, *then the sum* $T + S + K \in \Phi(X,Y)$ *and satisfy the following properties*

    (*i*) $\alpha(T + S + K) \leq \alpha(T)$, *and*
    (*ii*) $\overline{\beta}(T + S + K) \leq \overline{\beta}(T)$. $\diamond$

Proof. From Theorem 3.1.4, it follows that $T + S + K$ is a closed linear relation. Let T_1, S_1, and K_1 be the restrictions of the relation T, S, and K to X_T, respectively. Obviously, T is Fredholm linear relation and S_1, K_1 is a bounded linear relation. Moreover, it is easy to prove that

$$\|S_1 + K_1\| \leq (\beta_1 + \alpha_1(1 + \beta_1)) \leq \gamma(\widehat{T}).$$

Then, by using Corollary 2.16.1 (*vii*), we have $S_1 + K_1 + T_1$ is a Fredholm linear relation. The properties (*i*) and (*ii*) are straightforward consequences of Theorem 2.9.3.   Q.E.D.

Corollary 3.3.2 *Let* S, T, *and* K *be three operators such that* $\mathcal{D}(T) \subset \mathcal{D}(S) \subset \mathcal{D}(K)$ *and let* $\widehat{T}$ *be the bijection associated with* T. *Suppose,*

(*a*) *there exists a constant* α_1 *such that*

$$\|Sx\| \leq \alpha_1 (\|x\| + \|Tx\|), \quad x \in \mathcal{D}(T),$$

(*b*) *there exists a constant* β_1 *such that* $\alpha_1(1 + \beta_1) < 1$, $(1 + \beta_1) < \gamma(\widehat{T})$ *and*

$$\|Kx\| \leq \beta_1 (\|x\| + \|Sx\|), \quad x \in \mathcal{D}(T),$$

If $T \in \Phi(X,Y)$, then $T + S + K \in \Phi(X,Y)$ and also satisfies the following properties

    (i) $\alpha(T + S + K) \leq \alpha(T)$,
    (ii) $\beta(T + S + K) \leq \beta(T)$, and
    (ii) $i(T + S + K) = i(T)$. $\diamondsuit$

3.4   Atkinson Linear Relations

Theorem 3.4.1 *Let X and Y be two Banach spaces and $T \in C\mathcal{R}(X,Y)$. Then,*

(i) If $T \in \mathcal{A}_\alpha(X,Y)$, then $Q_T T \in \mathcal{A}_\alpha(X, Y/T(0))$.
(ii) If $Q_T T \in \mathcal{A}_\alpha(X, Y/T(0))$ and $T(0)$ is topologically complemented in Y, then $T \in \mathcal{A}_\alpha(X,Y)$.
(iii) If $T \in \mathcal{A}_\beta(X,Y)$, then $Q_T T \in \mathcal{A}_\beta(X, Y/T(0))$.
(iv) If $Q_T T \in \mathcal{A}_\beta(X, Y/T(0))$ and $T^*(0)$ is topologically complemented in Y^*, then $T \in \mathcal{A}_\beta(X,Y)$. $\diamondsuit$

Proof. (i) Let $T \in \mathcal{A}_\alpha(X,Y)$. Then, by using Lemma 2.14.3 (i), we deduce that $Q_T T \in \Phi_+(X, Y/T(0))$ with $R(Q_T T) = R(T)/\overline{T(0)} = R(T)/T(0)$. Since $R(T) \oplus N = Y$ for some closed subspace N of Y and $N(Q_T) = \overline{T(0)} = T(0) \subset R(T)$, then $R(Q_T T) \oplus Q_T N = Y/T(0)$. So, $R(Q_T T)$ is topologically complemented in $Y/T(0)$, as desired.

(ii) Let $Q_T T \in \mathcal{A}_\alpha(X, Y/T(0))$. By virtue of Lemma 2.14.3 (i), we have $T \in \Phi_+(X,Y)$. So, our purpose is to verify that $R(T)$ is topologically complemented in Y. In order to prove it, let us notice that $Y/T(0) = R(Q_T T) \oplus Q_T M$, for some closed subspace M of Y. The use of Lemma 2.3.1 allows us to ensure that $Y = R(T) + M$ and $T(0) = R(T) \bigcap M$. Since, by hypothesis, $T(0)$ is topologically complemented in Y, then there exists a closed subspace N of Y such that $T(0) \oplus N = Y$. Hence, $Y = R(T) \oplus (N \bigcap M)$. So, $R(T)$ is topologically complemented in Y, as desired.

The proof of (iii) (respectively, (iv)) follows immediately from (i) (respectively, (ii)) and Lemma 2.14.5.   Q.E.D.

As a direct consequence of Theorem 3.4.1, we have the following:

Corollary 3.4.1 *Let $T \in C\mathcal{R}(X,Y)$. Then,*

(i) If $T(0)$ is topologically complemented in Y, then

$$T \in \mathcal{A}_\alpha(X,Y) \quad \text{if and only if,} \quad Q_T T \in \mathcal{A}_\alpha(X,Y/T(0)).$$

(ii) If $T^*(0)$ is topologically complemented in Y^*, then

$$T \in \mathcal{A}_\beta(X,Y) \quad \text{if and only if,} \quad Q_T T \in \mathcal{A}_\beta(X,Y/T(0)). \qquad \diamond$$

Proposition 3.4.1 *Let $T \in C\mathcal{R}(X,Y)$ such that $T(0)$ is topologically complemented in Y. Then, the following properties are equivalent.*

(i) $T \in \mathcal{A}_\alpha(X,Y)$.
(ii) There exist $A \in \mathcal{L}(X,Y)$, a bounded finite rank projection $F \in \mathcal{L}(X)$ such that $N(A)$ is topologically complemented in Y, $R(A)$ is a closed subspace contained in $\mathcal{D}(T)$, $R(F) \subset \mathcal{D}(T)$ and $AT = (I - F)_{|\mathcal{D}(T)}$.
(iii) There exist $A \in \mathcal{L}(Y,X)$ and a continuous operator B in X such that $\mathcal{D}(T) \subset \mathcal{D}(B)$, $I - B \in \Phi(X)$, $R(B) \subset \mathcal{D}(T)$ and $AT = (I - B)_{|\mathcal{D}(T)}$. $\qquad \diamond$

Proof. $(i) \Longrightarrow (ii)$ Let us consider two cases for T:

<u>First case</u> : T is an operator. Since $T \in \Phi_+(X,Y)$, then there is a closed finite codimensional subspace M of X for which $N(T) \oplus M = X$. So, $N(T) \oplus (M \cap \mathcal{D}(T)) = \mathcal{D}(T)$. Since $R(T)$ is topologically complemented in Y, then there exists a closed subspace N of Y such that $R(T) \oplus N = Y$. Let P denote the bounded projection of Y onto $R(T)$ and let $A := (T_{|M})^{-1} P$. Then,

$$\begin{aligned}
\mathcal{D}(A) &= \{y \in \mathcal{D}(P) \text{ such that } Py \cap \mathcal{D}\left((T_{|M})^{-1}\right) \neq \emptyset\} \\
&= \{y \in Y \text{ such that } Py \in R(T_{|M}) = R(P)\} \\
&= Y.
\end{aligned}$$

By virtue of Theorem 2.5.4, $(T_{|M})^{-1}$ is continuous. Hence, A is continuous. Moreover,

$$\begin{aligned}
R(A) &= \left((T_{|M})^{-1}\right) PY \\
&= (T_{|M})^{-1}(T_{|M})(M \cap \mathcal{D}(T)) \\
&= (M \cap \mathcal{D}(T))
\end{aligned}$$

and

$$\begin{aligned}
N(A) &= P^{-1}N\left((T_{|M})^{-1}\right) \\
&= N(P) \\
&= N.
\end{aligned}$$

Let F be the bounded projection of X onto $N(T)$. Then, it is obvious that $R(F) \subset \mathcal{D}(T)$. For $x \in \mathcal{D}(T)$, we can write $x = x_1 + x_2$, where $x_1 \in N(T)$ and $x_2 \in M \bigcap \mathcal{D}(T)$. Then,

$$
\begin{aligned}
ATx &= (T_{|M})^{-1} P(T_{|M}) x_2 \\
&= x_2 \\
&= (I - F)x.
\end{aligned}
$$

Therefore, $(i) \implies (ii)$ whenever T is an operator.

<u>Second case</u> : T is a linear relation. Then, by using Lemma 2.14.3 (i), we have $Q_T T \in \Phi_+(X, Y/T(0))$ with $R(Q_T T) = R(T)/T(0)$. Since $R(T) \bigoplus N = Y$ for some closed subspace N of Y and $N(Q_T) = T(0) \subset R(T)$, then $R(Q_T T) \bigoplus Q_T N = Y/T(0)$. In this situation, it follows from the first case applied to $Q_T T$ that there exist $A \in \mathcal{L}(Y/T(0), X)$ and a bounded finite rank projection $F \in \mathcal{L}(X)$ such that $R(F) \subset \mathcal{D}(Q_T T) = \mathcal{D}(T)$. $N(A)$ is topologically complemented in $Y/T(0)$, $R(A)$ is a closed subspace contained in $\mathcal{D}(Q_T T)$ and $(AQ_T)T = (I - F)_{|\mathcal{D}(T)}$. Hence, it only remains to prove that $N(AQ_T)$ is topologically complemented in Y. To ensure this, we note that Q_T is a bounded lower semi-Fredholm operator from Y onto $Y/T(0)$ and $N(Q_T) = T(0)$ is topologically complemented in Y by hypothesis, then from Theorem3.1.5 that $N(AQ_T)$ is topologically complemented in Y, as required.

$(ii) \implies (iii)$ It is clear.

$(iii) \implies (i)$ Let us consider various cases for T :

<u>First case</u> : T is a densely defined operator. In such case the result was proved by V. Müller-Horrig in Theorem 2.14.5.

<u>Second case</u> : T is an operator. Let A and B satisfy the conditions in (iii). Then, it is clear that $ATG_T = (I - B)_{|X_T}$. Since $TG_T \in \mathcal{L}(X_T, Y)$ and X_T is complete, it follows from the first case applied to TG_T that $TG_T \in \mathcal{A}_\alpha(X_T, Y)$. Hence, $T \in \mathcal{A}_\alpha(X, Y)$, as desired.

<u>Third case</u> : T is a linear relation. Let A and B as in (iii). Recalling that a linear relation S is an operator if, and only if, $S(0) = \{0\}$. Hence, $AQ_T^{-1}(0) = AT(0) = (I - B)_{|\mathcal{D}(T)} = \{0\}$. So, AQ_T^{-1} is an operator. Clearly, it is everywhere defined and continuous. Thus, AQ_T^{-1} is bounded operator. So, $AQ_T^{-1} \in \mathcal{L}(Y/T(0), X)$. On the other hand,

$$
Q_T^{-1} Q_T T(\mathcal{D}(T)) = \Big(T(\mathcal{D}(T)) \bigcap \mathcal{D}(Q_T) \Big) + N(Q_T)
$$

which implies that $(AQ_T^{-1})Q_T T = AT$. Thus, it follows from the second case that $Q_T T \in \mathcal{A}_\alpha(X, Y/T(0))$. Hence, $Q_T T \in \Phi_+(X, Y/T(0))$ and $R(Q_T T)$ is topologically complemented in $Y/T(0)$. By virtue of Lemma 2.14.3 (i), $T \in \Phi_+(X, Y/T(0))$. So, it remains to see that $R(T)$ is topologically complemented in Y. To prove this, we note that $Y/T(0) = R(Q_T T) \oplus Q_T U$ for some closed subspace U of X containing $T(0)$, $Y = R(T) + U$ and $T(0) = R(T) \cap U$. Since by hypothesis $T(0) \oplus V = Y$ for some closed subspace V of Y, then $Y = R(T) \oplus (V \cap U)$. So, $R(T)$ is topologically complemented in Y, as desired. This completes the proof.   Q.E.D.

Definition 3.4.1 *An operator $T_l \in \mathcal{L}(Y, X)$ satisfying Proposition 3.4.1 (ii) is called a left regularizer (or a left generalized inverse) of T.* $\diamond$

Proposition 3.4.2 *Let $T \in \mathcal{A}_\alpha(X)$ such that $T(0)$ is topologically complemented in X and let $K \in \mathcal{K}_T(X)$. Then, $T + K \in \mathcal{A}_\alpha(X)$ and $i(T + K) = i(T)$.*

$\diamond$

Proof. Let A and B as in part (iii) of Proposition 3.4.1. Then,

$$A(T + K) = (I - (B - AK))_{|\mathcal{D}(T)} \, .$$

Further, $AK(0) \subset AT(0) = \{0\}$, so by Theorem 2.12.2, AK is compact single valued. Then, applying the implication $(iii) \Longrightarrow (i)$ in Proposition 3.4.1, we get $T + K \in \mathcal{A}_\alpha(X)$ and the equality $i(T + K) = i(T)$ follows immediately from Lemma 3.5.5.   Q.E.D.

Theorem 3.4.2 *Let $T \in BCR(X)$. Then,*

(i) If $T \in \mathcal{A}_\alpha(X)$ and $T(0)$ is topologically complemented in X, then $T_l T \in \Phi(X)$ and $i(T_l T) = 0$.

(ii) If $T \in \mathcal{A}_\beta(X)$, $\dim(T(0)) < \infty$ and $TT_r - TT_r \in \mathcal{KR}(X)$, then

   (ii_1) $TT_r \in \Phi(X)$ and $i(TT_r) = \dim(T(0))$.

   (ii_2) $T_r \in \Phi_+(X)$ and if T, T_r have finite indices, then $i(T) + i(T_r) - \dim(T(0)) = 0$. $\diamond$

Proof. (i) Follows immediately from Proposition 3.4.1 and Theorem 2.16.2.

(ii) (ii_1) By using Lemma 2.14.6, there exist $T_r \in \mathcal{L}(X)$ and $F \in \mathcal{L}(X)$, which is a finite rank projection such that

$$\begin{aligned} TT_r &= I - F + TT_r - TT_r \\ &= I - K, \end{aligned}$$

where $K = F - TT_r + TT_r$. Hence, $K(0) = T(0)$ and $K \in K\mathcal{R}(X)$. So, from Theorem 2.16.2, we infer that $TT_r \in \Phi(X)$ and $i(TT_r) = \dim(T(0))$.

(ii) (ii_2) By using (ii_1), we have $TT_r \in \Phi_+(X)$ and by applying Theorem 2.14.3, we infer that $T_r \in \Phi_+(X)$. Moreover, by using Lemma 2.4.2, we have $i(T) + i(T_r) = \dim(T(0))$. Therefore, (ii_2) holds.   Q.E.D.

Theorem 3.4.3 *Let* $T \in C\mathcal{R}(X,Y)$ *such that* $T(0)$ *is topologically complemented in* Y. *Then,*

(i) *If* A *is the left generalized inverse of* $Q_T T$, *then* AQ_T *is the left generalized inverse of* T.

(ii) *If* A *is the left generalized inverse of* T, *then* AQ_T^{-1} *is the left generalized inverse of* $Q_T T$.   ◇

Proof. (i) Let $Q_T T \in A_\alpha(X, Y/T(0))$. Then, by using Proposition 3.4.1, there exist $A \in \mathcal{L}(Y/T(0), X)$ and $F \in \mathcal{L}(X)$ which is, a bounded finite rank projection such that $R(F) \subset \mathcal{D}(Q_T T) = \mathcal{D}(T)$. Moreover, $N(A)$ is topologically complemented in $Y/T(0)$, $R(A)$ is a closed subspace contained in $\mathcal{D}(Q_T T)$, and $AQ_T T = (AQ_T)T = (I - F)_{|\mathcal{D}(T)}$. Finally, it is sufficient to prove that $N(AQ_T)$ is topologically complemented in Y. To ensure this, we notice that Q_T is a bounded lower semi-Fredholm operator from Y onto $Y/T(0)$ and $N(Q_T) = T(0)$ is, by hypothesis, topologically complemented in Y. Hence, according to Theorem 3.1.5, we deduce that $N(AQ_T)$ is topologically complemented in Y. So, $T_l = AQ_T$.

(ii) Let $T_l = A$. Then, by using Proposition 3.4.1, there exists $F \in \mathcal{L}(X)$, which is a bounded finite rank projection such that $N(A)$ is topologically complemented in Y, $R(A)$ is a closed subspace contained in $\mathcal{D}(T)$, $R(F) \subset \mathcal{D}(T)$, and $AT = (I - F)_{|\mathcal{D}(T)}$. Since $T(0)$ is closed, then

$$(AQ_T^{-1})Q_T T = AT = (I - F)_{|\mathcal{D}(T)}.$$

Knowing that $AQ_T^{-1}(0) = AT(0) = (I - F)_{|\mathcal{D}(T)} = \{0\}$, this allows us to deduce that AQ_T^{-1} is an operator and clearly, it is everywhere defined and continuous. Thus, $AQ_T^{-1} \in \mathcal{L}(Y/T(0), X)$. Now, by applying Proposition 3.4.1, we infer that $(Q_T T)_l = AQ_T^{-1}$.   Q.E.D.

3.5   The α-and β-Atkinson Perturbation Classes

The goal of this section is to study the class of α- and β-Atkinson perturbation of linear relations and to present the relationship between upper semi-Fredholm perturbations (respectively, lower semi-Fredholm perturbations) and the sets of α-Atkinson perturbations(respectively, β-Atkinson perturbations).

Theorem 3.5.1 *Let X be a complex Banach space. Then,*

(i) $\mathcal{KR}(X) \subset \mathcal{SSR}(X) \subset \mathcal{PR}\left(\Phi_+(X)\right) \subset \mathcal{PR}\left(\mathcal{A}_\alpha(X)\right).$
(ii) $\mathcal{KR}(X) \subset \mathcal{PR}\left(\Phi_-(X)\right) \subset \mathcal{PR}\left(\mathcal{A}_\beta(X)\right).$ $\diamond$

Proof. *(i)* First, let us notice that the inclusion $\mathcal{KR}(X) \subset \mathcal{SSR}(X)$ is a simple consequence of Proposition 2.12.1. Second, we have to prove that $\mathcal{SSR}(X) \subset \mathcal{PR}(\Phi_+(X))$. Let $A \in \mathcal{SSR}(X)$ and $B \in \Phi_+(X)$ such that $\mathcal{D}(B) \subset \mathcal{D}(A)$ and $A(0) \subset B(0)$. Since B is closed, then $\overline{B(0)} = B(0)$. By using Corollary 2.16.1 (iv), we deduce that $A + B \in \Phi_+(X)$. This shows that

$$\mathcal{SSR}(X) \subset \mathcal{PR}\left(\Phi_+(X)\right).$$

Finally, we have to prove that $\mathcal{PR}(\Phi_+(X)) \subset \mathcal{PR}\left(\mathcal{A}_\alpha(X)\right)$. For this purpose, let $A \in \mathcal{PR}\left(\Phi_+(X)\right)$ and let $B \in \mathcal{A}_\alpha(X)$ such that $\mathcal{D}(B) \subset \mathcal{D}(A)$ and $A(0) \subset B(0)$. By using Proposition 2.12.1, we conclude that

$$A + B \in \mathcal{CR}(X). \tag{3.53}$$

By applying Proposition 3.4.1 to B, there exist $B_l \in \mathcal{L}(X)$ and F, which is a finite rank linear operator, such that $\mathcal{D}(B) \subset \mathcal{D}(F)$, $\mathcal{R}(F) \subset \mathcal{D}(B)$, and $B_l B = (I - F)_{|\mathcal{D}(B)}$. We can easily deduce that

$$B_l(A + B) = B_l B + B_l A = (I - F + B_l A)_{|\mathcal{D}(B)}. \tag{3.54}$$

Now, we propose to show that

$$B_l A \in \mathcal{PR}\left(\Phi_+(X)\right). \tag{3.55}$$

Let us consider two cases of B_l:

<u>First case</u> : B_l is invertible single valued. Since $B_l A(0) \subset B_l B(0) = \{0\}$, then $B_l A$ is an operator. For this, we take $S \in \Phi_+(X)$ which is single valued. Then,

$B_l^{-1}S + A \in \Phi_+(X)$. Moreover, $B_l \in \Phi_+(X)$ and it is single valued. By using Theorem 2.14.3, we have

$$\Phi_+(X) \ni B_l \left(B_l^{-1}S + A \right) = S + B_l A.$$

This implies that $B_l A \in \mathcal{PR}(\Phi_+(X))$.

<u>Second case</u> : B_l is not invertible single valued. Since $B_l \in \mathcal{L}(X)$, then $\rho(B_l) \neq \emptyset$. Hence, there exists $\lambda \in \rho(B_l)$ such that B_l is the sum of two invertible operators

$$B_l = \lambda + (B_l - \lambda).$$

By applying the previous case and Proposition 2.17.1 (i), we have $B_l A \in \mathcal{PR}(\Phi_+(X))$. Consequently, (3.55) holds. So, by using (3.53), (3.54) and (3.55), we deduce that

$$B_l(A + B) \in \Phi_+(X). \tag{3.56}$$

Moreover, $R(A_l) \subset \mathcal{D}(A + B)$ and $R(F - B_l A) \subset \mathcal{D}(A + B)$ and clearly, we have $(A + B)(0) = B(0)$. Then, $(A + B)(0)$ is complemented in X. By using (3.53), (3.54), (3.56), Proposition 3.4.1, we infer that $A + B \in \mathcal{A}_\alpha(X, Y)$. So,

$$\mathcal{PR}(\Phi_+(X)) \subset \mathcal{PR}\left(\mathcal{A}_\alpha(X) \right).$$

(ii) In order to prove the inclusion $\mathcal{KR}(X) \subset \mathcal{PR}(\Phi_-(X))$, let $A \in \mathcal{KR}(X)$. Hence, we take $B \in \Phi_-(X)$ such that $\mathcal{D}(B) \subset \mathcal{D}(A)$ and $A(0) \subset B(0)$. Then, by using Proposition 2.8.1, we have

$$(A + B)^* = A^* + B^*. \tag{3.57}$$

We claim that $A^*(0) \subset B^*(0)$ and $\mathcal{D}(B^*) \subset \mathcal{D}(A^*)$. Indeed, since $\mathcal{D}(B) \subset \mathcal{D}(A)$, then by using Proposition 2.8.1, we have

$$A^*(0) = \mathcal{D}(A)^\perp \subset \mathcal{D}(B)^\perp = B^*(0).$$

Hence, $A^*(0) \subset B^*(0)$, and by applying

$$\mathcal{D}(B^*) \subset B(0)^\perp \subset A(0)^\perp \subset \mathcal{D}(A^*),$$

we have $\mathcal{D}(B^*) \subset \mathcal{D}(A^*)$ which proves our claim. By using Proposition 2.11.2, we have $A^* \in \mathcal{KR}(X^*)$ and by using Theorem 2.14.4, we have $B^* \in \Phi_+(X^*)$. Hence, we conclude that

$$A^* + B^* \in \Phi_+(X^*).$$

Then, from (3.57) and Theorem 2.14.4, we have $A + B \in \Phi_-(X)$. So, $A \in \mathcal{PR}(\Phi_-(X))$.

The proof of the the inclusion $\mathcal{PR}(\Phi_-(X)) \subset \mathcal{PR}(A_\beta(X))$ follows from the same reasoning as previously.   Q.E.D.

Theorem 3.5.2 *Let X and Y be two Banach spaces. Then,*

(i) If $T \in A_\alpha(X,Y)$ and $S \in \mathcal{PR}(A_\alpha(X,Y))$, then $T + S \in A_\alpha(X,Y)$ and $i(T + S) = i(T)$.
(ii) If $T \in A_\beta(X,Y)$ and $S \in \mathcal{PR}(A_\beta(X,Y))$, then $T + S \in A_\beta(X,Y)$ and $i(T + S) = i(T)$.
(iii) If $T \in \Phi_+(X,Y)$ and $S \in \mathcal{PR}(\Phi_+(X,Y))$, then $T + S \in \Phi_+(X,Y)$ and $i(T + S) = i(T)$.
(iv) If $T \in \Phi_-(X,Y)$ and $S \in \mathcal{PR}(\Phi_-(X,Y))$, then $T + S \in \Phi_-(X,Y)$ and $i(T + S) = i(T)$. $\diamond$

Proof. *(i)* Let $S \in \mathcal{PR}(A_\alpha(X,Y))$. Then, $S(0) \subset T(0)$, $\mathcal{D}(T) \subset \mathcal{D}(S)$ and S is continuous. By using Proposition 2.7.2, we have $T + S \in CR(X,Y)$. Thus, in view of Definition 2.17.1, we infer that $T + S \in A_\alpha(X,Y)$. Let $\lambda \in \mathbb{K}$, then $\lambda S \in \mathcal{PR}(A_\alpha(X,Y))$. So, $\lambda S + T \in A_\alpha(X,Y)$ for all $\lambda \in \mathbb{K}$. It follows that $\varphi(\lambda) = i(\lambda S + T)$ is continuous from $[0,1]$ into $\mathbb{Z}$ ($\mathbb{Z} :=$ integers), where $[0,1]$ has the usual topology and $\mathbb{Z}$ is the topological space consisting of the integers, and $\{-\infty\}$ with the discrete topology. Hence, φ is a constant function. In particular, $i(T) = \varphi(0) = \varphi(1) = i(T + S)$.

The proofs of *(ii)*, *(iii)* and *(iv)* may be achieved by using the same reasoning as *(i)*.   Q.E.D.

Corollary 3.5.1 *Let S, $T \in LR(X)$ such that $S(0) \subset T(0)$ and $\mathcal{D}(T) \subset \mathcal{D}(S)$.*

(i) Assume that $T(0)$ is topologically complemented in X. Then,

   (i_1) If $T \in A_\alpha(X)$ and $S \in \mathcal{SSR}(X)$, then $T + S \in A_\alpha(X)$ and $i(T+S) = i(T)$.

   (i_2) If $T \in W^l(X)$ and $S \in \mathcal{SSR}(X)$, then $T + S \in W^l(X)$.

(ii) Assume that $T^*(0)$ is topologically complemented in X^*. Then,

   (ii_1) If $T \in A_\beta(X)$ and $S \in \mathcal{KR}(X)$, then $T + S \in A_\beta(X)$ and $i(T+S) = i(T)$.

(ii_2) *If $T \in \mathcal{W}^r(X)$ and $S \in \mathcal{KR}(X)$, then $T + S \in \mathcal{W}^r(X)$.*

(iii) *If $T \in \Phi_+(X)$ and $S \in SSR(X)$, then $T + S \in \Phi_+(X)$ and $i(T + S) = i(T)$.*

(iv) *If $T \in \Phi_-(X)$ and $S \in \mathcal{KR}(X)$, then the linear relation $T + S \in \Phi_-(X)$ and $i(T + S) = i(T)$.* ◇

Theorem 3.5.3 *Let $T \in LR(X)$. Then,*

(i) *If $T \in \mathcal{A}_\alpha(X)$, $T(0)$ is topologically complemented in X, and $\alpha(T) \leq \beta(T)$, then there exists $K \in \mathcal{L}(X)$ with $\dim(R(K)) \leq \alpha(T)$, such that the linear relation $S = T - K$ satisfies $S \in \mathcal{GR}_l(X)$.*

(ii) *If $T \in \mathcal{A}_\beta(X)$, $T^*(0)$ is topologically complemented in X^*, and $\beta(T) \leq \alpha(T)$, then there exists $K \in \mathcal{L}(X)$ with $\dim(R(K)) \leq \alpha(T)$ such that the linear relation $S = T - K$ satisfies $S \in \mathcal{GR}_r(X)$.* ◇

Proof. (i) Let $T \in \mathcal{A}_\alpha(X)$ and $i(T) \leq 0$. If $\alpha(T) = 0$, then we can consider $K = 0$ and $S = T$. Suppose that $1 \leq n := \alpha(T)$ and let $\{x_1, \ldots, x_n\}$ be a basis of $N(T)$. Let us choose $x'_1, \ldots, x'_n \in X^*$ such that $x'_i(x_j) = \delta_{ij}$, where $\delta_{ij} = 0$ if $i \neq j$ and $\delta_{ij} = 1$ if $i = j$. Let us choose $y_1, \ldots, y_n \in X$ such that $[y_1], \ldots, [y_n] \in X/R(T)$ are linearly independent. The single valued linear operator K is defined by

$$K : X \ni x \longrightarrow Kx = \sum_{i=1}^{n} x'_i(x)y_i \in X.$$

Hence, K is an everywhere defined linear operator,

$$\|Kx\| \leq \left(\sum_{i=1}^{n} \|x'_i\| \|y_i\| \right) \|x\|,$$

with $\dim(R(K)) \leq n$ and $T - K$ is injective. By using Corollary 3.5.3 (i), we deduce that $R(T - K)$ is a closed and complemented subspace of X. So, $T - K \in \mathcal{GR}_l(X)$.

The proof of (ii) may be achieved in the same way as in the proof of (i). This completes the proof. Q.E.D.

As a direct consequence of Theorems 3.5.3 and 3.5.5, we infer the following result.

Corollary 3.5.2 *Let $T \in LR(X)$.*

(*i*) *Let us assume that $T(0)$ is topologically complemented in X. Then, the following conditions are equivalent*

(i_1) $T \in \mathcal{W}^l(X)$.

(i_2) *There exist $K \in \mathcal{K}_T(X)$ and $S \in \mathcal{GR}_l(X)$, such that $T = S + K$.*

(i_3) *There exist $K \in \mathcal{PR}_\alpha(\mathcal{A}(X))$ and $S \in \mathcal{GR}_l(X)$ such that $T = S + K$.*

(*ii*) *Let us assume that $T^*(0)$ is topologically complemented in X^*. Then, the following conditions are equivalent*

(ii_1) $T \in \mathcal{W}^r(X)$.

(ii_2) *There exist $K \in \mathcal{K}_T(X)$ and $S \in \mathcal{GR}_r(X)$ such that $T = S + K$.*

(ii_3) *There exist $K \in \mathcal{PR}_\beta(\mathcal{A}(X))$ and $S \in \mathcal{GR}_r(X)$ such that $T = S + K$.*

$\diamond$

Theorem 3.5.4 *Let $T \in CR(X,Y)$ and $S \in LR(X,Y)$ be continuous such that $S(0) \subset T(0)$ and S is T-bounded with T-bound $\delta < 1$. Then, the following statements hold*

(*i*) *If $T \in \mathcal{A}_\alpha(X,Y)$ and SG_T is α-Atkinson perturbation, then $T + S \in \mathcal{A}_\alpha(X,Y)$ and $i(T + S) = i(T)$.*

(*ii*) *If $T \in \mathcal{A}_\beta(X,Y)$ and SG_T is β-Atkinson perturbation, then $T + S \in \mathcal{A}_\beta(X,Y)$ and $i(T + S) = i(T)$.*

(*iii*) *If $T \in \Phi_+(X,Y)$ and SG_T is an upper semi-Fredholm perturbation, then $T + S \in \Phi_+(X,Y)$ and $i(T + S) = i(T)$.*

(*iv*) *If $T \in \Phi_-(X,Y)$ and SG_T is an lower semi-Fredholm perturbation, then $T + S \in \Phi_-(X,Y)$ and $i(T + S) = i(T)$.* $\diamond$

Proof. (*i*) Let $T \in \mathcal{A}_\alpha(X,Y)$. Since $S \in LR(X,Y)$ is continuous such that $S(0) \subset T(0)$ and S is T-bounded with T-bound $\delta < 1$, then by using Proposition 2.7.2, we have $(T + S)G_T \in CR(X_T,Y)$. Observe that $(T + S)G_T = TG_T + SG_T$ with $TG_T \in \mathcal{A}_\alpha(X_T,Y)$ and $SG_T \in \mathcal{PR}(\mathcal{A}_\alpha(X_T,Y))$. Hence, $(T + S)G_T \in \mathcal{A}_\alpha(X_T,Y)$. Thus, by applying Remark 2.7.2, we infer that $i(T + S) = i((T + S)G_T)$. Since $N(TG_T) = N(T)$ and $R(T) = R(TG_T)$, it follows from Theorem 3.5.4 that $i((T + S)G_T) = i(TG_T) = i(T)$.

The proofs of (*ii*), (*iii*) and (*iv*) may be achieved by the same reasoning as for (*i*).   Q.E.D.

Theorem 3.5.5 *Let X and Y be two complete spaces. Then, the following statements hold*

(i) $SSR(X,Y) \subset PR(\mathcal{A}_\alpha(X,Y))$.

(ii) $KR(X,Y) \subset PR(\mathcal{A}_\beta(X,Y))$. $\diamond$

Proof. (i) Let $A \in SSR(X,Y)$. We will prove that $A \in PR(\mathcal{A}_\alpha(X,Y))$. Let $B \in \mathcal{A}_\alpha(X,Y)$ such that $\mathcal{D}(B) \subset \mathcal{D}(A)$ and $A(0) \subset B(0)$. Since A is continuous, then by using Proposition 2.7.2, we conclude that $A + B \in CR(X,Y)$. Applying Proposition 3.4.1 to B, there exist $B_l \in \mathcal{L}(Y,X)$ and F, which is a finite rank linear operator, such that $\mathcal{D}(B) \subset \mathcal{D}(F)$, $\mathcal{R}(F) \subset \mathcal{D}(B)$, and $B_l B = (I - F)_{|\mathcal{D}(B)}$. We can easily deduce that

$$B_l(A + B) = B_l B + B_l A = (I - F + B_l A)_{|\mathcal{D}(B)}. \tag{3.58}$$

Now, in order to show that $B_l(A + B) \in \Phi_+(X)$, we will prove that $B_l A \in PR(\Phi_+(X))$. Since $A \in SSR(X,Y)$ and B_l is bounded single valued, then by using Proposition 2.3.4 (iii), we deduce that $B_l A \in SSR(X)$. It follows from Theorem 3.5.5 that $B_l A \in PR(\Phi_+(X))$. Now, by using (3.58) and $B_l A \in PR(\Phi_+(X))$, we conclude that $B_l(A + B) \in \Phi_+(X)$. Moreover, $R(B_l) \subset \mathcal{D}(A + B)$ and $R(F - B_l A) \subset \mathcal{D}(A + B)$ and clearly, we have $(A + B)(0) = B(0)$. Then, $(A + B)(0)$ is complemented in X. Now, the use of Proposition 3.4.1 allows us to conclude that $A + B \in \mathcal{A}_\alpha(X,Y)$. Hence, $SSR(X,Y) \subset PR(\mathcal{A}_\alpha(X,Y))$.

(ii) Let $A \in KR(X,Y)$. We are going to prove that $A \in PR(\mathcal{A}_\beta(X,Y))$. Indeed, we take $B \in \mathcal{A}_\beta(X,Y)$ such that $\mathcal{D}(B) \subset \mathcal{D}(A)$ and $A(0) \subset B(0)$. We prove that $A + B \in \mathcal{A}_\beta(X,Y)$. In view of Lemma 2.14.5 (ii), it is sufficient to prove that $(A+B)^* \in \mathcal{A}_\alpha(Y^*, X^*)$. Since, $\mathcal{D}(B) \subset \mathcal{D}(A)$ and A is continuous, then by using Proposition 2.8.1 (iii), we have $A^* + B^* = (A+B)^*$. By referring to Lemmas 2.3.4 (ii) and 2.14.5, we have $B^* \in \mathcal{A}_\alpha(Y^*, X^*)$, $A^*(0) \subset B^*(0)$, and $\mathcal{D}(B^*) \subset \mathcal{D}(A^*)$. Since $A^* \in PRK(Y^*, X^*)$, then in view of Propositions 2.12.1 and 2.8.1, implies that A^* is continuous strictly singular linear relation. Hence, $A^* + B^* \in \mathcal{A}_\alpha(Y^*, X^*)$. So, $A + B \in \mathcal{A}_\beta(X,Y)$. Q.E.D.

As an immediate consequence of Theorems 3.5.4 and 3.5.5:

Corollary 3.5.3 *Let $T \in CR(X,Y)$ and $S \in LR(X,Y)$ be continuous such that $S(0) \subset T(0)$ and S is T-bounded with T-bound $\delta < 1$.*

(i) *Assume that $T(0)$ is topologically complemented in X. Then,*

(i_1) *If $T \in \mathcal{A}_\alpha(X,Y)$ and $SG_T \in KR(X_T,Y)$ (respectively, $SG_T \in SSR(X_T,Y)$), then $T + S \in \mathcal{A}_\alpha(X,Y)$ and $i(T + S) = i(T)$.*

(i_2) *If $T \in \mathcal{W}^l(X,Y)$ and $SG_T \in KR(X_T,Y)$ (respectively, $SG_T \in SSR(X_T,Y)$), then $T + S \in \mathcal{W}^l(X,Y)$.*

(ii) Assume that $T^*(0)$ is topologically complemented in X^*. Then,

(ii_1) *If $T \in \mathcal{A}_\beta(X,Y)$ and $SG_T \in \mathcal{KR}(X_T,Y)$, then $T + S \in \mathcal{A}_\beta(X,Y)$ and $i(T+S) = i(T)$.*

(ii_2) *If $T \in \mathcal{W}^r(X,Y)$ and $SG_T \in \mathcal{KR}(X_T,Y)$, then $T+S \in \mathcal{W}^r(X,Y)$.*

$\diamond$

3.6   Index of a Linear Relations

Proposition 3.6.1 *Let n and m_i, $1 \le i \le n$ be some positive integers, and let $\lambda_i \in \mathbb{K}$, $1 \le i \le n$ be some distinct constants. Let $T \in LR(X)$ such that $\rho(T) \ne \emptyset$. Let $P(T)$ be defined in (2.47). If each factor $T - \lambda_k$ has finite index, then*

$$i(P(T) - \mu) = \sum_{k=1}^{n} i\left((T - \lambda_k)^{m_k}\right).$$

$\diamond$

Proof. We first show that each

$$\rho(T - \lambda_k) \ne \emptyset. \tag{3.59}$$

Since $T - \alpha = (T - \lambda_k) + (\lambda_k - \alpha)$ for all $\alpha \in \mathbb{C}$, we have $\eta - \lambda_k \in \rho(T - \lambda_k)$ whenever $\eta \in \rho(T)$. So, (3.59) holds. Now, combine (3.59) with Lemma 2.5.1, we have

$$P(T) - \mu = c \prod_{k=1}^{n} (T - \lambda_k)^{m_k}.$$

It only remains to see that

$$i\left(\prod_{k=1}^{n}(T - \lambda_k)^{m_k}\right) = \sum_{k=1}^{n} i\left((T - \lambda_k)^{m_k}\right).$$

We prove the result by induction. For $n = 1$, it is clear by Lemma 2.5.1 that

$$i\left((T - \lambda_k)^{m_k}\right) = m_k i(T - \lambda_k).$$

Assume that it is true for $n = r$, that is

$$i\left(\prod_{k=1}^{r}(T - \lambda_k)^{m_k}\right) = \sum_{k=1}^{r} i\left((T - \lambda_k)^{m_k}\right).$$

Let $A = \prod_{k=1}^{r}(T - \lambda_k)^{m_k}$ and $B = (T - \lambda_{r+1})^{m_{r+1}}$. Then, by the induction hypothesis, we infer that A and B have finite index. Thus, it follows from Theorem 2.14.3 that

$$i(AB) = i(A) + i(B) + \dim \left(\frac{X}{\mathcal{D}(A) + R(B)} \right) - \dim \left(N(A) \bigcap B(0) \right).$$

By applying Lemma 2.13.1, we have

$$\mathcal{D}(A) = \mathcal{D}(T^{m_1 + m_2 + \cdots + m_r}) = \mathcal{D}\left((T - \lambda_{r+1})^{m_1 + m_2 + \cdots + m_r} \right),$$

and

$$R(B) = R\left((T - \lambda_{r+1})^{m_{r+1}} \right).$$

Hence,

$$\mathcal{D}(A) + R(B) = \mathcal{D}\left((T - \lambda_{r+1})^{m_1 + m_2 + \cdots + m_r} \right) + R\left((T - \lambda_{r+1})^{m_{r+1}} \right) = X. \tag{3.60}$$

It is enough to observe that by Lemma 2.5.6 (ii), we have

$$
\begin{aligned}
\alpha(A) &\leq \alpha\left((T - \lambda_1)^{m_1} \right) + \alpha\left((T - \lambda_2)^{m_2} \right) + \cdots + \alpha\left((T - \lambda_r)^{m_r} \right) \\
&\leq m_1 \alpha(T - \lambda_1) + m_2 \alpha(T - \lambda_2) + \cdots + m_r \alpha(T - \lambda_r) \\
&< \infty.
\end{aligned}
$$

Hence,

$$0 < \dim \left(N(A) \bigcap B(0) \right) = \delta < \infty. \tag{3.61}$$

Now, it follows from (3.60) and (3.61) that $i(AB) = i(A) + i(B) - \delta$. Hence, $i(AB) \leq i(A) + i(B)$. It only remains to verify that $i(A) + i(B) \leq i(AB)$. We first note that $AB = BA$ and by virtue of the part (i) of Lemma 2.5.1 and, in view of Theorem 2.14.3, we get

$$i(BA) = i(AB) = i(A) + i(B) + \dim \left(\frac{X}{\mathcal{D}(B) + R(A)} \right) - \dim \left(N(B) \bigcap A(0) \right).$$

The use of Theorem 2.14.3, make us conclude that A is a Fredholm relation, in particular

$$\dim \left(\frac{X}{R(A)} \right) < \infty$$

which implies that

$$\dim \left(\frac{X}{\mathcal{D}(B) + R(A)} \right) < \infty. \tag{3.62}$$

Now, we prove

$$A(0) \bigcap N(B) = \{0\}. \tag{3.63}$$

Indeed, applying Lemmas 2.5.1 and 2.13.1, we get

$$A(0) = T^{m_1+m_2+\cdots+m_r}(0) = (T - \lambda_{r+1})^{m_1+m_2+\cdots+m_r}(0).$$

So,

$$A(0) \bigcap N(B) = \left((T - \lambda_{r+1})^{m_1+m_2+\cdots+m_r}(0)\right) \bigcap N\left((T - \lambda_{r+1})^{m_{r+1}}\right).$$

The last equality together with (3.59) and Lemma 2.5.1 ensures that $A(0) \bigcap N(B) = \{0\}$. We conclude from both (3.62) and (3.63) that $i(A) + i(B) \leq i(AB)$.   Q.E.D.

As an immediate consequence of Proposition 3.6.1:

Corollary 3.6.1 *Let $T \in LR(X)$ such that $\rho(T) \neq \emptyset$ and $P(T)$ defined in (2.47). Then,*

(i) If each $T - \lambda_k \in \Phi_+(X)$ and $i(T - \lambda_k) \in]-\infty, 0]$, for some $k \in \mathbb{N}$, then $P(T) - \mu \in \Phi_+(X)$ and $i(P(T) - \mu) \in]-\infty, 0]$.

(ii) If each $T - \lambda_k \in \Phi_-(X)$ and $i(T - \lambda_k) \in [0, +\infty[$, for some $k \in \mathbb{N}$, then $P(T) - \mu \in \Phi_-(X)$ and $i(P(T) - \mu) \in [0, +\infty[$.   ◇

Proposition 3.6.2 *Let $T \in LR(X)$ such that $\rho(T) \neq \emptyset$ and $P(T)$ defined in (2.47). Then,*

(i) If each $T - \lambda_k \in \Phi_+(X)$ and $i(T - \lambda_k) = -\infty$, for some $k \in \mathbb{N}$, then $P(T) - \mu \in \Phi_+(X)$ and $i(P(T) - \mu) = -\infty$.

(ii) If each $T - \lambda_k \in \Phi_-(X)$ and $i(T - \lambda_k) = +\infty$, for some $k \in \mathbb{N}$, then $P(T) - \mu \in \Phi_-(X)$ and $i(P(T) - \mu) = +\infty$.   ◇

Proof. (i) We note that $\rho(T - \lambda_k) \neq \emptyset$, by (3.59) and since $T - \lambda_k$ is closed, we deduce from Lemma 2.13.4 that $(T - \lambda_k)^{m_k}$ and $P(T) - \mu$ are closed linear relations. Using Theorem 2.14.4, we have $(T - \lambda_k)^{m_k} \in \Phi_+(X)$ and $P(T) - \mu \in \Phi_+(X)$. Furthermore, by Lemma 2.13.1, we get

$$R(P(T) - \mu) = \bigcap_{k=1}^{n} R\left((T - \lambda_k)^{m_k}\right).$$

Thus, it follows from Lemma 2.1.3 that

$$\left(\frac{\frac{X}{R(P(T)-\mu)}}{\frac{R(T-\lambda_k)}{R(P(T)-\mu)}}\right) \cong \frac{X}{R(T - \lambda_k)}.$$

The last property combined with both $\alpha(T - \lambda_k) < \infty$ and $i(T - \lambda_k) = -\infty$ leads to $\beta(P(T) - \mu) = +\infty$. Therefore, (i) holds.

(ii) We note that $\rho(T - \lambda_k) \neq \emptyset$, by (3.59) and since $T - \lambda_k$ is closed, we deduce from Lemma 2.13.4 that $(T - \lambda_k)^{m_k}$ and $P(T) - \mu$ are closed linear relations. Since $(T - \lambda_k)^{m_k} \in \Phi_-(X)$, it follows from Lemma 2.5.1 that $P(T) - \mu \in \Phi_-(X)$. It only remains to see that $i(P(T) - \mu) = +\infty$. To this end, we shall verify that

$$\alpha\left((T - \lambda_k)^2\right) + 2\beta(T - \lambda_k) = \beta\left((T - \lambda_k)^2\right) + 2\alpha(T - \lambda_k). \tag{3.64}$$

Indeed, from Theorem 2.14.3, it follows that

$$\alpha\left((T - \lambda_k)^2\right) + 2\beta(T - \lambda_k) + \dim\left((T - \lambda_k)(0) \bigcap N(T - \lambda_k)\right) =$$
$$\beta((T - \lambda_k)^2) + 2\alpha(T - \lambda_k) + \dim\left(\frac{X}{\mathcal{D}(T - \lambda_k) + R(T - \lambda_k)}\right).$$

This implies by the use of (3.59) and Lemma 2.5.1 the validity of (3.64). So,

$$\alpha\left((T - \lambda_k)^{m_k}\right) = +\infty. \tag{3.65}$$

Since $T - \lambda_k$ is a lower semi-Fredholm relation with nonempty resolvent set, then $(T - \lambda_k)^2 \in \Phi_-(X)$. So, $\beta(T - \lambda_k)$ and $\beta\left((T - \lambda_k)^2\right)$ are both finite. Furthermore, since $i(T - \lambda_k) = +\infty$, by hypothesis and $\beta(T - \lambda_k) < \infty$, we have $\alpha(T - \lambda_k) = +\infty$. In this situation, we infer from (3.64) that $\alpha\left((T - \lambda_k)^2\right) = +\infty$ and thus continuing in this way, we obtain the property (3.65). Hence,

$$\alpha(P(T) - \mu) = +\infty. \tag{3.66}$$

By using Lemma 2.13.1, we have

$$N(P(T) - \mu) = \sum_{i=1}^{n} N\left((T - \lambda_i)^{m_i}\right)$$
$$= \sum_{\substack{i=1 \\ i \neq k}}^{n} N\left((T - \lambda_i)^{m_i}\right) + N\left((T - \lambda_k)^{m_k}\right).$$

Thus, by (3.66), it follows that $\alpha(P(T) - \mu) = +\infty$. This property together with the fact $P(T) - \mu \in \Phi_-(X)$ proves that $i(P(T) - \mu) = +\infty$, as desired. This completes the proof.   Q.E.D.

3.6.1   Index of Upper Semi-Fredholm Relation Under Strictly Singular Perturbation

Throughout this section, we investigate the behavior of the index of upper and lower semi-Fredholm multivalued linear operators under strictly singular perturbation.

Lemma 3.6.1 *Let* T, $S \in LR(X,Y)$ *such that* $\dim(S(0)) < \infty$. *Then,*

(i) If $T \in \mathcal{F}_+(X,Y)$, *then* $Q_{T+S}T \in \mathcal{F}_+\left(X, Y/\overline{(S+T)(0)}\right)$.

(ii) If $T \in \mathcal{F}_-(X,Y)$, *then* $Q_{T+S}T \in \mathcal{F}_-\left(X, Y/\overline{(S+T)(0)}\right)$.

(iii) If S *is strictly singular, then* $Q_{T+S}S$ *is strictly singular.*   ◇

Proof. (i) From Propositions 2.14.1 and 2.14.2, $T \in \mathcal{F}_+(X,Y)$ if and only if, $Q_T T \in \mathcal{F}_+\left(X, Y/\overline{T(0)}\right)$. Since

$$
\begin{aligned}
\dim\left(\frac{\overline{T(0)+S(0)}}{\overline{T(0)}}\right) &= \dim\left(\frac{S(0)}{\overline{T(0)} \cap S(0)}\right) \\
&= \dim(S(0)) - \dim\left(\overline{T(0)} \cap S(0)\right) < \infty,
\end{aligned}
$$

then by using Corollary 2.4.1, we get

$$
Q_{T+S}T = Q_{\frac{\overline{T(0)+S(0)}}{\overline{T(0)}}}^{Y/\overline{T(0)}} Q_T^Y T \in \mathcal{F}_+\left(X, Y/\overline{(S+T)(0)}\right).
$$

(ii) In similar way, $T \in \mathcal{F}_-(X,Y)$ if and only if, $Q_T T \in \mathcal{F}_-\left(X, Y/\overline{T(0)}\right)$. Since $\dim\left(\frac{\overline{T(0)+S(0)}}{\overline{T(0)}}\right) < \infty$, then by using Corollary 2.4.1, we get

$$
Q_{T+S}T = Q_{\frac{\overline{T(0)+S(0)}}{\overline{T(0)}}}^{Y/\overline{T(0)}} Q_T^Y T \in \mathcal{F}_-\left(X, Y/\overline{(S+T)(0)}\right).
$$

(iii) From Corollary 2.4.1, it is clear that

$$
Q_{T+S}S = Q_{\frac{\overline{T(0)+S(0)}}{S(0)}}^{Y/S(0)} Q_S S.
$$

Since $Q_{\frac{\overline{T(0)+S(0)}}{S(0)}}^{Y/S(0)}$ is continuous and $Q_S S(0) = 0$, then by using Theorem 2.12.2, we deduce that $Q_{T+S}S = Q_{\frac{\overline{T(0)+S(0)}}{S(0)}}^{Y/S(0)} Q_S S$ is strictly singular.   Q.E.D.

Theorem 3.6.1 *Let* X *and* Y *be Banach spaces and let* $T \in CR(X,Y)$. *If* $T \in \Phi_+(X,Y)$ *and* $S \in SSR(X,Y)$ *is a continuous linear relation such that* $\dim(S(0)) < \infty$ *and* $\mathcal{D}(S) \supset \overline{\mathcal{D}(T)}$. *Then,* $T+S \in \Phi_+(X,Y)$ *and* $i(T+S) = i(T) + \dim(S(0)) - \dim\left(T(0) \cap S(0)\right)$.   ◇

Proof. Since T is closed, then by using Proposition 2.7.1, we have $T(0)$ is closed and $Q_T T$ is closed. Since $\dim(S(0)) < \infty$, then $T(0) + S(0)$ is closed. Since Y is a Banach space, then $Y/(T(0) + S(0))$ is a Banach space. By combining Proposition 2.9.4 and Corollary 2.4.1, we get $\dim\left(\frac{T(0)+S(0)}{T(0)}\right) < \infty$ and $Q_{T+S}T = Q^{Y/T(0)}_{\frac{T(0)+S(0)}{T(0)}} Q^Y_T T$ is closed. From Proposition 2.7.1 and Theorem 2.7.1, we obtain that $T + S$ and $Q_{T+S}(T + S)$ are closed. Furthermore, by using Lemma 3.6.1, we get $Q_{T+S}T \in \Phi_+(X, Y/(T(0) + S(0)))$ and $Q_{T+S}S$ is strictly singular. Finally, by using Corollary 2.16.1 (iv), we obtain that $Q_{T+S}(T + S) = Q_{T+S}T + Q_{T+S}S \in \Phi_+(X, Y/(T(0) + S(0)))$ and

$$i(Q_{T+S}(T + S)) = i(Q_{T+S}T + Q_{T+S}S) = i(Q_{T+S}T). \qquad (3.67)$$

Therefore, since $T + S$ and $Q_{T+S}(T + S)$ are closed, it follows from Lemma 2.14.3 that $T + S \in \Phi_+(X, Y)$. Moreover, it can be shown that

$$
\begin{aligned}
R(Q_{T+S}T) &= Q_{T+S}T(\mathcal{D}(T)) \\
&= Q_{T+S}(R(T)) \\
&= Q_{T+S}(R(T) + S(0)) \\
&= \frac{R(T) + S(0)}{T(0) + S(0)}.
\end{aligned}
$$

Therefore, using both Theorems 2.1.1 and 2.1.2, we get

$$
\begin{aligned}
\beta(T) &= \dim\left(\frac{Y}{R(T)}\right) \\
&= \dim\left(\frac{\frac{Y}{R(T)}}{\frac{R(T)+S(0)}{R(T)}}\right) + \dim\left(\frac{R(T) + S(0)}{R(T)}\right) \\
&= \dim\left(\frac{Y}{R(T) + S(0)}\right) + \dim\left(\frac{R(T) + S(0)}{R(T)}\right) \\
&= \dim\left(\frac{\frac{Y}{T(0)+S(0)}}{\frac{R(T)+S(0)}{T(0)+S(0)}}\right) + \dim\left(\frac{R(T) + S(0)}{R(T)}\right) \\
&= \beta(Q_{T+S}T) + \dim\left(\frac{S(0)}{R(T) \cap S(0)}\right).
\end{aligned}
$$

So,

$$\beta(T) = \beta(Q_{T+S}T) + \dim(S(0)) - \dim\left(R(T) \cap S(0)\right). \qquad (3.68)$$

On the other hand, since $T(0)$ is closed, then by using Proposition 2.9.2, we have

$$N(T) = N(Q_T T).$$

Now, it is clear that $N(T) = N(Q_T T)$. So, it follows that

$$
\begin{aligned}
\frac{N(Q_{T+S}T)}{N(T)} &\cong \frac{R(T) \cap N(Q_{T+S})}{T(0) \cap N(Q_{T+S})} \\
&\cong \frac{R(T) \cap (T+S)(0)}{T(0) \cap (T+S)(0)} \\
&\cong \frac{\mathcal{D}(T^{-1}) \cap (T+S)(0)}{N(T^{-1}) \cap (T+S)(0)} \\
&\cong \frac{T^{-1}(T+S)(0)}{T^{-1}(0)} \\
&\cong \frac{T^{-1}T(0) + T^{-1}S(0)}{T^{-1}(0)} \\
&\cong \frac{T^{-1}(0) + T^{-1}S(0)}{T^{-1}(0)}.
\end{aligned}
$$

Since $T^{-1}(0) \subset T^{-1}S(0)$, we get

$$
\frac{N(Q_{T+S}T)}{N(T)} \cong \frac{T^{-1}S(0)}{T^{-1}(0)}.
$$

By applying the same reasoning as above, we obtain that

$$
\frac{N(Q_{T+S}T)}{N(T)} \cong \frac{R(T) \cap S(0)}{T(0) \cap S(0)}.
$$

From Theorem 2.1.1, we have

$$
\dim N(Q_{T+S}T) = \dim N(T) + \dim \left(\frac{N(Q_{T+S}T)}{N(T)} \right).
$$

Since $\dim S(0) < \infty$, then

$$
\alpha(Q_{T+S}T) = \alpha(T) + \dim \left(R(T) \cap S(0) \right) - \dim \left(T(0) \cap S(0) \right). \qquad (3.69)
$$

The (3.67), (3.68) and (3.69) gives that

$$
\begin{aligned}
i(Q_{T+S}(T+S)) &= i(Q_{T+S}T) \\
&= i(T) + \dim(S(0)) - \dim(T(0) \cap S(0)).
\end{aligned}
$$

Finally, by using Lemma 2.14.3, we get

$$
i(Q_{T+S}(T+S)) = i(T+S).
$$

Therefore,

$$
i(T+S) = i(T) + \dim S(0) - \dim T(0) \cap S(0). \qquad \text{Q.E.D.}
$$

Corollary 3.6.2 *Let X and Y be Banach spaces, let $T \in CR(X,Y)$ such that $T \in \Phi_+(X,Y)$ and let $S \in SSR(X,Y)$ be a continuous linear relation such that $\dim(S(0)) < \infty$ and $\mathcal{D}(S) \supset \overline{\mathcal{D}(T)}$. Then,*

(i) If $S(0) \subset T(0)$, then $i(T + S) = i(T)$.
(ii) If $S(0) \bigcap T(0) = \{0\}$, then $i(T + S) = i(T) + \dim(S(0))$. ◇

3.6.2 Index of an Lower Semi-Fredholm Relation Under Strictly Singular Perturbations

In this section, we investigate the behavior of the index of an lower semi-Fredholm multivalued linear operators under strictly singular perturbation.

Theorem 3.6.2 *Let X and Y be Banach spaces, let $T \in CR(X,Y)$ such that $T \in \Phi_-(X,Y)$, and let $S \in LR(X,Y)$ such that $\dim S(0) < \infty$, $\mathcal{D}(S) \supset \overline{\mathcal{D}(T)}$ and $S^* \in SSR(Y^*,X^*)$ is a continuous linear relation. Then, $T + S \in \Phi_-(X,Y)$ and*

$$i(T + S) = i(T) + \dim(S(0)) - \dim\left(T(0) \bigcap S(0)\right).$$ ◇

Proof. In the same way of the proof of Theorem 3.6.1, it is clear that $T(0) + S(0)$ is closed. Therefore, $Y/(T(0) + S(0))$ is Banach space. Furthermore,

$$Q_{T+S}T = Q^{Y/T(0)}_{\frac{T(0)+S(0)}{T(0)}} Q^Y_T T$$

is closed. First, from Lemma 3.6.1, it follows that $Q_{T+S}T \in \Phi_-(X, Y/(T(0) + S(0)))$. Second, we show that $(Q_{T+S}S)^*$ is continuous. In fact, by combining both Lemma 2.8.1 and Proposition 2.5.5, we have

$$
\begin{aligned}
\|(Q_{T+S}S)^*\| &= \|S^* J^{Y^*}_{(T(0)+S(0))^\perp}\| \\
&\leq \|S^*\|\,\|J^{Y^*}_{(T(0)+S(0))^\perp}\| \\
&< \infty.
\end{aligned}
$$

Moreover, since $J^{Y^*}_{(T(0)+S(0))^\perp}$ is single valued and continuous, and by using Lemma 2.8.1, we have $(Q_{T+S}S)^* = S^* J^{Y^*}_{(T(0)+S(0))^\perp}$ is strictly singular. Finally, since $\overline{\mathcal{D}(Q_{T+S}T)} = \overline{\mathcal{D}(T)} \subset \mathcal{D}(Q_{T+S}S) = \mathcal{D}(S)$, then $Q_{T+S}T + Q_{T+S}S = Q_{T+S}(T + S) \in \Phi_-(X, Y/(T(0) + S(0)))$ and

$$
\begin{aligned}
i(Q_{T+S}(T + S)) &= i(Q_{T+S}T + Q_{T+S}S) \\
&= i(Q_{T+S}T).
\end{aligned}
$$

From Lemma 2.14.3, we have

$$i(Q_{T+S}(T + S)) = i(T + S),$$

and by using a similar reasoning as the proof of Theorem 3.6.1, we show that

$$\beta(Q_{T+S}T) = \beta(T) - \dim(S(0)) + \dim\left(R(T)\bigcap S(0)\right),$$

and

$$\alpha(Q_{T+S}T) = \alpha(T) + \dim\left(R(T)\bigcap S(0)\right) - \dim\left(T(0)\bigcap S(0)\right).$$

Hence, $T + S \in \Phi_-(X,Y)$ and

$$i(T + S) = i(T) + \dim S(0) - \dim\left(T(0)\bigcap S(0)\right). \qquad \text{Q.E.D.}$$

Corollary 3.6.3 *Let X and Y be Banach spaces, let $T \in C\mathcal{R}(X,Y)$ such that $T \in \Phi_-(X,Y)$, and let $S \in L\mathcal{R}(X,Y)$ such that $\dim S(0) < \infty$, $\mathcal{D}(S) \supset \overline{\mathcal{D}(T)}$ and $S^* \in SS\mathcal{R}(Y^*, X^*)$ is a continuous linear relation. Then,*

(i) If $S(0) \subset T(0)$, then
$$i(T + S) = i(T).$$

(ii) If $S(0) \bigcap T(0) = \{0\}$, then
$$i(T + S) = i(T) + \dim S(0). \qquad \diamondsuit$$

Theorem 3.6.3 *Let X be a Banach space and let $K \in L\mathcal{R}(X)$ be a precompact linear relation such that $\mathcal{D}(K)$ is closed and $\dim K(0) < \infty$. Then, $I_X - K \in \Phi_+(X)$ and*

$$i(I_X - K) = \dim K(0) - \dim\left(\frac{X}{\mathcal{D}(K)}\right). \qquad \diamondsuit$$

Proof. Since $\mathcal{D}(I_X - K) = \mathcal{D}(K)$, it is obvious that $I_X - K = I_{\mathcal{D}(K)} - K$. By using Lemma 2.7.4, $I_{\mathcal{D}(K)}$ is closed. Moreover, $I_{\mathcal{D}(K)} \in \Phi_+(X)$, in fact, $N(I_{\mathcal{D}(K)}) = \{0\}$ and $R(I_{\mathcal{D}(K)}) = \mathcal{D}(K)$. Now, applying both Lemma 2.12.1 and Theorem 3.6.1, we obtain $I_X - K \in \Phi_+(X)$ and

$$
\begin{aligned}
i(I_X - K) &= i(I_{\mathcal{D}(K)} - K) \\
&= i(I_{\mathcal{D}(K)}) + \dim K(0) \\
&= \dim K(0) - \dim\left(\frac{X}{\mathcal{D}(K)}\right).
\end{aligned}
$$

This completes the proof. 　　　　　　　　　　　　　　　　Q.E.D.

3.7 Demicompact Linear Relations

Throughout this section, we give some fundamental results about demicompact linear relations.

3.7.1 Auxiliary Results on Demicompact Linear Relations

Definition 3.7.1 *A linear relation* $T : \mathcal{D}(T) \subset X \longrightarrow Y$ *is said to be demicompact if for every bounded sequence* $(x_n)_n$ *in* $\mathcal{D}(T)$ *such that* $Q_{I-T}(I-T)x_n \rightarrow y \in Y/\overline{(I-T)(0)}$, *there is a convergent subsequence of* $(Q_T x_n)_n$. $\diamond$

The families of all demicompact linear relations will be denoted by $\mathcal{DC}(X,Y)$. If $X = Y$, then we denote $\mathcal{DC}(X) := \mathcal{DC}(X,X)$. For $\mu \in \mathbb{C}$, we denote by

$$\mathcal{DC}_\mu(X,Y) = \{T \in CR(X,Y) \; : \; \|Tx\| \leq |\mu|\|x\|, \text{ for all } x \in \mathcal{D}(T) \text{ and } |\mu| < 1\}.$$

If $X = Y$, then we denote $\mathcal{DC}_\mu(X) := \mathcal{DC}_\mu(X,X)$.

Remark 3.7.1 *A linear relation* $T : \mathcal{D}(T) \subset X \longrightarrow X$ *is demicompact if and only if, for every bounded sequence* $(x_n)_n$ *in* $\mathcal{D}(T)$ *such that* $Q_T x_n - Q_T T x_n \rightarrow y \in X/\overline{T(0)}$, *there is a convergent subsequence of* $(Q_T x_n)_n$. *Since* $T(0)$ *is a linear subspace of* X, *then* $(I-T)(0) = T(0)$. *Therefore*, $Q_{I-T} = Q_T$. *Moreover*, $\mathcal{D}(Q_T) = X$ *and* $Q_{I-T}(I-T) = Q_T(I-T) = Q_T - Q_T T$. $\diamond$

Theorem 3.7.1 *Let* $T \in LR(X)$. *If* T *is a compact linear relation, then* T *is a demicompact linear relation.* $\diamond$

Proof. Let $(x_n)_n$ be a bounded sequence of $\mathcal{D}(T)$ such that

$$Q_T(x_n - Tx_n) = Q_T x_n - Q_T T x_n \rightarrow y_0.$$

Since $(x_n)_n$ is bounded, then there exists $M > 0$ such that for all $n \in \mathbb{N}$ $\|x_n\| \leq M$. Hence, $\frac{1}{M} Q_T T x_n \in Q_T T B_X$. But T is compact, i.e., $\overline{Q_T T B_X}$ is compact, then there exists a subsequence $(\frac{1}{M} Q_T T x_{\varphi(n)})_n$ converging to y_1. Finally, it follows that $(Q_T x_{\varphi(n)})_n$ converges to $y_0 + M y_1$. Q.E.D.

Proposition 3.7.1 *Let* $T \in LR(X)$. *If* $I - Q_T$ *is compact, then* T *is a demicompact linear relation if and only if* $Q_T T$ *is a demicompact operator.*$\diamond$

Proof. We suppose that T is demicompact. Let $(x_n)_n$ be a bounded sequence of $\mathcal{D}(T)$ such that $x_n - Q_T T x_n \to y$. We have

$$x_n - Q_T T x_n = (I - Q_T) x_n + Q_T x_n - Q_T T x_n. \tag{3.70}$$

Since $I - Q_T$ is compact and $(x_n - Q_T T x_n)_n$ is a convergent sequence, then it follows from (3.70), $(Q_T x_n - Q_T T x_n)_n$ has a convergent subsequence. When the latter is added to demicompactness of T, we get $(Q_T x_n)_n$ has a convergent subsequence. On the other hand,

$$\begin{aligned} x_n &= x_n - Q_T x_n + Q_T x_n \\ &= (I - Q_T) x_n + Q_T x_n. \end{aligned}$$

Since $I - Q_T$ is compact and $(Q_T x_n)_n$ has a convergent subsequence, then $(x_n)_n$ has a convergent subsequence. So, $Q_T T$ is demicompact. Conversely, we suppose that $Q_T T$ is a demicompact operator. Let $(x_n)_n$ be a bounded sequence of $\mathcal{D}(T)$ such that $Q_T x_n - Q_T T x_n \to y$. Then,

$$Q_T x_n - Q_T T x_n = -(I - Q_T) x_n + x_n - Q_T T x_n. \tag{3.71}$$

According to (3.71), $I - Q_T$ is compact and $(Q_T x_n - Q_T T x_n)_n$ is a convergent sequence, then $(x_n - Q_T T x_n)_n$ has a convergent subsequence. Considering the fact that $Q_T T$ is demicompact and $(x_n - Q_T T x_n)_n$ has a convergent subsequence, we obtain that $(x_n)_n$ has a convergent subsequence. On the other hand,

$$Q_T x_n = Q_T x_n - x_n + x_n = -(I - Q_T) x_n + x_n,$$

also we have $I - Q_T$ is compact and $(x_n)_n$ has a convergent subsequence, thus $(Q_T x_n)_n$ has a convergent subsequence. Q.E.D.

Proposition 3.7.2 *Let $T \in C\mathcal{R}(X)$. If T is 1-set-contraction, then μT is demicompact for each $\mu \in [0, 1[$.* $\diamond$

Proof. Let $(x_n)_n$ be a bounded sequence of $\mathcal{D}(T)$ such that

$$y_n = Q_{\mu T}(I - \mu T) x_n \to y.$$

If $\delta(Q_{\mu T}(x_n)) \neq 0$, then since $Q_{\mu T}(x_n) \subset y_n + Q_{\mu T}(T x_n)$ and using the fact that μT is 1-set-contraction, we get

$$\begin{aligned} \delta(\{Q_{\mu T} x_n\}) &\leq \delta(\{y_n\}) + \delta(Q_{\mu T} \mu T x_n\}) \\ &\leq \delta(\{Q_{\mu T} \mu T x_n)\}) \\ &\leq \mu \delta(\{Q_{\mu T} x_n\}) \\ &< \delta(\{Q_{\mu T} x_n\}), \end{aligned}$$

which is impossible. Hence, $\delta(\{Q_{\mu T}x_n\}) = 0$. So, $(Q_{\mu T}x_n)_n$ is relatively compact, and we deduce that μT is demicompact. Q.E.D.

Theorem 3.7.2 *Let $T \in LR(X)$ such that $\overline{T(0)} \subset N(T)$. If T is $\overline{T(0)}$-condensing, then T is demicompact.* $\diamond$

Proof. Let $(x_n)_n$ be a bounded sequence of $\mathcal{D}(T)$ such that $y_n = Q_T(I - T)x_n \to y$. Suppose that $\delta(\{x_n\}) \neq 0$, since $\{Q_T x_n\} \subset \{y_n\} + \{Q_T T x_n\}$, it follows, using Proposition 2.10.1 (iv), that

$$\begin{aligned}
\delta(\{Q_T x_n\}) &\leq \delta(\{y_n\}) + \delta(\{Q_T T x_n\}) \\
&\leq \delta(\{Q_T T x_n\}) \\
&< \delta(\{Q_T x_n\}),
\end{aligned}$$

which is impossible. Hence, $\delta(\{Q_T x_n\}) = 0$. So, $(Q_T x_n)_n$ is relatively compact. Therefore, $(Q_T x_n)_n$ has a convergent subsequence. Q.E.D.

3.7.2 Demicompactness and Fredholm Linear Relations

In this subsection, we present some results on Fredholm and upper semi-Fredholm linear relations involving demicompact linear relations.

Theorem 3.7.3 *Let X be a Banach space and $T \in LR(X)$. Then,*

(i) *If $I - T \in \mathcal{F}_+(X)$, then T is a demicompact linear relation.*
(ii) *If $\dim T(0) < \infty$ and T is a demicompact linear relation, then $I - T \in \mathcal{F}_+(X)$.*
(iii) *If $T \in CR(X)$, $\dim T(0) < \infty$ and T is a demicompact linear relation, then $I - T \in \Phi_+(X)$.* $\diamond$

Proof. (i) If $I - T \in \mathcal{F}_+(X)$, then by using Propositions 2.14.1 and 2.14.2, we have $Q_{I-T}(I - T) = Q_T(I - T) \in \mathcal{F}_+(X, X/\overline{T(0)})$. Hence, in view of Theorem 2.14.1, there exists a bounded linear operator A and a bounded finite rank projection operator P such that $AQ_T(I - T) = I_{\mathcal{D}(T)} - P$. Let $(x_n)_n \in \mathcal{D}(T)$ be a bounded sequence such that $Q_T(I - T)x_n \to y \in X/\overline{T(0)}$. Then, $AQ_T(I - T)x_n \to Ay$. Hence, $x_n - Px_n \to Ay$. Since P is compact, then by using Theorem 3.7.1, we have P is demicompact. Therefore, $(x_n)_n$ has a convergent subsequence.

(ii) If $I - T \notin \mathcal{F}_+(X)$, then by using Theorem 2.14.2, there exists a sequence $(x_n)_n \in \mathcal{D}(T)$ of norm one such that $(x_n)_n$ has no Cauchy subsequence and

$\lim_{n \to \infty} \|(I - T)x_n\| = 0$. On the other hand,

$$\begin{aligned}
\|(I - T)x_n\| &= \|Q_{I-T}(I - T)x_n\| \\
&= \|Q_T(I - T)x_n\|.
\end{aligned}$$

Then, $(Q_T(I - T)x_n)_n$ converges to 0. Since T is demicompact, then $(Q_T x_n)_n$ has a convergent subsequence. In view of Theorem 2.1.3, $(x_n)_n$ has a convergent subsequence. This contradicts the fact that $(x_n)_n$ has no Cauchy subsequence. Then, $I - T \in \mathcal{F}_+(X)$.

(iii) is a simple consequence of (ii) and Proposition 2.7.2.   Q.E.D.

Theorem 3.7.4 *Let* X *be a Banach space and* $T \in LR(X)$ *such that* $\dim T(0) < \infty$, $\mathcal{D}(T)$ *is closed and* $\dim(X/\mathcal{D}(T)) < \infty$. *If* μT *is a demicompact linear relation for all* $\mu \in [0, 1]$, *then* $I - T \in \mathcal{F}_+(X)$ *and* $i(I - T) = \dim T(0) - \dim(X/\mathcal{D}(T))$. $\diamond$

Proof. Since G_T is closed and $\dim \mu T G_T(0) = \dim \mu T(0) < \infty$, then by using Theorem 2.7.1, we have $G_T - \mu T G_T$ is closed and $Q_T(G_T - \mu T G_T) = Q_T G_T - \mu Q_T T G_T$ is closed. On the other hand, by using Theorem 3.7.3, we have $I - \mu T \in \mathcal{F}_+(X)$. The fact that $N(G_T) = \{0\}$ and $R(G_T) = \mathcal{D}(T)$, we have $G_T \in \Phi_+(X_T, X)$. Since G_T is closed and X is a Banach space, then $G_T \in \mathcal{F}_+(X_T, X)$ and

$$(I - \mu T)G_T = G_T - \mu T G_T \in \mathcal{F}_+(X_T, X).$$

Now, since $\dim T(0) < \infty$, then $Q_T(I - \mu T)G_T = Q_T G_T - \mu Q_T T G_T \in \mathcal{F}_+(X_T, X/\overline{T(0)})$. Since $Q_T G_T - \mu Q_T T G_T$ is closed, then $Q_T(I - \mu T)G_T = Q_T G_T - \mu Q_T T G_T \in \Phi_+(X_T, X/\overline{T(0)})$ for all $\mu \in [0, 1]$. By Theorem 3.3.2, the index $i(Q_T G_T - \mu Q_T T G_T)$ is continuous in μ. Since it is an integer, including infinite value, it must be constant for every $\mu \in [0, 1]$. Showing that

$$\begin{aligned}
i(Q_T G_T - \mu Q_T T G_T) &= i(Q_T G_T - Q_T T G_T) \\
&= i(Q_T G_T).
\end{aligned}$$

Hence,

$$\begin{aligned}
i(Q_T G_T) &= i(Q_T) + i(G_T) - \dim\left(G_T(0) \bigcap N(Q_T)\right) \\
&= \dim T(0) - \dim(X/\mathcal{D}(T)).
\end{aligned}$$

On the other hand,

$$
\begin{aligned}
i(Q_T G_T - Q_T T G_T) &= i(Q_T(I-T)G_T) \\
&= i(G_T) + \dim\left(\frac{X}{R(G_T) + \mathcal{D}(Q_T(I-T))}\right) \\
&\quad + i(Q_T(I-T)) - \dim\left(G_T(0) \cap N(Q_T(I-T))\right) \\
&= i(Q_T(I-T)) - \dim\left(X/\mathcal{D}(T)\right) + \dim\left(X/\mathcal{D}(T)\right) \\
&= i(I-T).
\end{aligned}
$$

Finally, we obtain

$$
i(I-T) = \dim T(0) - \dim\left(X/\mathcal{D}(T)\right).
$$

This completes the proof. $\hspace{6cm}$ Q.E.D.

Corollary 3.7.1 (i) *If T satisfies the hypotheses of Theorem 3.7.4, then $I - \lambda T \in \mathcal{F}_+(X)$ and $i(I - \lambda T) = \dim T(0) - \dim\left(X/\mathcal{D}(T)\right)$ for every $\lambda \in (0,1]$.*
(ii) *If $\dim(T(0) < \infty$, then T is demicompact if and only if, $I - T \in \mathcal{F}_+(X)$.*
$\hspace{12cm}\Diamond$

Proposition 3.7.3 *Let $T \in C\mathcal{R}(X)$ such that $\dim(T(0)) < \infty$ and let S be a closed selection of T. If S is demicompact, 1-set-contraction, then $I - T$ is a Fredholm linear relation.* $\hspace{5cm}\Diamond$

Proof. Let S be a selection of T, then $T = S + T - T$. Hence, $I - T = (I-S)+(I-T)-(I-T)$. This implies that $I-S$ is a selection of $I-T$. Moreover, S is demicompact, 1-set-contraction, then it follows that $I - S \in \Phi(X)$. Since $\dim(I-T)(0) = \dim(T(0)) < \infty$, then $I-T$ is a Fredholm liner relation. This completes the proof. $\hspace{7cm}$ Q.E.D.

Theorem 3.7.5 *Let $T \in BC\mathcal{R}(X)$ such that $\dim T(0) < \infty$. If T is 1-set-contraction, then $I - T \in \Phi(X)$ and $i(I-T) = \dim(T(0))$.* $\hspace{2cm}\Diamond$

Proof. If $T \in B\mathcal{R}(X)$ such that T is 1-set-contraction, then by using Proposition 3.7.2, we get μT is demicompact for each $\mu \in [0,1[$. It follows from Theorem 3.7.4 that $I - \mu T \in \Phi_+(X)$. This implies that $I - T \in \Phi_+(X)$. It is sufficient to prove that $\beta(I-T) < \infty$. Consider the maps

$$
\begin{aligned}
\varphi : \quad [0,1] &\longrightarrow \mathbb{Z} \\
\lambda &\longrightarrow i(Q_T(I-\mu T)).
\end{aligned}
$$

φ is continuous in μ and constant. Hence, $i(Q_T(I-T)) = i(I-T)$ and $i(Q_T(I-\mu T)) = i(Q_T(I-T)) = i(Q_T(I)) = \dim T(0) < \infty$. It follows that $\beta(I-T) < \infty$, we deduce that $I - T \in \Phi(X)$ and $i(I-T) = \dim T(0)$. This completes the proof. $\hspace{7cm}$ Q.E.D.

3.8 Relatively Demicompact Linear Relations

3.8.1 Auxiliary Results on Relatively Demicompact Linear Relations

Definition 3.8.1 *Let X and Y be Banach spaces. If $T : \mathcal{D}(T) \subset X \longrightarrow Y$ and $S : \mathcal{D}(S) \subset X \longrightarrow Y$ are two linear relations with $S(0) \subset T(0)$ and $\mathcal{D}(S) \subset \mathcal{D}(T)$, then T is said to be S-demicompact (or relative demicompact with respect to S), if for every bounded sequence $(x_n)_n$ in $\mathcal{D}(T)$ such that*

$$Q_{S-T}(S - T)x_n = Q_T(S - T)x_n \to y \in Y/\overline{T(0)},$$

there is a convergent subsequence of $(Q_T S x_n)_n$. $\diamondsuit$

We denote by

$$\mathcal{D}\mathcal{C}_S(X, Y) = \{T \in C\mathcal{R}(X, Y) \text{ such that } T \text{ is } S\text{-demicompact}\}.$$

If $X = Y$, then we denote $\mathcal{D}\mathcal{C}_S(X, X) := \mathcal{D}\mathcal{C}_S(X)$.

Lemma 3.8.1 *Let $T : \mathcal{D}(T) \subseteq X \longrightarrow Y$ and $S : \mathcal{D}(S) \subseteq X \longrightarrow Y$ are two linear relations with $S(0) \subseteq T(0)$, $\mathcal{D}(T) \subseteq \mathcal{D}(S)$ and $Q_T(S - I)$ is compact. Then, T is S-demicompact if and only if, TG_S is demicompact.* $\diamondsuit$

Proof. Suppose that T is S-demicompact. Let $\{x_n\}$ be a bounded sequence of $\mathcal{D}(T) \subseteq \mathcal{D}(S)$ such that

$$Q_{TG_S}(I - TG_S)x_n = Q_T(I - T)x_n$$

converges. In other words,

$$Q_T(S - T)x_n = Q_T(I - T)x_n + Q_T(S - I)x_n,$$

then, we have $\{Q_T(S-T)x_n\}$ and we use T as an S-demicompact multivalued linear relation. We obtain $\{Q_T S x_n\}$ which has a convergent subsequence. Thus, we get

$$Q_T x_n = Q_T(I - S)x_n + Q_T S x_n.$$

Therefore, $\{Q_T x_n\}$ has a convergent subsequence. Conversely, let TG_S be a demicompact. Let $\{x_n\}$ be a bounded sequence of $\mathcal{D}(T) \subseteq \mathcal{D}(S)$ such that $\{Q_T(S - T)x_n\}$ converges. On the other side, assuming that

$$Q_T(I - TG_S)x_n = Q_T(I - T)x_n = Q_T(S - T)x_n - Q_T(S - I)x_n,$$

then, $\{Q_T(I - TG_S)x_n\}$ converges. Using the fact that $\{Q_T(I - TG_S)x_n\}$ is convergent and the fact that TG_S is demicompact, we obtain

$$Q_{TG_S}x_n = Q_T x_n$$

which has a convergent subsequence. Finally, we have

$$Q_T S x_n = Q_T x_n - Q_T(I - S)x_n.$$

As a matter of fact, $\{Q_T S x_n\}$ has a convergent subsequence.   Q.E.D.

Proposition 3.8.1 *Let $\mu \in \mathbb{C}^*$ and let $T : \mathcal{D}(T) \subset X \longrightarrow Y$ and $S : \mathcal{D}(S) \subset X \longrightarrow Y$ such that $\mu S N(\mu S - T)$ is compact, $R(\mu S - T)$ is closed, $\mu S - T$ is open and $Q_{\mu S N(\mu S - T)}\mu S$ is continuous. Then, T is μS-demicompact. If, in addition, $\mathcal{D}(T) = \mathcal{D}(S)$, then μS is T-compact.*   ◇

Proof. Since $\mu S - T$ is open, $R(\mu S - T)$ is closed, and $Q_{\mu S N(\mu S - T)}\mu S$ is continuous, then $Q_T(\mu S - T)$ is open and $Q_{\mu S N(\mu S - T)}\mu S$ is continuous. Hence, in view of Remark 2.5.3, $Q_{\mu S N(\mu S - T)}\mu S(Q_T(\mu S - T))^{-1}$ is continuous. Let $(x_n)_n$ be a bounded sequence of $\mathcal{D}(T)$ such that $(Q_T(\mu S - T)x_n)_n$ converges to $Q_T x$. Hence, $(Q_{\mu S N(\mu S - T)}\mu S(Q_T(\mu S - T))^{-1}Q_T(\mu S - T)x_n)_n$ converges to $Q_{\mu S N(\mu S - T)}\mu S(Q_T(\mu S - T))^{-1}Q_T x$. Since $(Q_{\mu S N(\mu S - T)}\mu S x_n)_n$ converges to $\{Q_{\mu S N(\mu S - T)}\mu S(\mu S - T)^{-1}x\}$, and $\mu S N(\mu S - T)$ is compact, then $(\mu S x_n)_n$ has a convergent subsequence. So, T is μS-demicompact. If, in addition, $\mathcal{D}(T) = \mathcal{D}(S)$, then μS is T-compact.   Q.E.D.

Lemma 3.8.2 *Let $T : \mathcal{D}(T) \subset X \longrightarrow Y$ and $S : \mathcal{D}(S) \subset X \longrightarrow Y$ be two linear relations with $S(0) \subset T(0)$, $\mathcal{D}(T) \subset \mathcal{D}(S)$ and $Q_T(I - S)$ is compact. Then, T is S-demicompact if and only if, TG_S is demicompact.*   ◇

Proof. Suppose that T is S-demicompact. Let $(x_n)_n$ be a bounded sequence of $\mathcal{D}(T) \subset \mathcal{D}(S)$ and $S(0) \subset T(0)$, then

$$Q_T(S - T)x_n = Q_T(I - T)x_n + Q_T(S - I)x_n. \tag{3.72}$$

If $Q_{TG_S}(I - TG_S)x_n := Q_T(I - T)x_n$ converges, then by using (3.72) and $Q_T(I - S)$ is compact, we have $(Q_T(S - T)x_n)_n$ has a convergent subsequence. We use T is S-demicompact, we obtain $(Q_T S x_n)_n$ has a convergent subsequence. On the other hand, since

$$Q_T x_n = Q_T(I - S)x_n + Q_T S x_n,$$

then $(Q_T x_n)_n$ has a convergent subsequence. Conversely, let TG_S be demicompact and let $(x_n)_n$ be a bounded sequence of $\mathcal{D}(T) \subset \mathcal{D}(S)$ and $S(0) \subset T(0)$, then

$$Q_T(I - TG_S)x_n = Q_T(I - T)x_n = Q_T(S - T)x_n - Q_T(S - I)x_n. \quad (3.73)$$

Now, if $(Q_T(S - T)x_n)_n$ converges, then by using (3.73), we have $(Q_T(I - TG_S)x_n)_n$ converges. The fact that TG_S is demicompact, we obtain $Q_{TG_S}x_n := Q_T x_n$ has a convergent subsequence. On the other hand, since

$$Q_T S x_n = Q_T x_n - Q_T(I - S)x_n,$$

then $(Q_T S x_n)_n$ has a convergent subsequence. $\hfill$ Q.E.D.

Proposition 3.8.2 *Let $\mu \in \mathbb{C}$ and let $T : \mathcal{D}(T) \subset X \longrightarrow X$. If $T \in \mathcal{DC}_\mu(X)$, then $T \in \mathcal{DC}(X)$.* $\hfill \Diamond$

Proof. Let $T \in \mathcal{DC}_\mu(X)$, then

$$\|Tx\| \leq |\mu| \|x\|.$$

Hence,

$$\|x\| - \|Tx\| \geq (1 - |\mu|) \|x\|.$$

So,

$$\|x\| \quad \leq \quad \frac{\|(I - T)x\|}{1 - |\mu|} = \frac{\|Q_T(I - T)x\|}{1 - |\mu|}. \quad (3.74)$$

Now, take $(x_n)_n$ a bounded sequence of $\mathcal{D}(T)$ such that $Q_T(I - T)x_n \to y$. Applying (3.74), we get $\|x_n - y\| \to 0$. So, $(x_n)_n$ has a convergent subsequence. Finally, by Lemma 2.14.1 (ii), we get $(Q_T x_n)_n$ which has a convergent subsequence. Thus, $T \in \mathcal{DC}(X)$. $\hfill$ Q.E.D.

Proposition 3.8.3 *Let μ ba a non-zero scalar of $\mathbb{C}$ and let $T : \mathcal{D}(T) \subset X \longrightarrow X$ be a linear relation such that $\|Tx\| \leq |\mu| \|Q_T x\|$ for all $x \in \mathcal{D}(T)$. If $|\mu| < 1$, then T is a demicompact linear relation.* $\hfill \Diamond$

Proof. Since $\|Tx\| \leq |\mu| \|Q_T x\|$, then $\|Q_T x\| - \|Tx\| \geq (1 - |\mu|) \|Q_T x\|$. Hence,

$$\|Q_T x\| \quad \leq \quad \frac{\|Q_T(I - T)x\|}{1 - |\mu|}. \quad (3.75)$$

Now, take $(x_n)_n$ a bounded sequence of $\mathcal{D}(T)$ such that $Q_T(I - T)x_n \to y$. Applying (3.75), we get $\|Q_T(x_n - y)\| \to 0$. Thus, $(Q_T x_n)_n$ has a convergent subsequence. $\hfill$ Q.E.D.

Proposition 3.8.4 *Let $T, T_0 : \mathcal{D}(T) = \mathcal{D}(T_0) \subset X \longrightarrow Y$ be a linear relation and $S : \mathcal{D}(S) \subset X \longrightarrow Y$ be a closed linear relation with $\mathcal{D}(T) \subset \mathcal{D}(S)$ and $S(0) \subset T_0(0) \subset T(0)$. If T_0 is S-demicompact and there exists $a, b \in \mathbb{C}$ such that $|b| < 1$ and for all $x \in \mathcal{D}(T)$*

$$\|Tx - T_0x\| \leq |a| \|Sx - Tx\| + |b| \|Sx - T_0x\|,$$

then T is an S-demicompact linear relation. $\diamondsuit$

Proof. Since

$$
\begin{aligned}
\|Sx - T_0x\| - \|Sx - Tx\| &\leq \|Tx - T_0x + Sx - Sx\| \\
&= \|Tx - T_0x\| \text{ (as Proposition 2.3.4)} \\
&\leq |a| \|Sx - Tx\| + |b| \|Sx - T_0x\|,
\end{aligned}
$$

then

$$(1 - |b|) \|Sx - T_0x\| \leq (1 + |a|) \|Sx - Tx\|.$$

Hence,

$$\|Sx - T_0x\| \leq \left(\frac{1 + |a|}{1 - |b|} \right) \|Sx - Tx\|. \tag{3.76}$$

Now, take $(x_n)_n$ a bounded sequence of $\mathcal{D}(T)$ such that $Q_T(S - T)x_n \to y$. Applying (3.76), we get $\|Q_{T_0}(S - T_0)(x_n - y)\| \to 0$. So, by using the fact that T_0 is S-demicompact, we have $Q_{T_0}(S - T_0)x_n \to y$. We obtain that $(Q_{T_0}Sx_n)_n$ has a convergent subsequence. On the other side, we have

$$
\begin{aligned}
\|Q_T Sx_n\| &= \operatorname{dist}(T(0), Sx_n) \\
&\leq \operatorname{dist}(T_0(0), Sx_n) \\
&= \|Q_{T_0} Sx_n\|.
\end{aligned}
$$

So, $(Q_T Sx_n)_n$ has a convergent subsequence.   Q.E.D.

Proposition 3.8.5 *Let $T, T_0 : \mathcal{D}(T) = \mathcal{D}(T_0) \subset X \longrightarrow Y$ be a linear relation and $S : \mathcal{D}(S) \subset X \longrightarrow Y$ be a continuous and a closed linear relation with $\mathcal{D}(T) \subset \mathcal{D}(S)$ and $S(0) \subset T_0(0) \subset T(0)$. If T is continuous and T_0 is S-demicompact and there exists $a, b \in \mathbb{C}$ such that $|a| > 1$ and for all $x \in \mathcal{D}(T)$*

$$\|Tx - T_0x\| \geq |a| \|Sx - T_0x\| - |b| \|Sx - Tx\|,$$

then T is an S-demicompact linear relation. $\diamondsuit$

Proof. Since

$$\begin{aligned} \|Tx - T_0x\| &= \|Tx - T_0x + Sx - Sx\| \\ &\leq \|Sx - T_0x\| + \|Sx - Tx\|, \end{aligned}$$

then

$$|a| \|Sx - T_0x\| - |b| \|Sx - T_0x\| \leq \|Sx - Tx\| + \|Sx - Tx\|.$$

Hence,

$$(|a| - 1)\|Sx - T_0x\| \leq (1 + |b|)\|Sx - Tx\|.$$

So,

$$\|Sx - T_0x\| \leq \left(\frac{1 + |b|}{|a| - 1}\right) \|Sx - Tx\|. \tag{3.77}$$

Now, take $(x_n)_n$ a bounded sequence of $\mathcal{D}(T)$ such that $Q_T(S - T)x_n \to y$. Applying (3.77), we get $\|Q_{T_0}(S - T_0)(x_n - y)\| \to 0$. So, by using the fact that T_0 is S-demicompact, we have $Q_{T_0}(S - T_0)x_n \to y$. Thus, we obtain $(Q_{T_0}Sx_n)_n$ has a convergent subsequence. On the other side, we have

$$\begin{aligned} \|Q_T Sx_n\| &= \operatorname{dist}(T(0), Sx_n) \\ &\leq \operatorname{dist}(T_0(0), Sx_n) \\ &= \|Q_{T_0}Sx_n\|. \end{aligned}$$

Finally, we get $(Q_T Sx_n)_n$ has a convergent subsequence.   Q.E.D.

Proposition 3.8.6 *Let α, μ, $k \in \mathbb{C}$ such that $\mu \neq 0$. Let $T : \mathcal{D}(T) \subset X \longrightarrow Y$ be a continuous linear relation and $S : \mathcal{D}(S) \subset X \longrightarrow Y$. If T is a $(|k| - \overline{T(0)})$-set-contraction, then $|\alpha|T$ is μS-demicompact for each $|\alpha k| < 1$.* ◇

Proof. Let $(x_n)_n$ be a bounded sequence of $\mathcal{D}(T)$ such that $Q_{|\alpha|T}\mu Sx_n - Q_{|\alpha|T}|\alpha|Tx_n \to y$. Then,

$$Q_{|\alpha|T}\mu Sx_n = Q_{|\alpha|T}(\mu Sx_n - |\alpha|Tx_n) + Q_{|\alpha|T}|\alpha|Tx_n. \tag{3.78}$$

If $\gamma(\{Q_{|\alpha|T}\mu Sx_n\}) \neq 0$, then by using (3.78), we have

$$\begin{aligned} \gamma(\{Q_{|\alpha|T}\mu Sx_n\}) &\leq \gamma(\{Q_{|\alpha|T}(\mu Sx_n - |\alpha|Tx_n)\}) + \gamma(\{Q_{|\alpha|T}|\alpha|Tx_n\}), \\ &\leq |\alpha k| \gamma(\{Q_{|\alpha|T}x_n\}), \\ &< \gamma(\{Q_{|\alpha|T}x_n\}). \end{aligned}$$

Hence, the result is not accurate. It follows that $\gamma(\{Q_{|\alpha|T}\mu Sx_n\}) = 0$. So, $(Q_{|\alpha|T}\mu Sx_n)_n$ is relatively compact.   Q.E.D.

An immediate consequence of Proposition 3.8.6 is the following corollary:

Corollary 3.8.1 *Let μ, $k \in \mathbb{C}$ such that $\mu \neq 0$ and $k \neq -1$. Let $T : \mathcal{D}(T) \subset X \longrightarrow Y$ be a continuous linear relation and $S : \mathcal{D}(S) \subset X \longrightarrow Y$. If T is a $(|k| - \overline{T(0)})$-set-contraction, then $\frac{1}{|1+k|}T$ is μS-demicompact.*   ◇

Proposition 3.8.7 *Let α, μ, $k \in \mathbb{C}$ such that $\mu \neq 0$. Let $T : \mathcal{D}(T) \subset X \longrightarrow Y$ be a continuous linear relation and $S : \mathcal{D}(S) \subset X \longrightarrow Y$. If T is a $(|k|-\overline{T(0)})$-set-contraction, then $|\alpha|T$ is μS-demicompact for each $|\alpha k| < 1$.*   ◇

Proof. Let $(x_n)_n$ be a bounded sequence of $\mathcal{D}(T)$ such that $Q_{|\alpha|T}\mu Sx_n - Q_{|\alpha|T}|\alpha|Tx_n \to y$. Then,

$$Q_{|\alpha|T}\mu Sx_n = Q_{|\alpha|T}(\mu Sx_n - |\alpha|Tx_n) + Q_{|\alpha|T}|\alpha|Tx_n. \quad (3.79)$$

If $\gamma(\{Q_{|\alpha|T}\mu Sx_n\}) \neq 0$, then by using (3.79), we have

$$\begin{aligned}
\gamma(\{Q_{|\alpha|T}\mu Sx_n\}) &\leq \gamma(\{Q_{|\alpha|T}(\mu Sx_n - |\alpha|Tx_n)\}) + \gamma(\{Q_{|\alpha|T}|\alpha|Tx_n\}), \\
&\leq |\alpha k|\gamma(\{Q_{|\alpha|T}x_n\}), \\
&< \gamma(\{Q_{|\alpha|T}x_n\}).
\end{aligned}$$

Hence, the result is not accurate. It follows that $\gamma(\{Q_{|\alpha|T}\mu Sx_n\}) = 0$. So, $(Q_{|\alpha|T}\mu Sx_n)_n$ is relatively compact.   Q.E.D.

Theorem 3.8.1 *Let $\mu \in \mathbb{C}^*$, $\alpha \in (0,1)$, let $T : \mathcal{D}(T) \subset X \longrightarrow X$ be a linear relation and $S : \mathcal{D}(S) \subset X \longrightarrow X$ be a continuous linear relations. If there is $m \in \mathbb{N}^*$ such that $(Q_T\alpha T)^m$ is $Q_T\mu S$-compact and $Q_S - Q_T$ is compact, then αT is μS-demicompact.*   ◇

Proof. Let us consider the various cases for m:

<u>First case</u> : For $m = 1$. Using Lemma 3.8.2, we noticed that αT is μS-demicompact for each $\alpha \in (0,1)$.

<u>Second case</u> : For $m \in \mathbb{N}^* \setminus \{1\}$. Let $(x_n)_n$ be a bounded sequence of $\mathcal{D}(T)$ such that

$$y_n = Q_T\mu Sx_n - Q_T\alpha Tx_n \to y.$$

Then,

$$\sum_{k=0}^{m-1}(Q_T\alpha T)^k Q_T\mu S x_n$$

$$= \sum_{k=0}^{m-1}(Q_T\alpha T)^k y_n + \sum_{k=0}^{m-1}(Q_T\alpha T)^{k+1} x_n,$$

$$= \sum_{k=0}^{m-1}(Q_T\alpha T)^k y_n + \sum_{k=0}^{m-2}(Q_T\alpha T)^{k+1} x_n + (Q_T\alpha T)^m x_n,$$

$$= \sum_{k=0}^{m-1}(Q_T\alpha T)^k y_n + \sum_{k=0}^{m-2}(Q_T\alpha T)^{k+1} x_n + (Q_T\alpha T)^m x_n. \quad (3.80)$$

On the other hand,

$$\sum_{k=0}^{m-1}(Q_T\alpha T)^k Q_T\mu S x_n = Q_T\mu S x_n + \sum_{k=0}^{m-2}(Q_T\alpha T)^{k+1} Q_T\mu S x_n. \quad (3.81)$$

Then, by using both (3.80) and (3.81), we have

$$Q_T\mu S x_n = \sum_{k=0}^{m-1}(Q_T\alpha T)^k y_n + \sum_{k=0}^{m-2}(Q_T\alpha T)^{k+1}(Q_S - Q_T)x_n + (Q_T\alpha T)^m x_n.$$

$$(3.82)$$

Since $Q_T T Q_T$ and $(Q_T T)^n Q_T\mu S$ are single valued linear operator for all $n \geq 1$. We get $\sum_{k=0}^{m-2}(Q_T\alpha T)^{k+1} Q_{\alpha T}\mu S x_n$ is single valued. Then, by using (3.82), we have

$$\gamma(Q_T\mu S x_n) \leq \sum_{k=0}^{m-1}\overline{\gamma}((Q_T\alpha T)^k)\gamma(\{y_n\}) + \overline{\gamma}((Q_T\alpha T)^m)\gamma(\{x_n\}) +$$

$$\sum_{k=0}^{m-2}\overline{\gamma}((Q_T\alpha T)^{k+1})\overline{\gamma}(Q_S - Q_T)\gamma(\{x_n\})$$

$$= 0.$$

We conclude that $\gamma(\{Q_T\mu S x_n\}) = 0$, Hence, $(Q_T\mu S x_n)_n$ is relatively compact, then there is a convergent subsequence of $(Q_T\mu S x_n)_n$.   Q.E.D.

An immediate consequence of Theorem 3.8.1.

Corollary 3.8.2 *Let α, μ, $k \in \mathbb{C}$ such that $\mu \neq 0$. Let $T : \mathcal{D}(T) \subset X \longrightarrow X$ and $S : \mathcal{D}(S) \subset X \longrightarrow X$ be a continuous linear relations. Then,*

(i) If $m \in \mathbb{N}^*$, $k \neq -1$, $(Q_T T)^m$ is μS-compact and $Q_S - Q_T$ is compact, then $\frac{1}{|1+k|}T$ is $Q_T\mu S$-demicompact.

(ii) If $(Q_T|\alpha|T)^m$ is $Q_T\mu S$-compact for each $\alpha \in (0,1)$ and $m > 0$, then $|\alpha|T$ is μS-demicompact for each $\alpha \in (0,1)$.

(iii) If $m \in \mathbb{N}^*$, $k \geq 0$, $\overline{\gamma}((Q_T|\alpha|T)^m) \leq k$ and $Q_S - Q_T$ is compact, then $|\alpha|T$ is μS-demicompact for each $0 < \alpha^m k < 1$.

(iv) If $m \in \mathbb{N}^*$, $k \geq 0$, $\overline{\gamma}(T^m) \leq k$ and $Q_S - Q_T$ is compact, then $\frac{1}{1+k}T$ is μS-demicompact. ◇

3.8.2 Fredholm Theory by Means Relatively Demicompact Linear Relations

In this subsection, we present some results on Fredholm and upper semi-Fredholm linear relations involving relatively demicompact linear relations.

Proposition 3.8.8 [70, Theorem 2.3] *Let $T : \mathcal{D}(T) \subset X \longrightarrow X$, $S : \mathcal{D}(S) \subset X \longrightarrow X$ be densely defined closed linear operators with $\mathcal{D}(T) \subset \mathcal{D}(S)$ such that $S - T$ is closed. If T is S-demicompact, then $S - T$ is an upper semi-Fredholm operator.* ◇

Theorem 3.8.2 *Let $T : \mathcal{D}(T) \subset X \longrightarrow Y$ be a closed linear relation and $S : \mathcal{D}(S) \subset X \longrightarrow Y$ be a continuous and a closed linear relation with $\mathcal{D}(T) \subset \mathcal{D}(S)$ and $S(0) \subset T(0)$. If $Q_S - Q_T$ is compact, then $S - T \in \Phi_+(X, Y)$ if and only if, T is S-demicompact.* ◇

Proof. By using Proposition 2.7.2, we have $S - T$ is closed. If $S - T \in \Phi_+(X, Y)$, then by using Lemma 2.14.3 (i), we get

$$Q_{S-T}(S - T) = Q_T(S - T) \in \Phi_+(X, Y/T(0)). \tag{3.83}$$

In view of Theorem 2.14.1, there exist a bounded linear operator A and a bounded finite rank projection operator P such that

$$AQ_T(S - T) = I_{\mathcal{D}(T)} - P. \tag{3.84}$$

Let $(x_n)_n$ be a bounded sequence of $\mathcal{D}(T)$ such that $(Q_T(S-T)x_n)_n$ converges to $Q_T x$. Then, $(AQ_T(S - T)x_n)_n$ converges to $AQ_T x$. So, by (3.84), we have $(x_n - Px_n)_n$ converges to $AQ_T x$. Since P is compact single valued, then $(x_n)_n$ has a convergent subsequence. Finally, $(Q_T Sx_n)_n$ has a convergent subsequence. Conversely, let T be S-demicompact and $Q_S - Q_T$ be a compact linear relation. Using Lemma 3.8.2, we find that $Q_T T$ is $Q_T S$-demicompact. The latter implies and using Proposition 3.8.8, we obtain $Q_T S - Q_T T$ which is an upper semi-Fredholm single valued. On the other side, since $S(0) \subset T(0)$, then

$$Q_T S - Q_T T = Q_T(S - T).$$

By using (3.83), we have $Q_{S-T}(S-T)$ is an upper single valued linear operator semi-Fredholm. Using Lemma 2.14.3 (i), we obtain $S - T$ is an upper semi-Fredholm relation. Q.E.D.

Proposition 3.8.9 *Let $\mu \in \mathbb{C}^*$ and let $T : \mathcal{D}(T) \subset X \longrightarrow Y$ and $S : \mathcal{D}(S) \subset X \longrightarrow Y$ be two densely defined linear relations with $S(0) \subset T(0)$ and $\mathcal{D}(T) \subset \mathcal{D}(S)$. If $\mu S - T \in \mathcal{F}_+(X,Y)$, then T is μS-demicompact.* ◇

Proof. Suppose that $\mu S - T \in \mathcal{F}_+(X,Y)$, then by using Propositions 2.14.1 and 2.14.2, we have

$$Q_{\mu S-T}(\mu S - T) = Q_T(\mu S - T) \in \mathcal{F}_+(X, Y/\overline{T(0)}).$$

Then, there exists a bounded linear operator A and a bounded finite rank projection operator P such that

$$A Q_T(\mu S - T) = I_{\mathcal{D}(T)} - P. \tag{3.85}$$

Let $(x_n)_n$ be a bounded sequence of $\mathcal{D}(T)$ such that $(Q_T(\mu S - T)x_n)_n$ converges to $Q_T x$. Then, $(A Q_T(\mu S - T)x_n)_n$ converges to $A Q_T x$. So, by using (3.85), we have $(x_n - P x_n)_n$ converges to $A Q_T x$. Since P is compact, then $(x_n)_n$ has a convergent subsequence. Finally, $(Q_T \mu S x_n)_n$ has a convergent subsequence. Q.E.D.

Proposition 3.8.10 *Let $\mu \in \mathbb{C}^*$ and let $T : \mathcal{D}(T) \subset X \longrightarrow Y$ and $S : \mathcal{D}(S) \subset X \longrightarrow Y$ be two densely defined linear relations with $S(0) \subset T(0)$, $\mathcal{D}(T) \subset \mathcal{D}(S)$ and $\overline{T(0)}$ is compact. If T is a μS-demicompact linear relation, then $\mu S - T \in \mathcal{F}_+(X,Y)$.* ◇

Proof. We suppose that $\mu S - T \notin \mathcal{F}_+(X,Y)$. Then, $\mu S - T$ has a singular sequence. Then, there exists a sequence $(x_n)_n \in \mathcal{D}(T)$ of norm one such that $(x_n)_n$ has no Cauchy subsequence and $\lim_{n\to\infty} \|(\mu S - T)x_n\| = 0$. Since

$$\begin{aligned}
\|(\mu S - T)x_n\| &= \|Q_{\mu S-T}(\mu S - T)x_n\|, \\
&= \|Q_T(\mu S - T)x_n\| \text{ (since } S(0) \subset T(0)),
\end{aligned}$$

then $(Q_T(\mu S - T)x_n)_n$ converge to 0. Since T is μS-demicompact, then $(Q_{T\mu}Sx_n)_n$ has a convergent subsequence. Using the fact that $\overline{T(0)}$ is compact, then $(x_n)_n$ has a convergent subsequence. This contradicts the fact $(x_n)_n$ has no Cauchy subsequence. Finally, $\mu S - T \in \mathcal{F}_+(X, Y)$.   Q.E.D.

Theorem 3.8.3 *Let $\mu \in \mathbb{C}^*$ and let $T : \mathcal{D}(T) \subset X \longrightarrow Y$ be a densely defined closed linear relation and $S : \mathcal{D}(S) \subset X \longrightarrow Y$ be a continuous and densely defined closed linear relations with $\mathcal{D}(T) \subset \mathcal{D}(S)$ and $S(0) \subset T(0)$ such that $|\mu|S - T$ is closed. If T is $|\mu|S$-demicompact and $Q_{|\mu|S} - Q_T$ is compact, then $|\mu|S - T$ is an upper semi-Fredholm relation.* ◇

Proof. Let T be a $|\mu|S$-demicompact and $Q_{|\mu|S} - Q_T$ be a compact linear relation. Using Lemma 3.8.2, we found that $Q_T T$ is $Q_T |\mu|S$-demicompact. The latter implies $Q_T |\mu|S - Q_T T$ is an upper semi-Fredholm single valued. On the other hand,

$$Q_{|\mu|S-T}(|\mu|S - T) = Q_T(|\mu|S - T).$$

We noticed that $Q_{|\mu|S-T}(|\mu|S - T)$ is an upper single valued linear operator semi-Fredholm. Hence, $|\mu|S - T$ is an upper semi-Fredholm relation. Q.E.D.

Proposition 3.8.11 *Let $\alpha \in (0, 1)$, $\mu \in \mathbb{C}^*$ and let $T : \mathcal{D}(T) \subset X \longrightarrow Y$ and $S : \mathcal{D}(S) \subset X \longrightarrow Y$ be a continuous and densely defined closed linear relations with $\mathcal{D}(T) \subset \mathcal{D}(S)$ and $S(0) \subset T(0)$ such that S is nonzero and is relatively bounded with respect to T with T-bound < 1. If $|\mu|S - \alpha T$ is closed, $Q_{|\mu|S} - Q_T$ is compact and αT is $|\mu|S$-demicompact, then $|\mu|S - T$ is a Fredholm linear relation and $i(|\mu|S - T) = i(|\mu|S)$.* ◇

Proof. Let αT be a $|\mu|S$-demicompact and $Q_{|\mu|S} - Q_T$ be a compact linear relation. Applying Lemma 3.8.2, we get $Q_T T$ is $Q_T |\mu|S$-demicompact. Hence, $Q_T |\mu|S - Q_T T$ is Fredholm linear relation and $i(Q_T |\mu|S - Q_T T) = i(Q_T |\mu|S) = i(|\mu|S)$. On the other hand,

$$Q_{|\mu|S-T}(|\mu|S - T) = Q_T(|\mu|S - T).$$

So, $Q_{|\mu|S-T}(|\mu|S - T)$ is a Fredholm linear relation. Hence, $|\mu|S - T$ is a Fredholm linear relation and $i(|\mu|S - T) = i(|\mu|S)$.   Q.E.D.

An immediate consequence of Proposition 3.8.6 is the following corollary:

Corollary 3.8.3 *Let μ, $k \in \mathbb{C}$ such that $\mu \neq 0$ and $k \neq -1$. Let $T : \mathcal{D}(T) \subset X \longrightarrow Y$ be a continuous linear relation and $S : \mathcal{D}(S) \subset X \longrightarrow Y$. If T is a $(|k| - \overline{T(0)})$-set-contraction, then $\frac{1}{|1+k|}T$ is μS-demicompact.* ◇

3.9   Essentially Semi Regular Linear Relations

The notion of essentially semi regularity operators amongst the various concepts of regularity originated by the classical treatment of perturbation theory owed to T. Kato and was studied by many authors, for instance, we cite [4, 5, 18, 65, 80] and [95]. We remark that all the above authors considered only the case of bounded linear operators. It is the purpose of this section to consider these class of essentially semi regularity in the more general setting of linear relations in Hilbert spaces. Many properties of essentially semi regularity for the case of linear operators remain be valid in the context of linear relations, sometimes under supplementary conditions.

3.9.1   Some Proprieties of Essentially Semi Regular Linear Relations

Lemma 3.9.1 *Let X be a Hilbert space and $A \in BR(X)$ with closed range such that $\rho(A) \neq \emptyset$. If for every $n \in \mathbb{N}$, there exists a finite dimensional subspace $F \subset X$ such that $N(A) \subset R(A^n) + F$, then $R(A^n)$ is closed for all $n \in \mathbb{N}$.*   ◇

Proof. Proceeding by induction. For $n = 1$, it is clear by the assumption. We may assume that $F \subset N(A)$. Let $R(A^n)$ is closed, then $R(A^n) + F$ is closed since $\dim F < \infty$. Since $\rho(A) \neq \emptyset$, then by Lemma 2.18.2, we have A is closed. Hence, $A_0 = A_{|R(A^n)+F}$ is also closed. Since $F \subset N(A)$, then $0 < \gamma(A) < \gamma(A_0)$. So, A_0 is open. Consequently, by using Theorem 2.7.2, we infer that $R(A_0)$ is closed. Moreover, since $F \subset N(A)$ and $N(A) \subset R(A^n)+F$, then

$$A(F) + R(A^{n+1}) \subset A(N(A)) + R(A^{n+1}) \subset A(F) + R(A^{n+1}).$$

Therefore, $A(N(A)) + R(A^{n+1}) = R(A^{n+1}) + A(F)$. Hence,

$$
\begin{aligned}
R(A_0) &= A(F + R(A^n)) \\
&= A(F) + A(R(A^n)) \\
&= A(F) + R(A^{n+1}) \\
&= A(N(A)) + R(A^{n+1}) \\
&= AA^{-1}(0) + R(A^{n+1}) \quad \text{since } N(A) = A^{-1}(0) \\
&= A(0) + R(A^{n+1}) \quad \text{by Remark 2.3.3} \\
&= R(A^{n+1}), \quad (\text{as } A(0) \subset R(A^{n+1})).
\end{aligned}
$$

So, $R(A^{n+1})$ is closed. Thus, $R(A^n)$ is closed for all $n \in \mathbb{N}$. Q.E.D.

Theorem 3.9.1 *Let X be a Hilbert space and $A \in B\mathcal{R}(X)$. If A is essentially semi regular, then A^* is essentially semi regular.* ◇

Proof. Let A be essentially semi regular, then $R(A)$ is closed. It follows from Proposition 2.8.1 (iv) and (v) that $A^* \in B\mathcal{R}(X^*)$ and $R(A^*)$ is closed. Hence, since A is essentially semi regular and by using Lemma 3.9.1, we have $R(A^n)$ is closed. So, $R(A^{n*})$ is closed (see Proposition 2.8.1 (v)). Let n, $m \in \mathbb{N}$. In view of both Proposition 2.8.1 and Lemma 2.8.2, we infer that

$$
\begin{aligned}
R^\infty(A^*) &= \bigcap_{n \in \mathbb{N}} R(A^{*n}) \\
&= \bigcap_{n \in \mathbb{N}} R(A^{n*}) \text{ by Lemma 2.8.2} \\
&= \left(\bigcap_{n \in \mathbb{N}} N(A^n) \right)^\perp \\
&= (N^\infty(A))^\perp.
\end{aligned}
$$

Since

$$
\begin{aligned}
(N^\infty(A))^\perp &\supset (R(A) + F)^\perp \\
&= R(A)^\perp \bigcap F^\perp \\
&= N(A^*) \bigcap F^\perp,
\end{aligned}
$$

and $\operatorname{codim}(F^\perp) < \infty$, then $N(A^*) \subset_e R^\infty(A^*)$. This implies that A^* is essentially semi regular. Q.E.D.

If X is a Hilbert space, then the class of semi regular coincides with the class of all quasi-Fredholm linear relations of degree 0, introduced in Ref. [74]. In order to study the Kato decomposable linear relations in Hilbert spaces, we shall establish an analogous result when A is essentially semi regular linear relation in a Hilbert space which will be crucial to prove the main theorems of the next.

Theorem 3.9.2 *Let X be an infinite dimensional Hilbert space and $A \in B\mathcal{R}(X)$ everywhere defined. Then, the following statements are equivalent :*

(i) $N^\infty(A) \subset_e R^\infty(A)$ *and $R(A)$ is closed.*
(ii) $N^\infty(A) \subset_e R(A)$ *and $R(A)$ is closed.*
(iii) $N(A) \subset_e R^\infty(A)$ *and $R(A)$ is closed.*
(iv) *There exists a decomposition $X = X_1 \oplus X_2$ with the properties that*

$A_{|X_1} \subset X_1$, $A_{|X_2} \subset X_2$, $\dim X_1 < \infty$, $A_{|X_2}$ *is a semi regular linear relation and $A_{|X_1}$ is a nilpotent bounded operator of degree d.* ◇

Proof. The implications $(i) \Longrightarrow (ii)$ and $(ii) \Longrightarrow (iii)$ are clear because $N(A) \subset N^\infty(A)$ and $R^\infty(A) \subset R(A)$.

$(iii) \Longrightarrow (iv)$ To prove this, we first show that there exists $d \in \mathbb{N}$ such that

(a) $N(A) \cap R(A^d) = N(A) \cap R(A^n) \ \forall \ n \geq d$.
(b) $N(A) \cap R(A^d)$ is closed.
(c) $N(A) + R(A^d)$ is closed in X.

Since $N(A) \subset_e R^\infty(A)$, then

$$\dim \left(\frac{N(A)}{N(A) \cap R^\infty(A)} \right) = \dim \left(\frac{R(A) + N^\infty(A)}{R(A)} \right) < \infty.$$

So, we can deduce the codimension of $N(A) \cap R(A^m)$ in $N(A)$ is finite independent of m. Therefore, $(N(A) \cap R(A^n))_n$ is a decreasing sequence and has therefore, limit. Hence, there is some smallest $d \in \mathbb{N}$ for which $N(A) \cap R(A^d) = N(A) \cap R(A^{d+m})$ for all nonnegative integer m, and thus (a) is satisfied. It is clear that $N(A) \cap R(A^d)$ and $N(A) + R(A^d)$ are closed and thus (b) and (c) are satisfied. Now, proceeding exactly as in the proof of Theorem 2.15.3, we can construct two closed subspaces X_1 and X_2 of X verifying the following properties:

(a1) $X = X_1 \oplus X_2$ with $\dim X_1 < \infty$.
(b1) $A = A_{|X_1} \oplus A_{|X_2}$.
(c1) $A_{|X_2}$ is a regular linear relation
(d1) $A_{|X_1}$ is a nilpotent bounded operator of degree d.

Then, $(iii) \Longrightarrow (iv)$.

We prove $(iv) \Longrightarrow (i)$. Since $A = A_{|X_1} \oplus A_{|X_2}$, then $A^n = (A_{|X_1})^n \oplus (A_{|X_2})^n$ and $R^\infty(A) = R^\infty(A_{|X_1}) \oplus R^\infty(A_{|X_2})$. Since $A_{|X_1}$ is nilpotent of degree d, we obtain that $R^\infty(A) = R^\infty(A_{|X_2})$. The semi regularity of $A_{|X_2}$ and the fact that $R(A) = R(A_{|X_1}) \oplus R(A_{|X_2})$ entails that $R(A)$ is closed and $N(A_{|X_2}^n) \subset R^\infty(A_{|X_2}) = R^\infty(A)$ for every $n \in \mathbb{N}$. By assumption there is $d \in \mathbb{N}$ such that $A_{|X_1}^d = 0$. Thus, for every $n \geq d$, we have

$$N(A^n) = N \oplus N((A_{|X_2})^n) \subset R^\infty(A) \oplus N.$$

Hence, $N^\infty(A) \subset_e R^\infty(A)$, since N is finite-dimensional. Then, $(iv) \Longrightarrow (i)$. This completes the proof. Q.E.D.

Theorem 3.9.3 *Let X be a Hilbert space, $A \in C\mathcal{R}(X)$ be essentially semi regular and $\lambda \in \mathbb{D}(0, \varepsilon)\backslash\{0\}$. Then,*

(i) $A \in \Phi_+(X)$ if and only if, $\lambda - A \in \Phi_+(X)$.
(ii) $A \in \Phi_-(X)$ if and only if, $\lambda - A \in \Phi_-(X)$.
(iii) $A \in \Phi(X)$ if and only if, $\lambda - A \in \Phi(X)$. $\diamondsuit$

Proof. (i) Let $\lambda \in \mathbb{D}(0, \varepsilon)\backslash\{0\}$ such that $\lambda - A \in \Phi_+(X)$ and let $A_0 = A_{|R^\infty(A)}$ be the restriction of A to $R^\infty(A)$ and $\lambda_0 \in R^\infty(A)$. Since A is essentially semi regular, then by using Lemma 3.9.1, we have $R(A^n)$ is closed for every $n \in \mathbb{N}$. Thus, $R^\infty(A)$ is closed and A_0 is closed. It follows from Lemma 2.5.5 that A_0 is surjective and therefore, $A_0 \in \Phi_-(X)$. Applying Theorem 2.14.4, we get $\lambda_0 - A_0 \in \Phi_-(X)$ with $\beta(\lambda_0 - A_0) \le \beta(A_0) = 0$ and $i(\lambda_0 - A_0) = i(A_0)$ for all $|\lambda| < \varepsilon$. Moreover, by using Lemma 2.5.1, we have

$$N(\lambda_0 - A_0) = N(\lambda - A) \bigcap R^\infty(A) = N(\lambda - A).$$

Hence,

$$
\begin{aligned}
\alpha(\lambda - A) &= i(\lambda_0 - A_0) \\
&= i(A_0) \\
&= \alpha(A_0) \\
&= \dim(N(A) \bigcap R^\infty(A))
\end{aligned}
$$

for all $|\lambda| < \varepsilon$. Since A is essentially semi regular and $\lambda - A \in \Phi_+(X)$, then

$$\dim\left(N(\lambda - A)\right) = \dim\left(N(A) \bigcap R^\infty(A)\right) < \infty.$$

On the other hand, by using Theorem 3.9.2, there exists a decomposition $X = X_1 \oplus X_2$ such that $A_{|X_2}$ is semi regular linear relation, $A_{|X_1}$ is a nilpotent bounded operator and $\dim X_1 < \infty$. Then,

$$N(A) = N(A_{|X_1}) + N(A_{|X_2}) = N(A_{|X_1}) + N(A) \bigcap R^\infty(A).$$

By assumption, $N(A_{|X_1})$ is finite-dimensional and since $\dim N(A) \bigcap R^\infty(A) < \infty$, then $N(A)$ is finite-dimensional. Furthermore, $R(A) = R(A_{|X_1}) + R(A_{|X_2})$ is closed since $A_{|X_2}$ is semi regular and $R(A_{|X_1})$ is finite dimensional. Hence, $A \in \Phi_+(X)$. The opposite implication follows from Theorem 2.14.4.

(ii) Assume that $\lambda - A \in \Phi_-(X)$, then $R(\lambda - A)$ is closed. By using Proposition 2.8.1, we have $R(\lambda - A) = N(\lambda - A^*)^\perp$ for all $\lambda \in \mathbb{D}(0, \varepsilon)\backslash\{0\}$. Thus, by part

(i), we have

$$
\begin{aligned}
\beta(\lambda - A) = \alpha(\lambda - A^*) \ &= \ \dim(N(A^*) \cap R(A^{*d})) \\
&= \ \operatorname{codim} \left[N(A^*) \cap R(A^{*d}) \right]^{\perp} \\
&= \ \operatorname{codim} \left[N(A^*)^{\perp} + R(A^{*d})^{\perp} \right] \\
&= \ \operatorname{codim} \left[N(A^d) + R(A) \right]
\end{aligned}
$$

for all $0 < |\lambda| < \varepsilon$. Hence, $\operatorname{codim}\left(R(A) + N(A^d)\right)$ is finite. In view of Proposition 2.15.1, we have

$$
R(A) + N(A^d) = R(A_{|X_2}) \bigoplus X_1,
$$

which implies that $R(A_{|X_2}) + X_1$ is finitely codimensional. Since $\dim X_1 < \infty$, then $R(A_{|X_2})$ is finitely codimensional. Thus, $\beta(A) = \operatorname{codim} R(A) = \operatorname{codim} R(A_{|X_2}) < \infty$. By using the same reasoning as part (i), we have $R(A)$ is closed. Hence, $A \in \Phi_-(X)$. The opposite implication follows from Theorem 2.14.4.

The proof of (iii) is an immediate consequence of both (i) and (ii). Q.E.D.

3.9.2 Perturbation Results of Essentially Semi Regular Linear Relations

Lemma 3.9.2 *Let X be a Banach space, $n \in \mathbb{N}$, and $A \in C\mathcal{R}(X)$. Then,*

(i) *If A is semi regular with $\mathfrak{R}_c(A) = \{0\}$ and $\alpha(A) < \infty$, then $\alpha(A^n) = n\alpha(A)$.*

(ii) *If A is semi regular with finite codimensional range, then $\beta(A^n) = n\beta(A)$.*

$\diamondsuit$

Proof. (i) Since $\mathfrak{R}_c(A) = \{0\}$, then by using Lemma 2.5.6, we have

$$
\frac{N(A^n)}{N(A^{n-1})} \cong N(A) \cap R(A^{n-1}).
$$

Since A is semi regular, then it follows that $\dim N(A^n) = \dim N(A) + \dim N(A^{n-1})$. Thus, a successive repetition of this argument leads to $\alpha(A^n) = n\alpha(A)$.

(ii) Let n be a positive integer, then

$$
\frac{R(A^n)}{R(A^{n+1})} \cong \frac{X}{N(A^n) + R(A)}
$$

for every $n \in \mathbb{N}$. Since A is semi regular, then for every $n \in \mathbb{N}$, $N(A) \subset R(A^n)$, or equivalently, $N(A^n) \subset R(A)$. On the other hand, it follows from Lemma 2.3.5 (iv) that

$$\left(\frac{X}{R(A^{n-1})} \times \frac{R(A^{n-1})}{R(A^n)} \right) \cong \frac{X}{R(A^n)}.$$

Hence, $\mathrm{codim}(R(A^n)) = \mathrm{codim}(R(A^{n-1})) + \mathrm{codim}(R(A))$. By induction, we have $\beta(A^n) = n\beta(A)$ for all $n \in \mathbb{N}$.   Q.E.D.

Lemma 3.9.3 *Let X be a Hilbert space and $A \in C\mathcal{R}(X)$ be essentially semi regular. Then, there exists $\delta > 0$ such that*

(i) $\alpha(\lambda - A) = \alpha(A_{|X_2}) \leq \alpha(A)$ *for all* $0 < |\lambda| < \delta$.
(ii) $\beta(\lambda - A) = \beta(A_{|X_2}) \leq \beta(A)$ *for all* $0 < |\lambda| < \delta$.   ◇

Proof. (i) Since A is essentially semi regular, then there exists a decomposition $X = X_1 \oplus X_2$ with the properties that $\dim X_1 < \infty$, $A_{|X_2}$ is a semi regular linear relation and $A_{|X_1}$ is a nilpotent bounded operator of degree d $(A^d = 0)$. Then, $\lambda - A = (\lambda - A)_{|X_1} \oplus (\lambda - A)_{|X_2}$ for all $\lambda \in \mathbb{C}$. Therefore, $\alpha(\lambda - A) = \alpha((\lambda - A)_{|X_1}) + \alpha((\lambda - A)_{|X_2})$. Since $A_{|X_1}$ is a nilpotent bounded operator of degree d, then $(\lambda - A)_{|X_1}$ is invertible for all $\lambda \neq 0$ which implies that $\alpha((\lambda - A)_{|X_1}) = 0$. Since $A_{|X_2}$ is a semi regular linear relation, it follows from Lemma 3.9.2 that there exists a positive constant $\delta > 0$ such that $(\lambda - A)_{|X_2}$ is semi regular for all $0 < |\lambda| < \delta$. In view of Theorem 2.15.1, we have $\alpha((\lambda - A)_{|X_2}) = \alpha(A_{|X_2})$ for all $0 < |\lambda| < \delta$. Hence, $\alpha(\lambda - A) = \alpha((\lambda - A)_{|X_2}) = \alpha(A_{|X_2}) \leq \alpha(A)$ for all $0 < |\lambda| < \delta$.

(ii) Let A be essentially semi regular. Then, there exists a decomposition $X = X_1 \oplus X_2$ with the properties that $A_{|X_1} \subset X_1$, $A_{|X_2} \subset X_2$, $\dim X_1 < \infty$, $A_{|X_2}$ is a semi regular linear relation and $A_{|X_1}$ is a nilpotent bounded operator of degree d. Hence, $\lambda - A = (\lambda - A)_{|X_1} \oplus (\lambda - A)_{|X_2}$. Therefore, $\beta(\lambda - A) = \beta((\lambda - A)_{|X_1}) + \beta((\lambda - A)_{|X_2})$. Now, a same reasoning as the proof of (i) leads to the result.   Q.E.D.

Theorem 3.9.4 *Let X be a Hilbert space and $A \in C\mathcal{R}(X)$ be essentially semi regular linear relation. Then,*

(i) *If $\alpha(A) < \infty$ and $\mathfrak{R}_c(A) = 0$, then there exists a positive constant $\delta > 0$ such that*

$$\alpha(\lambda - A) = b.s.mult(A)$$

and if $\beta(A) < \infty$, then there exists a positive constant $\delta > 0$ such that

$$\beta(\lambda - A) = s.mult(A)$$

for all $0 < |\lambda| < \delta$. In particular, the following two functions

$$\lambda \longrightarrow s.mult(\lambda - A)$$

and

$$\lambda \longrightarrow b.s.mult(\lambda - A)$$

are constant on a neighborhood of the origin $\mathbb{D}(0, \delta) = \{\lambda \in \mathbb{C} \text{ such that } |\lambda| < \delta\}$.

(ii) When k is large enough, if $\alpha(A) < \infty$ and $\mathfrak{R}_c(A) = 0$, then

$$s.mult(A) = \dim\left(\frac{X}{R(A) + N(A^k)}\right) = \dim\left(\frac{X}{R(A) + \overline{N(A)^\infty}}\right)$$

and if $\beta(A) < \infty$, then

$$b.s.mult(A) = \dim\left(N(A)\bigcap R(A^k)\right) = \dim\left(N(A)\bigcap R(A)^\infty\right).$$

(iii) A is upper semi-Fredholm (respectively, lower semi-Fredholm) if and only if, $b.s.mult(A) < \infty$ (respectively, $s.mult(A) < \infty$). Moreover, if $b.s.mult(A) < \infty$ and $s.mult(A) < \infty$, then A is semi-Fredholm and

$$i(A) = b.s.mult(A) - s.mult(A). \qquad \diamond$$

Proof. *(i)* Since A is essentially semi regular, then by applying Theorem 3.9.2, there exists a decomposition $X = X_1 \oplus X_2$ with the properties that $A_{|X_1} \subset X_1$, $A_{|X_2} \subset X_2$, $\dim X_1 < \infty$, $A_{|X_2}$ is a semi regular linear relation and $A_{|X_1}$ is a nilpotent bounded operator of degree d. Let $\dim X_1 = n_0$, then by using Lemma 3.9.2, we have

$$
\begin{aligned}
b.s.mult(A) &= \lim_{k\to\infty}\left(\frac{\alpha(A^k)}{k}\right) \\
&= \lim_{k\to\infty}\left(\frac{\alpha(A^k_{|X_2})}{k}\right) + \lim_{k\to\infty}\left(\frac{\alpha(A^k_{|X_1})}{k}\right) \\
&= \lim_{k\to\infty}\left(\frac{\alpha(A^k_{|X_2})}{k}\right) + \lim_{k\to\infty}\left(\frac{n_0}{k}\right) \\
&= \lim_{k\to\infty}\left(\frac{k\alpha(A_{|X_2})}{k}\right) \\
&= \alpha(A_{|X_2}),
\end{aligned}
$$

and

$$\begin{aligned}
s.mult(A) &= \lim_{k\to\infty}\left(\frac{\beta(A^k)}{k}\right)\\
&= \lim_{k\to\infty}\left(\frac{\beta(A^k_{|X_1})}{k}\right) + \lim_{k\to\infty}\left(\frac{\beta(A^k_{|X_2})}{k}\right)\\
&= \lim_{k\to\infty}\left(\frac{\beta(A^k_{|X_2})}{k}\right) + \lim_{k\to\infty}\left(\frac{n_0}{k}\right)\\
&= \lim_{k\to\infty}\left(\frac{k\beta(A_{|X_2})}{k}\right)\\
&= \beta(A_{|X_2}).
\end{aligned}$$

By using Lemma 3.9.3, there exists a positive constant $\delta > 0$ such that

$$\alpha(\lambda - A) = \alpha(A_{|X_2}) = b.s.mult(A)$$

and

$$\beta(\lambda - A) = \beta(A_{|X_2}) = s.mult(A)$$

for all $0 < |\lambda| < \delta$.

(ii) Since A is essentially semi regular, then by applying Theorem 3.9.2, there exists a decomposition $X = X_1 \oplus X_2$ with the properties that $A_{|X_1} \subset X_1$, $A_{|X_2} \subset X_2$, $\dim X_1 < \infty$, $A_{|X_2}$ is a semi regular linear relation and $A_{|X_1}$ is a nilpotent bounded operator of degree d. By using Proposition 2.15.1 and in view of (i), we have

$$b.s.mult(A) = \alpha(A_{|X_2}) = \dim\left(N(A)\bigcap R(A^k)\right) = \dim\left(N(A)\bigcap R(A)^\infty\right).$$

Let $k \geq \dim X_1$, then

$$\begin{aligned}
R(A) + \overline{N(A)^\infty} &= R(A_{|X_1}) + R(A_{|X_2}) + \overline{N^\infty(A_{|X_1}) + N^\infty(A_{|X_2})}\\
&= R(A_{|X_1}) + R(A_{|X_2}) + \overline{N^\infty(A_{|X_2}) + X_1}\\
&= R(A_{|X_1}) + R(A_{|X_2}) + \overline{N^\infty(A_{|X_2})} + X_1\\
&= R(A_{|X_1}) + R(A_{|X_2}) + X_1\\
&= R(A_{|X_1}) + R(A_{|X_2}) + N(A^k_{|X_1}) + N(A^k_{|X_2})\\
&= R(A) + N(A^k).
\end{aligned}$$

Thus,

$$\frac{X}{R(A) + N(A^k)} = \frac{X}{R(A) + \overline{N(A)^\infty}}$$

for all $k \geq \dim X_1$. Consequently,

$$
\begin{aligned}
s.mult(A) &= \beta(A_{|X_2}) \\
&= \dim\left(\frac{X_2}{A(X_2)}\right) \\
&= \dim\left(\frac{X_2 \oplus X_1}{A(X_2) \oplus X_1}\right) \\
&= \dim\left(\frac{X}{R(A) + N(A^k)}\right) \\
&= \dim\left(\frac{X}{R(A) + \overline{N(A)^\infty}}\right).
\end{aligned}
$$

(iii) If $A \in \Phi_+(X)$, then $\dim N(A) < \infty$. This implies that $\dim N(A) \cap R(A) < \infty$. By using (ii), we have $b.s.mult(A) < \infty$. If $A \in \Phi_-(X)$, then $\beta(A) < \infty$. This implies that $\mathrm{codim}\,(N^\infty(A) + R(A)) < \infty$. Hence, by using (ii), we have $s.mult(A) < \infty$. To prove the converse, let A be essentially semi regular, then $R(A)$ is closed. By applying Theorem 3.9.2, there exists a decomposition $X = X_1 \oplus X_2$ such that $A_{|X_2}$ is semi regular linear relation, $A_{|X_1}$ is a nilpotent bounded operator and $\dim X_1 < \infty$. Suppose that $b.s.mult(A) < \infty$, then $\alpha(A_{|X_2}) < \infty$ and consequently, $\alpha(A) = \alpha(A_{|X_1}) + \alpha(A_{|X_2}) < \infty$. Since $R(A)$ is closed, then $A \in \Phi_+(X)$. Let $s.mult(A) < \infty$, then $\beta(A_{|X_2}) < \infty$. Hence, $\beta(A) = \beta(A_{|X_1}) + \beta(A_{|X_2}) < \infty$. Since $R(A)$ is closed, then $A \in \Phi_-(X)$. On the other hand, if $\alpha(A)$ and $\beta(A)$ are finite and $\mathfrak{R}_c(A) = 0$, then by using Lemma 3.9.2, we have $i(A^k) = \alpha(A^k) - \beta(A^k) = k\, i(A)$ and

$$
\begin{aligned}
i(A) &= \lim_{k\to\infty}\left(\frac{\alpha(A^k)}{k}\right) - \lim_{k\to\infty}\left(\frac{\beta(A^k)}{k}\right) \\
&= b.s.mult(A) - s.mult(A).
\end{aligned}
$$

This completes the proof.   Q.E.D.

Chapter 4

Essential Spectra and Essential Pseudospectra of a Linear Relations

In this chapter, we emphasize the strong connection between the spectral theory of closed linear relations and that of some closed linear operators. As a matter of fact, we develop a spectral theory for a certain class of linear operators, obtaining as consequences most of the main spectral properties of linear relations. Furthermore, we introduce and study the pseudospectra and the essential pseudospectra of linear relations. We start by giving the definition and we investigate the characterization and some properties of these pseudospectra. We end this chapter by gives some new results related to the pseudospectra and the essential pseudospectra of linear relations. We start by studying the stability of these pseudospectra and some characterization.

4.1 Characterization of the Essential Spectrum of a Linear Relations

Theorem 4.1.1 *Let X be a Banach space and let $T \in C\mathcal{R}(X)$. Then,*

(i) $\sigma_{e5}(T) = \sigma_{e4}(T) \bigcup \{\lambda \in \mathbb{C} \text{ such that } i(T - \lambda) \neq 0\}$.
(ii) $\sigma_{eap}(T) = \sigma_{e1}(T) \bigcup \{\lambda \in \mathbb{C} \text{ such that } i(T - \lambda) > 0\}$.
(iii) $\sigma_{e\delta}(T) = \sigma_{e2}(T) \bigcup \{\lambda \in \mathbb{C} \text{ such that } i(T - \lambda) < 0\}$. $\diamondsuit$

Proof. *(i)* Let $\lambda \notin \sigma_{e4}(T) \bigcup \{\lambda \in \mathbb{C} \text{ such that } i(T - \lambda) \neq 0\}$. Then, $T - \lambda \in \Phi(X)$ and $i(T - \lambda) = 0$. Hence, $n = \alpha(T - \lambda) = \beta(T - \lambda) < \infty$. So, there exists an everywhere defined single valued linear operator K with $\dim R(K) \leq \alpha(T - \lambda)$ such that $T - \lambda - K$ is bijective. Furthermore, see the proof of Lemma 3.5.3 (iii), K is defined by

$$Kx := \sum_{i=1}^{n} x_i'(x)y_i, \quad x \in X,$$

where $\{x_1, \ldots, x_n\}$ is a basis of $N(T - \lambda)$ and $x_1', \ldots, x_n'$ are linear functionals such that $x_i'(x_j) = \delta_{ij}$. Choose $y_1, \ldots, y_n \in X$ such that $[y_1], \ldots, [y_n] \in X/R(\lambda - T)$ are linearly independent $(n = \alpha(T - \lambda) = \beta(T - \lambda))$. Hence, it is clear that K is a bounded finite rank operator. So, $K \in \mathcal{KR}(X)$ and $\{0\} = K(0) \subset (T - \lambda)(0) = T(0) - \lambda(0) = T(0)$. Thus, it is clear that $K \in \mathcal{K}_T(X)$ and $(\lambda - T) - K = \lambda - (T + K)$ is bijective. Therefore, $\lambda \in \rho(T + K)$. This shows that $\lambda \notin \bigcap_{K \in \mathcal{K}_T(X)} \sigma(T + K)$ and

$$\sigma_{e5}(T) \subset \sigma_{e4}(T) \bigcup \{\lambda \in \mathbb{C} \text{ such that } i(\lambda - T) \neq 0\}. \tag{4.1}$$

To prove the inverse inclusion of (4.1). Suppose $\lambda \notin \sigma_{e5}(T)$, then there exists $K \in \mathcal{K}_T(X)$ such that $\lambda \in \rho(T + K)$. Hence, $\lambda - (T + K) \in \Phi(X)$ and $i(\lambda - (T + K)) = 0$. Using Remark 2.3.4, we get $\lambda - T \in \Phi(X)$ and $i(\lambda - T) = i(\lambda - (T + K)) = 0$. We conclude $\lambda \notin \sigma_{e4}(T) \bigcup \{\lambda \in \mathbb{C} \text{ such that } i(\lambda - T) \neq 0\}$. So,

$$\sigma_{e4}(T) \bigcup \{\lambda \in \mathbb{C} \text{ such that } i(\lambda - T) \neq 0\} \subset \sigma_{e5}(T).$$

Therefore,

$$\sigma_{e5}(T) = \sigma_{e4}(T) \bigcup \{\lambda \in \mathbb{C} \text{ such that } i(\lambda - T) \neq 0\}.$$

The proof of other cases is analogous. Q.E.D.

Corollary 4.1.1 *Let X be a Banach space and $T \in CR(X)$. Then,*

(i) $\lambda \notin \sigma_{e5}(T)$ *if and only if,* $\lambda - T \in \Phi(X)$ *and* $i(\lambda - T) = 0$.
(ii) $\lambda \notin \sigma_{eap}(T)$ *if and only if,* $\lambda - T \in \Phi_+(X)$ *and* $i(\lambda - T) \leq 0$.
(iii) $\lambda \notin \sigma_{e\delta}(T)$ *if and only if,* $\lambda - T \in \Phi_-(X)$ *and* $i(\lambda - T) \geq 0$. ◇

Theorem 4.1.2 *Let X be Banach spaces and $T \in CR(X)$. Then,*

(i) $C\sigma(T) \subset \bigcap_{i \in \Lambda} \sigma_{ei}(T)$, *where* $\Lambda := \{1, 2, 3, 4, 5, ap, \delta\}$, *and*
(ii) $R\sigma(T) \subset \sigma_{e\delta}(T)$. ◇

Proof. (i) Let $\lambda \in C\sigma(T)$, then $R(\lambda - T)$ is not closed (otherwise $\lambda \in \rho(T)$). Therefore, $\lambda \in \sigma_{ei}(T)$, $i = 1, \ldots, 5, ap, \delta$. Consequently,

$$C\sigma(T) \subset \bigcap_{i \in \Lambda} \sigma_{ei}(T), \quad \text{where } \Lambda := \{1, 2, 3, 4, 5, ap, \delta\}.$$

(*ii*) Let $\lambda \in R\sigma(T)$, then $\beta(\lambda - T) \neq 0$. Hence, $i(\lambda - T) < 0$ and $\lambda - T$ is one to one. This implies, by the use of Corollary 4.1.1 (*iii*) that $\lambda \in \sigma_{e\delta}(T)$. This completes the proof.   Q.E.D.

Theorem 4.1.3 *Let $T \in C\mathcal{R}(X)$. If $0 \in \rho(T)$, then for $\lambda \neq 0$, $\lambda - T \in \Phi(X)$ if and only if, $\lambda^{-1} - T^{-1} \in \Phi(X)$ and $i(\lambda - T) = i(\lambda^{-1} - T^{-1})$.*   ◇

Proof. Let $\lambda \in \mathbb{C}^*$. By using Remark 2.18.2 (with $\mu = 0$), we have

$$\lambda - T = (0 - \lambda)[(0 - \lambda)^{-1} - (0 - T)^{-1}](0 - T).$$

Then, for all $\lambda \neq 0$, we have

$$\lambda - T = -\lambda(\lambda^{-1} - T^{-1})T. \tag{4.2}$$

Now, let us prove that

$$R(\lambda - T) = R(\lambda^{-1} - T^{-1}). \tag{4.3}$$

Indeed, by (4.2), we have

$$R(\lambda - T) = R\left((\lambda^{-1} - T^{-1})T\right) = (\lambda^{-1} - T^{-1})R(T).$$

The fact that $R(T) = X$, we have

$$R(\lambda - T) = R(\lambda^{-1} - T^{-1}). \tag{4.4}$$

Now, by using Corollary 2.19.1 (with $\mu = 0$), we infer that

$$N(\lambda - T) = N(\lambda^{-1} - T^{-1}). \tag{4.5}$$

Applying (4.4) and (4.5), we have $\lambda - T \in \Phi(X)$ if and only if, $\lambda^{-1} - T^{-1} \in \Phi(X)$. Further, in view of (4.4) and (4.5), we have

$$\begin{aligned}
i(\lambda - T) &= \alpha(\lambda - T) - \beta(\lambda - T) \\
&= \alpha(\lambda^{-1} - T^{-1}) - \beta(\lambda^{-1} - T^{-1}) \\
&= i(\lambda^{-1} - T^{-1}).
\end{aligned}$$

This completes the proof.   Q.E.D.

The following corollary follows immediately from Theorem 4.1.3 and Lemma 2.13.5 (*ii*) (*b*).

Corollary 4.1.2 *Let $\alpha \in \rho(T)$ and $\lambda \in \mathbb{C}$ such that $\lambda \neq \alpha$. Then, $\lambda \in \sigma_{ei}(T)$ if and only if, $(\alpha - \lambda)^{-1} \in \sigma_{ei}((T - \alpha)^{-1})$, $i \in \{1, 2, 3, 4, 5, ap, \delta, q\phi_d\}$.*   ◇

Theorem 4.1.4 *Let X be a Banach space and $T \in LR(X)$ such that $0 \in \rho(T)$. Then, for $0 \neq \lambda \in \mathbb{C}$, we have*

(i) $des(T - \lambda) = des(T^{-1} - \lambda^{-1})$, and
(ii) $asc(T - \lambda) = asc(T^{-1} - \lambda^{-1})$. $\Diamond$

Proof. *(i)* For all $n \in \mathbb{N}$. We prove by induction the following

$$(\lambda - T)^n T = T(\lambda - T)^n. \tag{4.6}$$

Indeed, $(\lambda - T)T = -(\lambda - T)(0 - T) = -(0 - T)(\lambda - T)$. Hence, $(\lambda - T)T = T(\lambda - T)$. Let $n \in \mathbb{N}^*$, we suppose that

$$(\lambda - T)^n T = T(\lambda - T)^n.$$

Then,

$$(\lambda - T)^{n+1} T = (\lambda - T)^n (\lambda - T)T = (\lambda - T)^n T(\lambda - T) = T(\lambda - T)^n (\lambda - T).$$

So, (4.6) holds. Now, we prove that $R\left((\lambda - T)^n\right) = R\left((\lambda^{-1} - T^{-1})^n\right)$ for all $n \in \mathbb{N}$. Let $\lambda \in \mathbb{C}^*$. By using (4.4), we have $R(\lambda - T) = R(\lambda^{-1} - T^{-1})$. Let $n \in \mathbb{N}^*$, we assume that $R\left((\lambda - T)^n\right) = R\left((\lambda^{-1} - T^{-1})^n\right)$. Then,

$$
\begin{aligned}
R\left((\lambda - T)^{n+1}\right) &= (\lambda - T)R\left((\lambda - T)^n\right) \\
&= -\lambda(\lambda^{-1} - T^{-1})TR\left((\lambda - T)^n\right) \\
&= -\lambda(\lambda^{-1} - T^{-1})R\left(T(\lambda - T)^n\right) \\
&= -\lambda(\lambda^{-1} - T^{-1})R\left((\lambda - T)^n T\right) \quad \text{(by (4.6))} \\
&= -\lambda(\lambda^{-1} - T^{-1})R((\lambda - T)^n) \\
&= -\lambda(\lambda^{-1} - T^{-1})R((\lambda^{-1} - T^{-1})^n)) \\
&= R((\lambda^{-1} - T^{-1})^{n+1}). \tag{4.7}
\end{aligned}
$$

So, (4.7) gives *(i)*.

(ii) First, we show that for all $\lambda \in \mathbb{C}^*$, we have

$$(\lambda^{-1} - T^{-1})T = T(\lambda^{-1} - T^{-1}). \tag{4.8}$$

Let $\lambda \in \mathbb{C}^*$ and $(x, x) \in G(I_{\mathcal{D}(T)})$. Since $G(I_{\mathcal{D}(T)}) \subset G(TT^{-1})$, then $x \in Ty$ for some $y \in \mathcal{D}(T)$ (as T is surjective). Hence, $(y, x) \in G(T)$ and $(x, y) \in G(T^{-1})$. So, $(x, x) \in G(TT^{-1})$. On the other hand,

$$
\begin{aligned}
\lambda^{-1}T - TT^{-1} &= T\lambda^{-1} - TT^{-1} \\
&\subset T(\lambda^{-1} - T^{-1}),
\end{aligned}
$$

and

$$(\lambda^{-1} - T^{-1})T \; \subset \; \lambda^{-1}T - T^{-1}T$$
$$= \; \lambda^{-1}T - I_{\mathcal{D}(T)}$$
$$= \; \lambda^{-1}T - TT^{-1}.$$

Hence,

$$(\lambda^{-1} - T^{-1})T \subset T(\lambda^{-1} - T^{-1}). \tag{4.9}$$

Now, we prove that

$$N(T(\lambda^{-1} - T^{-1})) = N((\lambda^{-1} - T^{-1})T). \tag{4.10}$$

Indeed,

$$N(T(\lambda^{-1} - T^{-1})) \; = \; (\lambda^{-1} - T^{-1})^{-1}N(T)$$
$$= \; (\lambda^{-1} - T^{-1})^{-1}(0) \quad (\text{as } N(T) = \{0\})$$
$$= \; N(\lambda^{-1} - T^{-1})$$
$$= \; N(\lambda - T) \quad (\text{as Eq. (4.5)})$$
$$= \; N(-\lambda(\lambda^{-1} - T^{-1})T) \quad (\text{as Eq. (4.2)})$$
$$= \; N((\lambda^{-1} - T^{-1})T).$$

Hence, (4.10) holds. Now, we claim

$$R(T(\lambda^{-1} - T^{-1})) = R((\lambda^{-1} - T^{-1})T). \tag{4.11}$$

Indeed,

$$R(T(\lambda^{-1} - T^{-1})) \; = \; TR(\lambda^{-1} - T^{-1})$$
$$= \; TR(\lambda - T)$$
$$= \; R(T(\lambda - T))$$
$$= \; R((\lambda - T)T)$$
$$= \; (\lambda - T)R(T)$$
$$= \; R(\lambda - T)$$
$$= \; R(\lambda^{-1} - T^{-1})$$
$$= \; R((\lambda^{-1} - T^{-1})T) \quad (\text{since } T \text{ is surjective}),$$

which proves our claim. Now, (4.9), (4.10), and (4.11) ensures that (4.8) is true. Now, we prove by induction that $N((\lambda - T)^n) = N((\lambda^{-1} - T^{-1})^n)$ for all $n \in \mathbb{N}$ and $\lambda \in \mathbb{C}^*$. By using (4.5), we have $N(\lambda - T) = N(\lambda^{-1} - T^{-1})$. Let $n \in \mathbb{N}^*$, we assume $N((\lambda - T)^n) = N((\lambda^{-1} - T^{-1})^n)$. Then,

$$
\begin{aligned}
N((\lambda^{-1} - T^{-1})^{n+1}) &= N((\lambda^{-1} - T^{-1})^n(\lambda^{-1} - T^{-1})) \\
&= (\lambda^{-1} - T^{-1})N((\lambda^{-1} - T^{-1})^n) \\
&= (\lambda^{-1} - T^{-1})N((\lambda - T)^n) \\
&= (\lambda^{-1} - T^{-1})N((\lambda - T)^nT) \quad \text{(by (4.2) and } T \text{ is injective)} \\
&= (\lambda^{-1} - T^{-1})T^{-1}N((\lambda - T)^n) \\
&= (T(\lambda^{-1} - T))^{-1}N((\lambda - T)^n) \quad \text{(by (4.3))} \\
&= ((\lambda^{-1} - T)T)^{-1}N((\lambda - T)^n) \\
&= \left(-\frac{1}{\lambda}(-\lambda)(\lambda^{-1} - T)T\right)^{-1}N((\lambda - T)^n) \\
&= \left(-\frac{1}{\lambda}(\lambda - T)\right)^{-1}N((\lambda - T)^nT) \\
&= N((\lambda - T)^{n+1}) \quad \text{(by Eq. (4.2)).} \tag{4.12}
\end{aligned}
$$

So, (4.12) gives (ii).   Q.E.D.

A consequence of Theorem 4.1.4 is the following:

Corollary 4.1.3 *Let $\alpha \in \rho(T)$ and $\lambda \in \mathbb{C}$ such that $\lambda \neq \alpha$. Then, $\lambda \in \sigma_{eb}(T)$ if and only if, $(\alpha - \lambda)^{-1} \in \sigma_{eb}((T - \alpha)^{-1})$.* ◇

Proposition 4.1.1 *Let X be a Banach space and let T, $S \in LR(X)$. If $ST \subset TS$ and for some $\lambda \in \rho(T) \bigcap \rho(S)$, we have $(\lambda - T)^{-1} - (\lambda - S)^{-1} \in \mathcal{PR}(\Phi(X))$, then*

$$
\sigma_{eb}(T) = \sigma_{eb}(S). \qquad ◇
$$

Proof. We may assume $\lambda = 0$. Let $J = T^{-1} - S^{-1}$. Then,

$$
JS^{-1} = (T^{-1} - S^{-1})S^{-1} \subset (T^{-1}S^{-1} - S^{-1}S^{-1}) = (ST)^{-1} - S^{-1}S^{-1}.
$$

Since $((ST)^{-1} - S^{-1}S^{-1}) \subset ((TS)^{-1} - S^{-1}S^{-1}) = (S^{-1}T^{-1} - S^{-1}S^{-1})$ and $(S^{-1}T^{-1} - S^{-1}S^{-1}) \subset S^{-1}(T^{-1} - S^{-1})$, we have $JS^{-1} \subset S^{-1}J$. Now, using both $\mathfrak{R}_c(S^{-1}) = \{0\}$ and $J \in \mathcal{PR}(\Phi(X))$, we infer that $\sigma_{eb}(T^{-1}) = \sigma_{eb}(S^{-1})$. Finally, in view of Corollary 4.1.3, we get $\sigma_{eb}(T) = \sigma_{eb}(S)$.   Q.E.D.

Theorem 4.1.5 *Let X be a Banach space and let T, $S \in LR(X)$. Then,*

(i) If for some $\lambda \in \rho(T) \bigcap \rho(S)$, we have $(\lambda - T)^{-1} - (\lambda - S)^{-1} \in \mathcal{PR}(\Phi(X))$, then

$$\sigma_{ei}(T) = \sigma_{ei}(S), \ i = 4, 5.$$

(ii) If for some $\lambda \in \rho(T) \bigcap \rho(S)$, we have $(\lambda - T)^{-1} - (\lambda - S)^{-1} \in \mathcal{PR}(\Phi_+(X))$, then

$$\sigma_{ei}(T) = \sigma_{ei}(S), \ i = 1, ap.$$

(iii) If for some $\lambda \in \rho(T) \bigcap \rho(S)$, we have $(\lambda - T)^{-1} - (\lambda - S)^{-1} \in \mathcal{PR}(\Phi_-(X))$, then

$$\sigma_{ei}(T) = \sigma_{ei}(S), \ i = 2, \delta.$$

(iv) If for some $\lambda \in \rho(T) \bigcap \rho(S)$, we have $(\lambda - T)^{-1} - (\lambda - S)^{-1} \in \mathcal{PR}(\Phi_+(X)) \bigcap \mathcal{PR}(\Phi_-(X))$, then

$$\sigma_{e3}(T) = \sigma_{e3}(S). \qquad \diamond$$

Proof. (i) We may assume $\lambda = 0$. From, the fact that $T^{-1} - S^{-1}$ is a Fredholm perturbation, it follows that $\Phi_{T^{-1}} = \Phi_{S^{-1}}$ and $i(\eta - T^{-1}) = i(\eta - S^{-1})$ for all $\eta \in \Phi_{T^{-1}}$. If we apply Corollary 4.1.2 to both T and S, we can see that $\Phi_T = \Phi_S$ and $i(\lambda - T) = i(\lambda - S)$ for $\lambda \in \Phi_T$.

The proof of (ii), (iii) and (iv) may be checked in the same way as the proof of (i). $\qquad$ Q.E.D.

Theorem 4.1.6 *Let X be a Banach space and $T \in CR(X)$. If $\dim T(0) < \infty$, then $\sigma_{e5}(T) = \sigma_l(T)$.* $\qquad \diamond$

Proof. Let $\lambda \notin \sigma_l(T)$, then there exists $K \in \Psi_T(X)$ such that $\lambda \in \rho(T + K)$. This implies that $\lambda - T - K$ is injective, open and $R(\lambda - T - K) = X$. So, $(\lambda - T - K)^{-1}$ is a bounded linear operator on X. Since $R(\lambda - T - K) = X$, then

$$
\begin{aligned}
\mathcal{D}(K(\lambda - T - K)^{-1}) &= (\lambda - T - K)(\mathcal{D}(K)) \\
&\supset (\lambda - T - K)(\mathcal{D}(K) \bigcap \mathcal{D}(T)) \\
&= (\lambda - T - K)(\mathcal{D}(\lambda - T - K)) \\
&= R(\lambda - T - K) \\
&= X,
\end{aligned}
$$

and according to Theorem 3.7.4, we have $I + K(\lambda - T - K)^{-1} \in \Phi_+(X)$ and $i(I + K(\lambda - T - K)^{-1}) = \dim K(\lambda - T - K)^{-1}(0) = \dim K(0) < \infty$. Since $\lambda - T - K$ is injective and open, then $\lambda - T - K$ is in $\Phi_+(X)$. In view of Proposition 2.14.2, we have

$$(I + K(\lambda - T - K)^{-1})(\lambda - T - K) \in \Phi_+(X). \qquad (4.13)$$

The use of Remark 2.3.3, leads to $0 \in K^{-1}(0)$. So, $(\lambda - T - K)(0) \subset (\lambda - T - K)K^{-1}(0) = N(K(\lambda - T - K)^{-1})$. Hence,

$$
\begin{aligned}
(I + K(\lambda - T - K)^{-1})(\lambda - T - K) &= \lambda - T - K + K(\lambda - T - K)^{-1}(\lambda - T - K) \\
&= \lambda - T - K + K. \quad\quad (4.14)
\end{aligned}
$$

As (4.14) and Proposition 2.3.4, we have

$$
(I + K(\lambda - T - K)^{-1})(\lambda - T - K) = \lambda - T. \quad\quad (4.15)
$$

By using (4.13) and (4.15), we have $\lambda - T \in \Phi_+(X)$. Therefore,

$$
\begin{aligned}
i(\lambda - T) &= i((I + K(\lambda - T - K)^{-1})(\lambda - T - K)) \\
&= i(I + K(\lambda - T - K)^{-1}) + i(\lambda - T - K) + \\
&\quad \dim\left(\frac{X}{R(\lambda - T - K) + \mathcal{D}(I + K(\lambda - T - K)^{-1}}\right) - \\
&\quad \dim\left((\lambda - T - K)(0) \bigcap N(I + K(\lambda - T - K)^{-1})\right) \\
&= \dim K(0) - \\
&\quad \dim\left((\lambda - T - K)(0) \bigcap N(I + K(\lambda - T - K)^{-1})\right) .(4.16)
\end{aligned}
$$

On the other hand,

$$
\begin{aligned}
\frac{(I + K(\lambda - T - K)^{-1})(\lambda - T - K)(0)}{(I + K(\lambda - T - K)^{-1})(0)} \\
= \frac{\mathcal{D}(I + K(\lambda - T - K)^{-1}) \bigcap (\lambda - T - K)(0)}{N(I + K(\lambda - T - K)^{-1}) \bigcap (\lambda - T - K)(0)} \\
= \frac{\mathcal{D}(K(\lambda - T - K)^{-1}) \bigcap (\lambda - T - K)(0)}{N(I + K(\lambda - T - K)^{-1}) \bigcap (\lambda - T - K)(0)} \\
= \frac{(\lambda - T - K)\mathcal{D}(K) \bigcap (\lambda - T - K)(0)}{N(I + K(\lambda - T - K)^{-1}) \bigcap (\lambda - T - K)(0)} \\
= \frac{(\lambda - T - K)(0)}{N(I + K(\lambda - T - K)^{-1})(\lambda - T - K)(0)}.
\end{aligned}
$$

But

$$
\frac{(I + K(\lambda - T - K)^{-1})(\lambda - T - K)(0)}{(I + K(\lambda - T - K)^{-1})(0)} = \frac{(\lambda - T - K)(0)}{K(0)}.
$$

Since $(\lambda - T - K)(0) = -T(0) - K(0) = T(0) + K(0) = T(0)$ has finite dimensional space, then

$$
\dim K(0) = \dim N(I + K(\lambda - T - K)^{-1}) \bigcap (\lambda - T - K)(0). \quad\quad (4.17)
$$

Hence, by (4.16) and (4.17), we have $i(\lambda - T) = 0$. So, $\sigma_{e5}(T) \subset \sigma_l(T)$. To prove the opposite inclusion. Let $K \in \mathcal{K}_T(X)$ and $\lambda \in \rho(T + K)$, then

$(\lambda - T - K)^{-1}$ is single valued and continuous. Hence, $-K(\lambda - T - K)^{-1}$ is precompact. Since X is a Banach space, $-K(\lambda - T - K)^{-1}$ is compact, and by using Theorem 3.7.1, we have $-K(\lambda - T - K)^{-1} \in \Lambda_X$. So, $K \in \Psi_T(X)$. Thus, $\sigma_l(T) \subset \sigma_{e5}(T)$.   Q.E.D.

Corollary 4.1.4 *Let X be a Banach space and $T \in CR(X)$ such that $\dim T(0) < \infty$. If $\Sigma(X)$ is a subset of $\Psi_T(X)$ containing $\mathcal{K}_T(X)$, then*

$$\sigma_{e5}(T) = \bigcap_{K \in \Sigma(X)} \sigma(T + K). \qquad \diamondsuit$$

Proof. Since $\Sigma(X) \subset \Psi_T(X)$, we obtain

$$\bigcap_{K \in \Psi_T(X)} \sigma(T + K) \subset \bigcap_{K \in \Sigma(X)} \sigma(T + K).$$

Applying Theorem 4.1.6, we get

$$\sigma_{e5}(T) \subset \bigcap_{K \in \Sigma(X)} \sigma(T + K).$$

On the other hand, since $\mathcal{K}_T(X) \subset \Psi_T(X)$, then

$$\bigcap_{K \in \Sigma(X)} \sigma(T + K) \subset \sigma_{e5}(T). \qquad \text{Q.E.D.}$$

Corollary 4.1.5 *Let X be a Banach space, $T \in CR(X)$ such that $\dim T(0) < \infty$, and $\mathcal{H}_T(X)$ be a subset of $\Psi_T(X)$, containing $\mathcal{K}_T(X)$. If for all $K_1, K_2 \in \mathcal{H}_T(X)$, $K_1 \pm K_2 \in \mathcal{H}_T(X)$, then for every $K \in \mathcal{H}_T(X)$,*

$$\sigma_{e5}(T) = \sigma_{e5}(T + K). \qquad \diamondsuit$$

Proof. We denote

$$\sigma^*(T) = \bigcap_{K \in \mathcal{H}_T(X)} \sigma(T + K).$$

From Corollary 4.1.5, we have $\sigma_{e5}(T) = \sigma^*(T)$. Furthermore, for each $K \in \mathcal{H}_T(X)$, we have $\mathcal{H}_T(X) + K = \mathcal{H}_T(X)$. Then, $\sigma^*(T + K) = \sigma^*(T)$. Hence, we get the desired result.   Q.E.D.

4.1.1 α- and β-Essential Spectra of Linear Relations

Proposition 4.1.2 *Let $T \in CR(X)$. Then, $\sigma_{e\alpha}(T)$ and $\sigma_{e\beta}(T)$ are closed.* $\diamondsuit$

Proof. Using Lemma 2.14.5, we deduce that $\sigma_{e\alpha}(T) = \sigma_{e\beta}(T^*)$. It is sufficient to show that $\sigma_{e\alpha}(T)$ is closed. Let $\lambda \notin \sigma_{e\alpha}(T)$. Then, $\lambda - T \in \mathcal{A}_\alpha(X)$. By using Proposition 3.4.1, we deduce that $T_l(\lambda - T) = (I - F)_{|\mathcal{D}(T)}$, where $T_l \in \mathcal{L}(X)$, F is a bounded finite rank operator such that $N(T_l)$ is topologically complemented in X, $R(T_l)$ is a closed subspace contained in $\mathcal{D}(T)$, and $R(F) \subset \mathcal{D}(T)$. Therefore,

$$
\begin{aligned}
0 \ < \gamma(I - F) \ &\le \ \gamma\left((I - F)_{|N(I-F)+\mathcal{D}(T)}\right) \\
&\le \ \gamma\left((I - F)_{|R(F)+\mathcal{D}(T)}\right) \\
&\le \ \gamma\left((I - F)_{|\mathcal{D}(T)}\right) \quad (\text{as } R(F) \subset \mathcal{D}(T)).
\end{aligned}
$$

Thus, $\gamma(T_l(\lambda - T)) > 0$. Now, we take $\eta \in \mathbb{C}$ such that

$$
|\eta - \lambda| < \frac{\gamma(T_l(\lambda - T))}{\|T_l\|}. \tag{4.18}
$$

Moreover,

$$
T_l(\eta - T) = T_l(\eta - \lambda + \lambda - T) = (I - F + (\eta - \lambda)T_l)_{|\mathcal{D}(T)}. \tag{4.19}
$$

Using (4.18), we have $\|(\eta - \lambda)T_l\| \le \gamma(T_l(\lambda - T))$, and from Corollary 2.16.1 (iv) and (4.19), it follows that $T_l(\eta - T) \in \Phi_+(X)$. Moreover, from Proposition 3.4.1, it is clear that $R(T_l) \subset \mathcal{D}(T) = \mathcal{D}(\eta - T)$, $R(F + (\eta - \lambda)T_l) \subset \mathcal{D}(T) = \mathcal{D}(\eta - T)$ and $(\eta - T)(0) = T(0)$ is complemented in X. Hence, by applying Proposition 3.4.1, we have $\eta - T \in \mathcal{A}_\alpha(X)$. Then, $\{\lambda \in \mathbb{C} \text{ such that } \lambda - T \in \mathcal{A}_\alpha(X)\}$ is open. Hence, $\sigma_{e\alpha}(T)$ is closed. $\qquad$ Q.E.D.

Theorem 4.1.7 *Let $T \in LR(X)$. Then,*

(i) *If $T(0)$ is topologically complemented in X, then*

$$
\sigma_{el}(T) \ = \sigma_{e\alpha}(T) \bigcup \{\lambda \in \mathbb{C} \text{ such that } i(\lambda - T) > 0\}.
$$

(ii) *If $T^*(0)$ is topologically complemented in X^*, then*

$$
\sigma_{er}(T) \ = \sigma_{e\beta}(T) \bigcup \{\lambda \in \mathbb{C} \text{ such that } i(\lambda - T) < 0\}. \qquad \diamond
$$

Proof. (i) Let $\lambda \notin (\sigma_{e\alpha}(T) \bigcup \{\lambda \in \mathbb{C} \text{ such that } i(\lambda - T) > 0\})$. Then, $\lambda - T \in \mathcal{A}_\alpha(X)$ and $i(\lambda - T) \le 0$. Hence, $\lambda - T \in \Phi_+(X)$ with $i(\lambda - T) \le 0$. Thus, by virtue of Theorem 3.5.3 (i), $\lambda - T$ can be expressed in the form

$$
\lambda - T = U + K, \tag{4.20}
$$

where $K \in \mathcal{K}_{\lambda-T}(X) = \mathcal{K}_T(X)$ and U is injective linear operator. Moreover, $\lambda - T \in \mathcal{A}_\alpha(X)$ and $(\lambda - T)(0) = T(0)$ is topologically complemented in X. By using Proposition 3.4.1, we obtain $T_l(\lambda - T) = (I - F)_{|\mathcal{D}(T)}$, where $T_l \in \mathcal{L}(X)$, F is a bounded finite rank operator such that $N(T_l)$ is topologically complemented in X, $R(T_l)$ is a closed subspace contained in $\mathcal{D}(T)$, and $R(F) \subset \mathcal{D}(T)$. Now, let $L = F + T_l K$. Then,

$$R(L) \subset R(F) + R(T_l) \subset \mathcal{D}(T) = \mathcal{D}(\lambda - T - K) = \mathcal{D}(T) \bigcap \mathcal{D}(K),$$

and in view $K(0) \subset T(0)$, we have

$$(\lambda - T - K)(0) = T(0).$$

Hence, $(\lambda - T - K)(0)$ is topologically complemented in X. Moreover,

$$T_l(\lambda - T - K) = (I - L)_{|\mathcal{D}(T)}.$$

So, $\lambda - T - K \in \mathcal{A}_\alpha(X)$. Thus, $R(\lambda - T - K)$ is closed and topologically complemented in X. By using (4.20), we have $\lambda - T - K$ is injective and by applying Proposition 3.4.1, we infer that $\lambda - T - K \in \mathcal{GR}_l(X)$. Hence, $\lambda \notin \sigma_l(T + K)$ and we can deduce that

$$\lambda \notin \bigcap_{K \in \mathcal{K}_T(X)} \sigma_l(T + K).$$

Conversely, let us assume that $\lambda \notin \bigcap_{K \in \mathcal{K}_T(X)} \sigma_l(T + K)$. Then, there exists $K \in \mathcal{K}_T(X)$ such that $\lambda - T - K \in \mathcal{GR}_l(X)$. Since $K(0) \subset T(0) = (\lambda - T)(0)$ and $\mathcal{D}(\lambda - T) = \mathcal{D}(T) \subset \mathcal{D}(K)$, and by using Proposition 2.3.4, we have $\lambda - T = (\lambda - T - K) + K$. Now, in view of Corollary 3.5.3 (i), we infer that $\lambda - T \in \mathcal{A}_\alpha(X)$ and $i(\lambda - T) = i(\lambda - T - K) = -\beta(\lambda - T - K) \le 0$. This is equivalent to say that $\lambda \notin (\sigma_{e\alpha}(T) \bigcup \{\lambda \in \mathbb{C} \; : i(\lambda - T) > 0\})$.

(ii) Let $\lambda \notin (\sigma_{e\beta}(T) \bigcup \{\lambda \in \mathbb{C} \; : i(\lambda - T) < 0\})$. Then, $\lambda - T \in \mathcal{A}_\beta(X)$ and $i(\lambda - T) \ge 0$ which implies that $\lambda - T \in \Phi_-(X)$ and $i(\lambda - T) \ge 0$. By using Lemma 3.5.3, $\lambda - T$ can be expressed in the form

$$\lambda - T = S + F, \tag{4.21}$$

where S is a surjective linear relation and $F \in \mathcal{K}_{\lambda-T}(X) = \mathcal{K}_T(X)$. Moreover, $\lambda - T \in \mathcal{A}_\beta(X)$, and by applying Lemma 2.14.6, we deduce that

$$(\lambda - T)T_r = I - K + (\lambda - T)T_r - (\lambda - T)T_r \tag{4.22}$$

for some bounded operators T_r and K such that $R(T_r) \subset \mathcal{D}(T)$, $T_r(\lambda - T)$ and $K(\lambda - T)$ are continuous operators and $I - K \in \Phi(X)$. Then, one finds out from (4.22) that

$$(\lambda - T - F)T_r = I - K - FT_r + (\lambda - T - F)T_r - (\lambda - T - F)T_r.$$

Clearly, $K + FT_r$ is a bounded operator, $I - K - FT_r \in \Phi(X)$ and, $T_r(\lambda - T - F)$ and $(K + FT_r)(\lambda - T - F)$ are continuous operators. By applying Lemma 2.14.6, we deduce that $N(\lambda - T - F)$ is topologically complemented in X. By using (4.21), we infer that $\lambda - T - F$ is surjective. So, $\lambda - T - F \in \mathcal{GR}_r(X)$. Hence, $\lambda \notin \sigma_r(T + F)$. We can deduce that

$$\lambda \notin \bigcap_{F \in \mathcal{K}_T(X)} \sigma_r(T + F).$$

Conversely, let us assume that $\lambda \notin \bigcap_{F \in \mathcal{K}_T(X)} \sigma_r(T + F)$. Then, there exists $F \in \mathcal{K}_T(X)$ for which $\lambda - T - F \in \mathcal{GR}_r(X)$. Since $F(0) \subset T(0) = (\lambda - T)(0)$ and $\mathcal{D}(\lambda - T) = \mathcal{D}(T) \subset \mathcal{D}(F)$, and by using Proposition 2.3.4, we infer that $\lambda - T = (\lambda - T - F) + F$. Since $((\lambda - T - F)(0))' = T^*(0)$, then $((\lambda - T - F)(0))^*$ is topologically complemented in X^*. By using Corollary 3.5.3 (ii), we infer that $\lambda - T \in \mathcal{A}_\beta(X)$ and $i(\lambda - T) = i(\lambda - T - F) = \alpha(\lambda - T - F) \geq 0$. This is equivalent to say that $\lambda \notin (\sigma_{e\beta}(T) \bigcup \{\lambda \in \mathbb{C} \text{ such that } i(\lambda - T) < 0\})$. This completes the proof.Q.E.D.

Corollary 4.1.6 (i) *Assume that $T(0)$ is topologically complemented in X. Then,*

(i_1) $\sigma_{el}(T) = \{\lambda \in \mathbb{C} \text{ such that } \lambda - T \notin \mathcal{W}^l(X)\}$, *and*

(i_2) $\lambda \notin \sigma_{el}(T)$ *if and only if,* $\lambda - T \in \mathcal{A}_\alpha(X)$ *and* $i(\lambda - T) \leq 0$.

(ii) *Assume that $T^*(0)$ is topologically complemented in X^*. Then,*

(ii_1) $\sigma_{er}(T) = \{\lambda \in \mathbb{C} \text{ such that } \lambda - T \notin \mathcal{W}^r(X)\}$, *and*

(ii_2) $\lambda \notin \sigma_{er}(T)$ *if and only if,* $\lambda - T \in \mathcal{A}_\beta(X)$ *and* $i(\lambda - T) \geq 0$.

(iii) *Assume that the conditions of (i) and (ii) are satisfied. Then,*

(iii_1) $\sigma_{el}(T)$ *and* $\sigma_{er}(T)$ *are closed,*

(iii_2) $\sigma_{el}(T) = \sigma_{er}(T^*)$, *and*

(iii_3) $\sigma_{er}(T^*) = \sigma_{el}(T)$. ◇

The aim of the following theorem is to give a refinement of the definitions of $\sigma_{el}(\cdot)$ and $\sigma_{er}(\cdot)$.

Theorem 4.1.8 *Let $T \in LR(X)$. Then,*

(i) If $T(0)$ is topologically complemented in X, then

$$\sigma_{el}(T) = \bigcap_{K \in \mathcal{PR}(\mathcal{A}_\alpha(X))} \sigma_l(T + K).$$

(ii) If $T^*(0)$ is topologically complemented in X^*, then

$$\sigma_{er}(T) = \bigcap_{K \in \mathcal{PR}(\mathcal{A}_\beta(X))} \sigma_r(T + K). \qquad \diamondsuit$$

Proof. *(i)* Let $\mathcal{U} := \bigcap_{K \in \mathcal{PR}(\mathcal{A}_\alpha(X))} \sigma_l(T + K)$. Since $\mathcal{K}_T(X) \subset \mathcal{PR}(\mathcal{A}_\alpha(X))$,
we have $\mathcal{U} \subset \sigma_{el}(T)$. Conversely, assume that $\lambda \notin \mathcal{U}$. Then, there exists
$K \in \mathcal{PR}(\mathcal{A}_\alpha(X))$ such that $\lambda \notin \sigma_l(T+K)$, that is $\lambda-(T+K) = (\lambda-T)-K \in$
$\mathcal{GR}_l(X)$. By applying Corollary 3.5.2 (i_1), we have $\lambda - T - K + K \in \mathcal{A}_\alpha(X)$
and $i(\lambda - T - K + K) \leq 0$. By using Proposition 2.3.4, we have

$$\lambda - T \in \mathcal{A}_\alpha(X) \quad \text{and } i(\lambda - T) \leq 0.$$

Now, the use of Corollary 4.1.6 *(i)* allows us to conclude that $\lambda \notin \sigma_{el}(T)$.

The proof of *(ii)* is analogous to the previous one.   Q.E.D.

Corollary 4.1.7 *Let X be a Banach space, and let $\mathfrak{N}(X)$ and $\mathfrak{M}(X)$ be any
subsets of $LR(X)$. Then,*

*(i) If $\mathcal{KR}(X) \subset \mathfrak{N}(X) \subset \mathcal{PR}(\mathcal{A}_\alpha(X))$ and if $T(0)$ is topologically
complemented in X, then*

$$\sigma_{el}(T) = \bigcap_{F \in \mathfrak{N}(X)} \sigma_l(T + F).$$

Moreover, if for all F_1, $F_2 \in \mathfrak{N}(X)$, we have $F_1 \pm F_2 \in \mathfrak{N}(X)$, then

$$\sigma_{el}(T + F) = \sigma_{el}(T) \text{ for all } F \in \mathfrak{N}(X).$$

*(ii) If $\mathcal{KR}(X) \subset \mathfrak{M}(X) \subset \mathcal{PR}(\mathcal{A}_\beta(X))$ and if $T^*(0)$ is topologically
complemented in X^*, then*

$$\sigma_{er}(T) = \bigcap_{F \in \mathfrak{M}(X)} \sigma_r(T + F).$$

Moreover, if for all F_1, $F_2 \in \mathfrak{M}(X)$, we have $F_1 \pm F_2 \in \mathfrak{M}(X)$, then

$$\sigma_{el}(T + F) = \sigma_{el}(T) \text{ for all } F \in \mathfrak{M}(X). \qquad \diamondsuit$$

Remark 4.1.1 (*i*) *If $T(0)$ is topologically complemented in X, then from Corollary 4.1.8 (i) and Proposition 2.17.1 (iii), it follows that*

$$\sigma_{el}(T + F) = \sigma_{el}(T) \text{ for all } F \in \mathcal{PR}\left(\mathcal{A}_\alpha(X)\right).$$

(*ii*) *If $T^*(0)$ is topologically complemented in X^*, then from Corollary 4.1.8 (ii) and Proposition 2.17.1 (iv), it follows that*

$$\sigma_{er}(T + F) = \sigma_{er}(T) \text{ for all } F \in \mathcal{PR}\left(\mathcal{A}_\beta(X)\right).$$

(*iii*) *By using both Proposition 2.3.4 and Definition 2.17.1, we have*

$$\sigma_{e\alpha}(T + F) = \sigma_{e\alpha}(T) \text{ for all } F \in \mathcal{PR}\left(\mathcal{A}_\alpha(X)\right)$$

and

$$\sigma_{e\beta}(T + F) = \sigma_{e\beta}(T) \text{ for all } F \in \mathcal{PR}\left(\mathcal{A}_\beta(X)\right). \qquad \Diamond$$

Theorem 4.1.9 *Let $\alpha \in \rho(T)$ and $\lambda \in \mathbb{C}$ such that $\lambda \neq \alpha$. Then, $\lambda \in \sigma_{ei}(T)$ if and only if, $(\alpha - \lambda)^{-1} \in \sigma_{ei}((T - \alpha)^{-1})$, $i = \alpha, \beta, l, r$.* $\qquad \Diamond$

Proof.   The proof of Theorem 4.1.9 is inspired from the proof of Theorem 4.1.3.   Q.E.D.

Remark 4.1.2 *Theorem 4.1.9 allows a very important relationship between the essential spectra of a linear relation and the essential spectra of a linear operator. Then, we can transfer some known results about the essential spectrum of a linear operator to the case of a linear relation.* $\qquad \Diamond$

As an illustration, let us see the following theorem:

Theorem 4.1.10 *Let $T, S \in LR(X)$. Then,*

(*i*) *If, for some $\lambda \in \rho(T) \bigcap \rho(S)$, we have $(\lambda - T)^{-1} - (\lambda - S)^{-1} \in \mathcal{PR}(\mathcal{A}_\alpha(X))$, then*

$$\sigma_{ei}(T) = \sigma_{ei}(S), i = \alpha, l.$$

(*ii*) *If, for some $\lambda \in \rho(T) \bigcap \rho(S)$, we have $(\lambda - T)^{-1} - (\lambda - S)^{-1} \in \mathcal{PR}\left(\mathcal{A}_\beta(X)\right)$, then*

$$\sigma_{ei}(T) = \sigma_{ei}(S), \ i = \beta, r.$$ $\Diamond$

Proof. (*i*) We may assume that $\lambda = 0$. The fact that $T^{-1} - S^{-1} \in \mathcal{PR}(\mathcal{A}_\alpha(X))$, allows us to deduce the existence of $F \in \mathcal{PR}(\mathcal{A}_\alpha(X))$ such that $T^{-1} = S^{-1} + F$. Then, $T^{-1}(0) = S^{-1}(0) = F(0) = \{0\}$. Hence, by applying Remark 4.1.1, we infer that

$$\sigma_{ei}(T^{-1}) = \sigma_{ei}(S^{-1}), \ i = \alpha, \ l.$$

Now, if we apply Theorem 4.1.9 to both T and S, we get

$$\sigma_{ei}(T^{-1}) = \sigma_{ei}(S^{-1}), \ i = \beta, \ r.$$

(*ii*) The proof of (*ii*) may be checked in the same way as in the proof of (*i*), which completes the proof. Q.E.D.

4.1.2 Invariance of Essential Spectra of Linear Relations

In this subsection we apply the reached findings to study the invariance and the characterization of the essential spectra of a closed multivalued linear operator.

Theorem 4.1.11 *Let X be complete space, $T \in \mathcal{CR}(X)$ with $\dim \mathcal{D}(T) = \infty$, and $S \in \mathcal{LR}(X)$ such that $\overline{\mathcal{D}(T)} \subset \mathcal{D}(S)$ and $S(0) \subset T(0)$. If S is T-precompact, then*

(*i*) $\sigma_{eap}(T + S) = \sigma_{eap}(T)$, *and*
(*ii*) $\sigma_{e\delta}(T + S) = \sigma_{e\delta}(T)$. ◇

Proof. (*i*) Let S be T-precompact, then SG_T is precompact. Since X and X_T are complete, then SG_T is compact. Suppose that $\lambda \notin \sigma_{eap}(T)$, then by using Corollary 4.1.1, we have $\lambda - T \in \Phi_+(X)$. Hence, $(\lambda - T)G_{\lambda-T} \in \Phi_+(X_T, X)$, which gives $(\lambda - T)G_T \in \Phi_+(X_T, X)$. Since SG_T is compact, then $(\lambda - T + S)G_T \in \Phi_+(X_T, X)$. So, $(\lambda - (T + S))G_{\lambda-(T+S)} \in \Phi_+(X_T, X)$. Thus, $\lambda - (T + S) \in \Phi_+(X)$. Moreover, by using Proposition 2.11.3, it is clear that $R((\lambda - (T + S))G_T) = R(\lambda - (T + S))$ and $N((\lambda - (T + S))G_T) = N(\lambda - (T + S))$. These relations imply that $i(\lambda - (T + S)) = i((\lambda - T)G_T + SG_T)$. Now, the use of Lemma 3.5.4 allows us to conclude that $i((\lambda - (T + S))) = i(\lambda - T)$, which implies that $\lambda \notin \sigma_{eap}(T + S)$. So, $\sigma_{eap}(T + S) \subset \sigma_{eap}(T)$. Conversely, let $\lambda \notin \sigma_{eap}(T + S)$. Then, by using Corollary 4.1.1, we have $\lambda - (T + S) \in \Phi_+(X)$. Arguing as before, it follows that $\lambda - (T + S - S) \in \Phi_+(X)$. Then, $\lambda - T \in \Phi_+(X)$ and, we have $i(\lambda - T) = i(\lambda - (T + S))$. Hence, $\lambda \notin \sigma_{eap}(T)$ see Corollary 4.1.1. Thus, $\sigma_{eap}(T + S) = \sigma_{eap}(T)$.

(ii) Now, suppose that $\lambda \notin \sigma_{e\delta}(T)$, then by Corollary 4.1.1, $\lambda - T \in \Phi_-(X)$. Hence, $(\lambda - T)G_{\lambda - T} \in \Phi_-(X_T, X)$. This implies that $(\lambda - T)G_T \in \Phi_-(X_T, X)$. Since SG_T is precompact, then $(\lambda - (T + S))G_{(\lambda - T)} \in \Phi_-(X_T, X)$. So, $(\lambda - (T + S))G_{\lambda - (T + S)} \in \Phi_-(X_T, X)$. Therefore, $\lambda - (T + S) \in \Phi_-(X)$ and, we have $i(\lambda - T) = i(\lambda - (T + S))$. Hence, $\lambda \notin \sigma_{e\delta}(T + S)$. Thus, $\sigma_{e\delta}(T + S) \subset \sigma_{e\delta}(T)$. Conversely, let $\lambda \notin \sigma_{e\delta}(T + S)$, then $\lambda - (T + S) \in \Phi_-(X)$. So, $(\lambda - (T + S))G_{\lambda - (T + S)} \in \Phi_-(X_T, X)$. Hence, $(\lambda - (T + S))G_T \in \Phi_-(X_T, X)$. The latter holds if and only if, $((\lambda - (T + S))G_T)^* \in \Phi_+(X^*, X_T^*)$ and $((\lambda - T)G_T)^* + (SG_T)^* \in \Phi_+(X^*, X_T^*)$. Since SG_T is precompact, then $(SG_T)^*$ is compact. This implies that $((\lambda - T)G_T)^* \in \Phi_+(X^*, X_T^*)$. Hence, $(\lambda - T)G_T \in \Phi_-(X_T, X)$. So, $\lambda - T \in \Phi_-(X)$ and $i(\lambda - T) = i(\lambda - (T + S))$. Thus, $\lambda \notin \sigma_{e\delta}(T)$. We infer that $\sigma_{e\delta}(T + S) = \sigma_{e\delta}(T)$, which completes the proof.   Q.E.D.

Theorem 4.1.12 *Let* $T \in C\mathcal{R}(X)$, *then*

(i) $\quad \bigcap_{K \in \mathcal{PR}(\mathcal{A}_\alpha(X))} \sigma_{ap}(T + K) \subset \sigma_{eap}(T) = \bigcap_{K \in \mathcal{SSR}(X)} \sigma_{ap}(T + K).$

(ii) $\quad \bigcap_{K \in \mathcal{PR}(\mathcal{A}_\beta(X))} \sigma_\delta(T + K) \subset \sigma_{e\delta}(T).$

(iii) $\quad \bigcap_{K \in \mathcal{PR}(\mathcal{A}_\alpha(X)) \cup \mathcal{PR}(\mathcal{A}_\beta(X))} \sigma(T + K) \subset \sigma_{e5}(T).$

(iv) *If* $T(0)$ *is topologically complemented in* X, *then*

$$\sigma_{el}(T) = \bigcap_{K \in \mathcal{PR}(\Phi_+(X))} \sigma_l(T + K) = \bigcap_{K \in \mathcal{PR}(\mathcal{A}_\alpha(X))} \sigma_l(T + K).$$

(v) *If* $T^*(0)$ *is topologically complemented in* X^*, *then*

$$\sigma_{er}(T) = \bigcap_{K \in \mathcal{PR}(\Phi_-(X))} \sigma_r(T + K) = \bigcap_{K \in \mathcal{PR}(\mathcal{A}_\beta(X))} \sigma_r(T + K). \qquad \diamondsuit$$

Proof. (i) Let $\mathcal{U} := \bigcap_{K \in \mathcal{PR}(\mathcal{A}_\alpha(X))} \sigma_{ap}(T + K)$. Since $\mathcal{K}_T(X) \subset \mathcal{PR}(\mathcal{A}_\alpha(X))$, we have $\mathcal{U} \subset \sigma_{eap}(T)$. Conversely, assume that $\lambda \notin \mathcal{U}$. Then, there exists $K \in \mathcal{SSR}(X)$ such that $\lambda \notin \sigma_{ap}(T + K)$. Moreover, since $\lambda - (T + K)$ is bounded below, then $(\lambda - T) - K \in \Phi_+(X)$ and $i((\lambda - T) - K) \leq 0$. Also, since $\mathcal{SSR}(X) \subset \mathcal{PR}(\Phi_+(X))$, then $(\lambda - T) - K + K \in \Phi_+(X)$ and $i((\lambda - T) - K + K) \leq 0$. Again, by applying Proposition 2.3.4, we infer that

$\lambda - T \in \mathcal{A}_\alpha(X)$ and $i(\lambda - T) \leq 0$. Finally, the use of Corollary 4.1.6 gives the wanted inclusion and achieves the proof of (i).

Statement (ii), (iii), (iv) and (v) can be checked in the same way as the assertion (i). Q.E.D.

As a direct consequence of Theorem 4.1.12, we infer the following result.

Corollary 4.1.8 *Let $\mathfrak{U}(X)$ and $\mathfrak{V}(X)$ be any subsets of $LR(X)$. Then,*

(i) *If $\mathcal{KR}(X) \subset \mathfrak{U}(X) \subset \mathcal{PR}(\Phi_+(X))$, then*

$$\sigma_{eap}(T) = \bigcap_{K \in \mathfrak{U}(X)} \sigma_{ap}(T + K).$$

If, for all $F_1, F_2 \in \mathfrak{U}(X)$, $F_1 \pm F_2 \in \mathfrak{U}(X)$, then for every $F \in \mathfrak{U}(X)$, we have

$$\sigma_{eap}(T + F) = \sigma_{eap}(T).$$

(ii) *If $\mathcal{KR}(X) \subset \mathfrak{U}(X) \subset \mathcal{A}_\alpha(X))$ and $T(0)$ is topologically complemented in X, then*

$$\sigma_{el}(T) = \bigcap_{F \in \mathfrak{U}(X)} \sigma_l(T + F).$$

If, for all $F_1, F_2 \in \mathfrak{U}(X)$, $F_1 \pm F_2 \in \mathfrak{U}(X)$, then for every $F \in \mathfrak{U}(X)$, we have

$$\sigma_{el}(T + F) = \sigma_{el}(T).$$

(iii) *If $\mathcal{KR}(X) \subset \mathfrak{U}(X) \subset \mathcal{PR}(\Phi_+(X))$, then*

$$\sigma_{e\delta}(T) = \bigcap_{K \in \mathfrak{V}(X)} \sigma_\delta(T + K).$$

If, for all $F_1, F_2 \in \mathfrak{U}(X)$, $F_1 \pm F_2 \in \mathfrak{U}(X)$, then for every $F \in \mathfrak{U}(X)$, we have

$$\sigma_{e\delta}(T + F) = \sigma_{e\delta}(T).$$

(iv) *If $\mathcal{KR}(X) \subset \mathfrak{V}(X) \subset \mathcal{PR}(\mathcal{A}_\beta(X))$ and $T^*(0)$ is topologically complemented in X^*, then*

$$\sigma_{er}(T) = \bigcap_{F \in \mathfrak{V}(X)} \sigma_r(T + F).$$

If, for all $F_1, F_2 \in \mathfrak{V}(X)$, $F_1 \pm F_2 \in \mathfrak{V}(X)$, then for every $F \in \mathfrak{V}(X)$, we have

$$\sigma_{er}(T + F) = \sigma_{er}(T).$$ ◇

Remark 4.1.3 *It follows immediately, from Corollary 4.1.8, that*

(i) $\sigma_{ei}(T + F) = \sigma_{ei}(T)$ *for all* $F \in \mathcal{PR}(\Phi_+(X))$ *and* $i \in \{1, ap\}$.

(ii) $\sigma_{ei}(T + F) = \sigma_{ei}(T)$ *for all* $F \in \mathcal{PR}(\Phi_-(X))$ *and* $i \in \{2, \delta\}$.

(iii) $\sigma_{e3}(T + F) = \sigma_{e3}(T)$ *for all* $F \in \mathcal{PR}(\Phi_+(X)) \bigcap \mathcal{PR}(\Phi_-(X))$.

(iv) $\sigma_{ei}(T + F) = \sigma_{ei}(T)$ *for all* $F \in \mathcal{PR}(\Phi_+(X)) \bigcup \mathcal{PR}(\Phi_-(X))$ *and* $i \in \{4, 5\}$.

(v) $\sigma_{e\alpha}(T + F) = \sigma_{e\alpha}(T)$ *for all* $F \in \mathcal{PR}(\mathcal{A}_\alpha(X))$.

(vi) $\sigma_{e\beta}(T + F) = \sigma_{e\beta}(T)$ *for all* $F \in \mathcal{PR}(\mathcal{A}_\beta(X))$.

(vii) If $T^*(0)$ *is topologically complemented in* X^*, *then* $\sigma_{er}(T + F) = \sigma_{er}(T)$ *for all* $F \in \mathcal{PR}(\mathcal{A}_\beta(X))$.

(vii) If $T(0)$ *is topologically complemented in* X, *then* $\sigma_{el}(T + F) = \sigma_{el}(T)$ *for all* $F \in \mathcal{PR}(\mathcal{A}_\alpha(X))$. ◇

Proposition 4.1.3 *Let* $X = l^p(\mathbb{N})$ *be the space of sequences* $x : \mathbb{N} \longrightarrow \mathbb{C}$ *summable with a power* $p \in [1, \infty)$ *and the standard norm. Consider the linear relation* T *defined by*

$$G(T) = \{(x, y) \in X \times X \text{ such that } x(n) = y(n - 1), n \geq 2\}$$

and let $F = K + F - F$, *where* K *is a compact single valued linear operator given by*

$$K(x(n)) = \begin{cases} \dfrac{x(n - 1)}{n - 1}, & x \in X \text{ and } n \geq 2 \\ 0, & n = 1, \end{cases}$$

and $F(0) = span\{e_1\}$ *(the subspace generated by* e_1*). Then,*

$$\{\lambda \in \mathbb{C} \text{ such that } |\lambda| = 1\} \subset \sigma(T) \subset \{\lambda \in \mathbb{C} \text{ such that } |\lambda| \geq 1\},$$

and

$$\sigma_{ei}(T + F) = \{\lambda \in \mathbb{C} \text{ such that } |\lambda| = 1\} \ i = 1, \ldots, 4. \quad ◇$$

Proof. It is clear that $T = L^{-1}$, where L is the left shift single valued operator defined by

$$L(x(n)) = x(n + 1), \quad n \geq 1 \text{ and } x \in X.$$

Then, T is closed and $\mathcal{D}(T) = R(L) = X$. Hence, T is everywhere defined closed linear relation. So,

$$\begin{aligned} \sigma(T) \bigcup \{\infty\} &= \left\{\frac{1}{\lambda} \in \mathbb{C} \text{ such that } \lambda \in \sigma(T)\right\} \\ &= \left\{\frac{1}{\lambda} \in \mathbb{C} \text{ such that } 0 \leq |\lambda| \leq 1\right\} \\ &= \{\lambda \in \mathbb{C} \text{ such that } |\lambda| \geq 1\}. \end{aligned}$$

On the other hand, we will check that $\sigma_{ei}(T) = \{\lambda \in \mathbb{C} \text{ such that } |\lambda| = 1\}$, $i = 1, \ldots, 4$. Indeed, $N(T) = L(0) = \{0\}$ and $R(T) = X$. Hence, $0 \in \rho(T)$. By using Remark 2.18.3, we have $\sigma_{ei}(T) = \{\lambda \in \mathbb{C} \text{ such that } |\lambda| = 1\}$, $i = 1, \ldots, 4$. So, $\{\lambda \in \mathbb{C} \text{ such that } |\lambda| = 1\} \subset \sigma(T) \subset \{\lambda \in \mathbb{C} \text{ such that } |\lambda| \geq 1\}$. Observe that $\|F\| = \|K + F - F\| \leq \|K\| + \|F - F\| = \|K\| < \infty$. Hence, F is continuous, $\mathcal{D}(F) = \mathcal{D}(K) = X$ and $F(0) = T(0)$. Therefore, by using Proposition 2.7.2, we have $T + F \in CR(X)$. Since F is bounded compact linear relation, then by using Theorem 3.5.5, we have $F \in \mathcal{PR}(\mathcal{A}_\alpha(X))$, and $F \in \mathcal{PR}(\mathcal{A}_\beta(X))$. Hence, by virtue of Remark 4.1.3, we have

$$\sigma_{ei}(T + F) = \sigma_{ei}(T), \ i = 1, \ldots, 5, ap, \delta, \alpha, \beta.$$

In particulary, $\sigma_{ei}(T + F) = \sigma_{ei}(T) = \{\lambda \in \mathbb{C} \text{ such that } |\lambda| = 1\}$ $i = 1, \ldots, 4$. This completes the proof.   Q.E.D.

We close this section by the following result:

Theorem 4.1.13 *Let $T \in CR(X)$ and $S \in LR(X)$ be continuous such that $S(0) \subset T(0)$ and S is T-bounded with T-bound < 1. Then,*

(i) If SG_T is α-Atkinson perturbation, then

$$\sigma_{e\alpha}(T + S) = \sigma_{e\alpha}(T).$$

Moreover, if $T(0)$ is topologically complemented in X, then

$$\sigma_{el}(T + F) = \sigma_{el}(T).$$

(ii) If SG_T is β-Atkinson perturbation, then

$$\sigma_{e\beta}(T + S) = \sigma_{e\beta}(T).$$

Moreover, if $T^*(0)$ is topologically complemented in X^*, then

$$\sigma_{er}(T + F) = \sigma_{er}(T). \qquad \diamond$$

Proof. (i) Let $\lambda \in \mathbb{C}$ such that $\lambda - T \in \mathcal{A}_\alpha(X)$. Then, $(\lambda - T)G_{\lambda-T} \in \mathcal{A}_\alpha(X_T, X)$. Hence, by the equivalence of the norms $\|\cdot\|_T$ and $\|\cdot\|_{\lambda-T}$, we have $(\lambda - T)G_T \in \mathcal{A}_\alpha(X_T, X)$. Moreover, $SG_T \in \mathcal{PR}(\mathcal{A}_\alpha(X_T, X))$. We can deduce that $(\lambda - (T + S))G_T \in \mathcal{A}_\alpha(X_T, X)$. Again, by the equivalence of the norms $\|\cdot\|_T$ and $\|\cdot\|_{T+S}$ (see Theorem 2.6.1), we have $(\lambda - (T+S))G_{T+S} \in \Phi_+(X)$. Thus, by using the equivalence of the norms $\|\cdot\|_{T+S}$ and $\|\cdot\|_{\lambda-(T+S)}$, we have $\lambda - (T + S)G_{\lambda-(T+S)} \in \mathcal{A}_\alpha(X_T, X)$. Hence, $\lambda - (T + S) \in \mathcal{A}_\alpha(X)$.

Observing that each step in the above argument is reversible, so we obtain the equivalence

$$\lambda - T \in \mathcal{A}_\alpha(X) \text{ if and only if, } \lambda - (T + S) \in \mathcal{A}_\alpha(X). \tag{4.23}$$

It follows from (4.23) that $\sigma_{ea}(T + S) = \sigma_{ea}(T)$. Moreover, by using Proposition 2.11.3, it is clear that $R((\lambda - (T + S))G_T) = R(\lambda - (T + S))$ and $N((\lambda - (T + S))G_T) = N(\lambda - (T + S))$. These relations imply that $i(\lambda - (T + S)) = i((\lambda - T)G_T + SG_T)$. Now, the use of Lemma 3.5.4 allows us to conclude that

$$i(\lambda - (T + S)) = i(\lambda - T). \tag{4.24}$$

Then, it follows from both (4.24) and Corollary 4.1.6 that $\sigma_{el}(T+S) = \sigma_{el}(T)$.

(ii) The proofs of (ii) may be achieved by using the same reasoning as (i), which completes the proof.   Q.E.D.

4.2   The Essential Spectrum of a Sequence of Linear Relations

The goal of this section is to examine the invariance of Weyl essential spectrum of closed linear relation T by sequence T_n, that converges in the generalized sense to T.

Lemma 4.2.1 *Let $(T_n)_n$ be a sequence of linear closed relations, $T \in \mathcal{CR}(X)$ such that $0 \in \rho(T)$ and $T_n \xrightarrow{g} T$. If $\mathcal{O} \subset \mathbb{C}$ is open and $0 \in \mathcal{O}$, then there exists $n_0 \in \mathbb{N}$ such that for every $n \geq n_0$, we have*

$$\sigma_{e5}(T_n^{-1}) \subset \sigma_{e5}(T^{-1}) + \mathcal{O}. \tag{4.25}$$

Proof. Let us assume that $0 \in \rho(T)$ and $T_n \xrightarrow{g} T$, then by using Theorem 3.2.8 (iii), we infer that $T_n^{-1} \in \mathcal{L}(X)$. Since T_n^{-1} is a bounded linear operator on X, then $\sigma_{e5}(T_n^{-1})$ is a compact subset of $\mathbb{C}$ (see [65]). Now, let us assume that (4.25) fails. Then, by passing to subsequence, we suppose that for each $n \in \mathbb{N}$, there exists $\lambda_n \in \sigma_{e5}(T_n^{-1})$ such that $\lambda_n \notin \sigma_{e5}(T^{-1}) + \mathcal{O}$. Assume that $\lim_{n \to +\infty} \lambda_n = \lambda$, which implies that $\lambda \notin \sigma_{e5}(T^{-1}) + \mathcal{O}$. Since $0 \in \mathcal{O}$, then we obtain $\lambda \notin \sigma_{e5}(T^{-1})$. This equivalent to saying that $\lambda - T^{-1} \in \Phi(X)$ and $i(\lambda - T^{-1}) = 0$. The fact that $\lambda_n - T_n^{-1} \in \mathcal{L}(X)$ for sufficiently large n and $\lambda_n - T_n^{-1} \to \lambda - T^{-1}$ as $n \to \infty$ implies from (ii) of Theorem 3.2.8 that

$$\widehat{\delta}(\lambda_n - T_n^{-1}, \lambda - T^{-1}) \to 0 \quad \text{as} \quad n \to \infty.$$

Since $R(\lambda - T^{-1})$ is closed, then by Proposition 2.9.3 (i) and Theorem 2.5.4 (ii), we deduce that $\gamma(\lambda - T^{-1}) > 0$. Hence, there exists $N \in \mathbb{N}$ such that, for all $n \geq N$, we have

$$\widehat{\delta}(\lambda_n - T_n^{-1}, \lambda - T^{-1}) \leq \frac{\gamma(\lambda - T^{-1})}{\sqrt{1 + [\gamma(\lambda - T^{-1})]^2}}.$$

Thus, allows us to conclude that $\lambda_n - T_n^{-1} \in \Phi(X)$ and $i(\lambda_n - T_n^{-1}) = i(\lambda - T^{-1}) = 0$. This implies that $\lambda_n \notin \sigma_{e5}(T_n^{-1})$, which is a contradiction. Q.E.D.

Theorem 4.2.1 *Let $(T_n)_n$ be a sequence of linear closed relations, $T \in CR(X)$ such that $0 \in \rho(T)$ and $T_n \xrightarrow{g} T$. Then,*

(i) *If $\mathcal{O} \subset \mathbb{C}$ is open and $0 \in \mathcal{O}$, then there exists $n_0 \in \mathbb{N}$ such that for every $n \geq n_0$, we have*

$$\sigma_{e5}(T_n) \subset \sigma_{e5}(T) + \mathcal{O}. \tag{4.26}$$

(ii) *Let $S \in BR(X)$ satisfy that $\mathcal{D}(S) \supset \mathcal{D}(T_n) \bigcup \mathcal{D}(T)$, $S(0) \subset T_n(0) \bigcap T(0)$ and $\|S\| < \gamma(T)$, for all $n \geq 1$. Then,*

$$\sigma_{e5}(T_n + S) \subset \sigma_{e5}(T) + \mathcal{O}, \text{ for all } n \geq n_0. \qquad \diamondsuit$$

Proof. (i) Since $T_n \xrightarrow{g} T$, then by using Lemma 4.2.1, there exists $n_0 \in \mathbb{N}$ such that

$$\sigma_{e5}(T_n^{-1}) \subset \sigma_{e5}(T^{-1}) + \mathcal{O} \text{ for all } n \geq n_0. \tag{4.27}$$

Now, take $\gamma_n \in \sigma_{e5}(T_n)$ such that $\gamma_n \notin \sigma_{e5}(T) + \mathcal{O}$. From $\sigma_{e5}(.)$ is upper semi continuous at T^{-1}(since T^{-1} is a single valued), there exists $k > 0$ such that $k^{-1} \leq |\gamma_n - \lambda_0|^{-1}$, so (γ_n) is bounded. Therefore it can be assumed that $\gamma_n \to \gamma$. Then, $\gamma \notin \sigma_{e5}(T) + \mathcal{O}$ and hence $\gamma \notin \sigma_{e5}(T)$. This implies that $(\gamma)^{-1} \notin \sigma_{e5}(T^{-1})$. We set $\lambda_n = (\gamma_n)^{-1}$, $\lambda = (\gamma)^{-1}$. Then by using (4.27), there exists $n_0 \in \mathbb{N}$ such that for all $n \geq n_0$, $\lambda_n \notin \sigma_{e5}((T_n)^{-1})$, which implies $\gamma_n \notin \sigma_{e5}((T_n))$, is a contradiction.

(ii) Let $S \in BR(X)$ such that $\|S\| < \gamma(T)$. Let $A_n = T_n + S$ and $A = T + S$, then by using Theorem 3.2.8 (i), we deduce that $T_n + S = A_n \xrightarrow{g} A = T + S$. The fact that $0 \in \rho(T)$ implies that T is open and injective with dense range.

Hence, it follows from Lemma 2.9.3 that $T + S$ is open and injective with dense range. In view of the above, $0 \in \rho(T + S)$. Now, by applying (i) to A_n and A, we conclude that there exists $n_0 \in \mathbb{N}$ such that

$$\sigma_{e5}(T_n + S) \subset \sigma_{e5}(T + S) + \mathcal{O}, \text{ for all } n \geq n_0.$$

Let us assume that $0 < |\lambda| < \gamma(T) - \|S\|$. Then, for any $\lambda \in \mathbb{C}$, we have

$$\|S\| < \gamma(T) - |\lambda|. \tag{4.28}$$

Suppose that $\lambda \notin \sigma_{e5}(T)$, then

$$\lambda - T \in \Phi(X) \text{ and } i(\lambda - T) = 0.$$

Since T is open and injective, then it follows from (4.28) and Lemma 2.9.2 that

$$\|S\| < \gamma(T) - |\lambda| \leq \gamma(\lambda - T), \text{ for any } \lambda \in \mathbb{C}.$$

Thus, by applying Corollary 2.16.1 (vii), we conclude that

$$\lambda - (T + S) \in \Phi(X) \text{ and } i(\lambda - (T + S)) = i(\lambda - T) = 0.$$

This is equivalent to saying that $\lambda \notin \sigma_{e5}(T + S)$. Therefore, $\sigma_{e5}(T + S) \subset \sigma_{e5}(T)$. Finally, we have proved the following result

$$\sigma_{e5}(T_n + S) \subset \sigma_{e5}(T) + \mathcal{O}, \text{ for all } n \geq n_0. \qquad \text{Q.E.D.}$$

Remark 4.2.1 *The compactness of the Weyl spectrum is not valid for the case of bounded linear relations. For this reason we are not able to investigate directly $\sigma_{e5}(T)$. Accordingly, we added the condition that $0 \in \rho(T)$ to work on the Weyl spectrum of a bounded linear operator T^{-1}.* $\diamond$

Proposition 4.2.1 *Let $X = l^p(\mathbb{N})$ be the space of sequences $x : \mathbb{N} \longrightarrow \mathbb{C}$ summable with a power $p \in [1, \infty)$ with the standard norm. Consider the linear relation T defined by*

$$G(T) := \{(x, y) \in X \times X : x(n) = y(n - 1), n \geq 2\}.$$

For each positive integer element m such that $n \geq 2$, let $T_n = (\frac{n}{n-1})T$. Then,

(i) T_n converges in the generalized sense to T.
(ii) $\sigma_{e4}(T_n) \neq \sigma_{e4}(T)$. $\diamond$

Proof. (i) It is clear that $T = L^{-1}$, where L is the left shift single valued operator defined by

$$L(x(n)) = x(n+1), \text{ for all } n \geq 1 \text{ and } x \in X.$$

Then, T is closed and $\mathcal{D}(T) = R(L) = X$. Hence, T is an everywhere defined closed linear relation. So, $T \in \mathcal{BR}(X)$, and according to Proposition 2.5.4 (iii), we deduce that $(T_n)_n \in \mathcal{BR}(X)$ for sufficiently large n. Now, we have to prove that $T_n \xrightarrow{g} T$. Indeed, since T is a linear relation, then for all $n \geq 2$, we get

$$T_n(0) = \left(\frac{n}{n-1}\right) T(0) = T\left(\frac{n}{n-1} \times 0\right) = T(0),$$

and $\|T_n - T\| \to 0$ when $m \to \infty$. Then, by applying Theorem 3.2.8 (ii), we conclude that T_n converges in the generalized sense to T.

(ii) Since $N(T) = L(0) = \{0\}$ and $R(T) = X$, then $0 \in \rho(T)$. It is clear that $\sigma_{e4}(T) = \{\lambda \in \mathbb{C} : |\lambda| = 1\}$ and $\sigma_{e4}(T_n) = \{\lambda \in \mathbb{C} : |\lambda| = 1 - \frac{1}{n}\}$, for all $n \geq 2$. This implies that

$$\sigma_{e4}(T_n) \neq \sigma_{e4}(T). \hspace{3cm} \text{Q.E.D.}$$

4.3   Spectral Mapping Theorem of Essential Spectra

Theorem 4.3.1 *Assume that $T \in \mathcal{CR}(X)$ has a nonempty resolvent set and $P(T) = \prod\limits_{i=1}^{n}(T - \lambda_i)^{m_i}$ as in Definition 2.13.1. Then, for any complex polynomial P, we have*

(i) $\sigma_{eap}(P(T)) \subset P(\sigma_{eap}(T))$, *and*
(ii) $\sigma_{e\delta}(P(T)) \subset P(\sigma_{e\delta}(T))$. $\Diamond$

Proof. (i) Let $P(T)$ defined in (2.47). Assume that $\mu \in \sigma_{eap}(P(T))$. If $\lambda_k \notin \sigma_{eap}(T)$ for all $k = 1, 2, \ldots, n$, then it follows from the characterization of $\sigma_{eap}(T)$, established in Theorem 2.14.3, that $T - \lambda_k \in \Phi_+(X)$ with $i(T - \lambda_k) \leq 0$, for all $k \in \{1, 2, \ldots, n\}$. Thus, we deduce from Theorem 2.14.3 that $P(T) - \mu$ is an upper semi-Fredholm relation. Let us consider two cases for the index:

<u>First case</u> : $i(T - \lambda_k) \in]-\infty, 0]$, for all $k \in \{1, 2, \ldots, n\}$. Then, $i(P(T) - \mu) \in]-\infty; 0]$, by Proposition 3.6.1 (i). Hence, $\mu \notin \sigma_{eap}(P(T))$ which contradicts the fact that $\mu \in \sigma_{eap}(P(T))$.

<u>Second case</u> : $i(T - \lambda_k) = -\infty$, for some $k \in \{1, 2, \ldots, n\}$. Then, $i(P(T) - \mu)$ would be $-\infty$, applying Proposition 3.6.2 (i). So, $\mu \notin \sigma_{eap}(P(T))$. Consequently, there exists $j \in \{1, 2, \ldots, n\}$ for which $\lambda_j \in \sigma_{eap}(T)$. Since $\mu = P(\lambda_j)$, then we conclude that $\sigma_{eap}(P(T)) \subset P(\sigma_{eap}(T))$.

(ii) This assertion may be proved with a similar scheme by using Propositions 3.6.1 (ii) and 3.6.2 (ii). Q.E.D.

4.3.1 Essential Spectra of the Sum of Two Linear Relations

In this subsection we investigate the essential spectra of the sum of two closed linear relations defined on a Banach space by means of essential spectra of these two linear relations.

Theorem 4.3.2 *Let X be a Banach space and let $A, B \in BCR(X)$ and $AB \subset BA$. Then,*

(i) *If A is a single valued linear operator and $AB \in PR(\Phi(X))$, then*

$$\sigma_{e4}(A + B) \backslash \{0\} \subset \left[\sigma_{e4}(A) \bigcup \sigma_{e4}(B)\right] \backslash \{0\}.$$

If, further, $BA \in PR(\Phi(X))$, then

$$\sigma_{e4}(A + B) \backslash \{0\} = \left[\sigma_{e4}(A) \bigcup \sigma_{e4}(B)\right] \backslash \{0\}.$$

(ii) *If A is single valued and $BA \in PR(\Phi(X))$, then*

$$\sigma_{e5}(A + B) \backslash \{0\} \subset \left[\sigma_{e5}(A) \bigcup \sigma_{e5}(B)\right] \backslash \{0\}.$$

Moreover, if $AB \in PR(\Phi(X))$ and Φ_A is connected, then

$$\sigma_{e5}(A + B) \backslash \{0\} = \left[\sigma_{e5}(A) \bigcup \sigma_{e5}(B)\right] \backslash \{0\}. \tag{4.29}$$

(iii) *If $AB \in PR(\Phi_+(X))$, then*

$$\sigma_{e1}(A + B) \backslash \{0\} \subset \left[\sigma_{e1}(A) \bigcup \sigma_{e1}(B)\right] \backslash \{0\}.$$

If, further, $BA \in PR(\Phi_+(X))$ then

$$\sigma_{e1}(A + B) \backslash \{0\} = \left[\sigma_{e1}(A) \bigcup \sigma_{e1}(B)\right] \backslash \{0\}. \tag{4.30}$$

(iv) If the hypothesis of (iii) is satisfied, then

$$\sigma_{eap}(A+B)\backslash\{0\} \subset \left[\sigma_{eap}(A)\bigcup\sigma_{eap}(B)\right]\backslash\{0\}.$$

If, further, Φ_A is connected, then

$$\sigma_{eap}(A+B)\backslash\{0\} = \left[\sigma_{eap}(A)\bigcup\sigma_{eap}(B)\right]\backslash\{0\}.$$

(vi) If the hypothesis of (iv) is satisfied, then

$$\sigma_{e\delta}(A+B)\backslash\{0\} \subset \left[\sigma_{e\delta}(A)\bigcup\sigma_{e\delta}(B)\right]\backslash\{0\}.$$

If, further, Φ_A is connected, then

$$\sigma_{e\delta}(A+B)\backslash\{0\} = \left[\sigma_{e\delta}(A)\bigcup\sigma_{e\delta}(B)\right]\backslash\{0\}.$$

(vii) If $AB \in \mathcal{PR}\left(\Phi_+(X)\right)\bigcap\mathcal{PR}\left(\Phi_-(X)\right)$, then

$$\sigma_{e3}(A+B)\backslash\{0\} \subset$$
$$([\sigma_{e3}(A)\bigcup\sigma_{e3}(B)]\bigcup[\sigma_{e1}(A)\bigcap\sigma_{e2}(B)]\bigcup[\sigma_{e2}(A)\bigcap\sigma_{e1}(B)])\backslash\{0\}.$$

Moreover, if $BA \in \mathcal{PR}\left(\Phi_+(X)\right)\bigcap\mathcal{PR}\left(\Phi_-(X)\right)$, then

$$\sigma_{e3}(A+B)\backslash\{0\} =$$
$$([\sigma_{e3}(A)\bigcup\sigma_{e3}(B)]\bigcup[\sigma_{e1}(A)\bigcap\sigma_{e2}(B)]\bigcup[\sigma_{e2}(A)\bigcap\sigma_{e1}(B)])\backslash\{0\}. \quad \Diamond$$

Proof. Let $A, B \in BCR(X)$ and $\lambda \in \mathbb{C}$. By using Proposition 2.3.6, it is easy to see that

$$(\lambda - A)(\lambda - B) = AB + \lambda(\lambda - A - B) \tag{4.31}$$

and

$$(\lambda - B)(\lambda - A) = BA + \lambda(\lambda - A - B). \tag{4.32}$$

(i) Let $\lambda \notin \sigma_{e4}(A)\bigcup\sigma_{e4}(B)\bigcup\{0\}$. Then, $\lambda - A \in \Phi(X)$ and $\lambda - B \in \Phi(X)$. By using Theorem 2.16.1, we have $(\lambda - A)(\lambda - B) \in \Phi(X)$. So, (4.31) gives

$$AB + \lambda(\lambda - A - B) \in \Phi(X).$$

Since $AB \in \mathcal{PR}(\Phi(X))$, then

$$AB - (\lambda - A)(\lambda - B) \in \Phi(X).$$

This implies that

$$AB - AB - \lambda(\lambda - A - B) \in \Phi(X). \tag{4.33}$$

Since $AB(0) \subset BA(0) = B(0) \subset \lambda(A + B - \lambda)(0) = (A + B)(0) = A(0) + B(0)$, then in view of (4.33), we have $\lambda - A - B \in \Phi(X)$. Hence, $\lambda \notin \sigma_{e4}(A + B)$. Therefore,

$$\sigma_{e4}(A + B)\backslash\{0\} \subset \left[\sigma_{e4}(A) \bigcup \sigma_{e4}(B)\right]\backslash\{0\}. \tag{4.34}$$

Prove the inverse inclusion of (4.34). Suppose $\lambda \notin \sigma_{e4}(A + B)\bigcup\{0\}$, then $\lambda - A - B \in \Phi(X)$. Since AB and BA are in $\mathcal{PR}(\Phi(X))$ and $AB(0) \subset BA(0) \subset (A + B)(0) = A(0) + B(0)$, then by using both (4.31) and (4.32), we have

$$(\lambda - A)(\lambda - B) \in \Phi(X) \text{ and } (\lambda - B)(\lambda - A) \in \Phi(X). \tag{4.35}$$

Applying Theorem 2.14.3, it is clear that $\lambda - A \in \Phi(X)$ and $\lambda - B \in \Phi(X)$. Therefore, $\lambda \notin \sigma_{e4}(A) \bigcup \sigma_{e4}(B)$. This proves that

$$\left[\sigma_{e4}(A) \bigcup \sigma_{e4}(B)\right]\backslash\{0\} \subset \sigma_{e4}(A + B)\backslash\{0\}.$$

(ii) Let $\lambda \notin [\sigma_{e5}(A) \bigcup \sigma_{e5}(B)]\backslash\{0\}$. Then, $A - \lambda \in \Phi(X)$, $i(A - \lambda) = 0$, $B - \lambda \in \Phi(X)$ and $i(B - \lambda) = 0$. Using Theorem 2.14.3, we infer that $(B - \lambda)(A - \lambda) \in \Phi(X)$ and $i((A - \lambda)(B - \lambda)) = 0$. Since

$$i\left((B - \lambda)(A - \lambda)\right) = i(B - \lambda) + i(A - \lambda) + \dim\left(\frac{X}{R(A - \lambda) + \mathcal{D}(B - \lambda)}\right)$$
$$- \dim\left(A(0) \bigcap N(B - \lambda)\right),$$

$R(A - \lambda) + \mathcal{D}(T - \lambda) = X$, and $A(0) \bigcap N(B - \lambda) = \{0\} \bigcap N(B - \lambda) = \{0\}$, then

$$i\left((B - \lambda)(A - \lambda)\right) = i(B - \lambda) + i(A - \lambda) = 0. \tag{4.36}$$

This implies that $i(BA + \lambda(\lambda - A - B)) = 0$. Moreover, $BA \in \mathcal{PR}(\Phi(X))$ and $BA(0) = B(0) \subset \lambda(A + B - \lambda)(0) = (A + B)(0) = A(0) + B(0)$. Hence, by using Proposition 3.3.2, we infer that $\lambda - A - B \in \Phi(X)$ and $i(\lambda - A - B) = 0$. Therefore, $\lambda \notin \sigma_{e5}(A + B)$, whence

$$\sigma_{e5}(A + B)\backslash\{0\} \subset \left[\sigma_{e5}(A) \bigcup \sigma_{e5}(B)\right]\backslash\{0\}. \tag{4.37}$$

To prove the inverse inclusion of (4.37). Let $\lambda \notin \sigma_{e5}(A + B)\backslash\{0\}$, then $A + B - \lambda \in \Phi(X)$ and $i(A + B - \lambda) = 0$. Since AB and BA are in $\mathcal{PR}(\Phi(X))$, then it is easy to show that $A - \lambda \in \Phi(X)$ and $B - \lambda \in \Phi(X)$. On the other hand, applying (4.31), (4.35), (4.36) and Proposition 3.3.2 (iii), we have

$$i[(B - \lambda)(A - \lambda)] = i(A - \lambda) + i(B - \lambda)$$
$$= i(A + B - \lambda) \tag{4.38}$$
$$= 0.$$

Since A is bounded single valued, we get $\rho(A) \neq \emptyset$. Besides, Φ_A is connected, this together with Proposition 3.6.2 (i), allow us to deduce that

$$\sigma_{e4}(A) = \sigma_{e5}(A).$$

Using the last equality and the fact that $A - \lambda \in \Phi(X)$, we deduce that $i(A - \lambda) = 0$. It follows from (4.38) that $i(B - \lambda) = 0$. We conclude that $\lambda \notin \sigma_{e5}(A) \bigcup \sigma_{e5}(B)$. Hence,

$$\left[\sigma_{e5}(A) \bigcup \sigma_{e5}(B)\right] \backslash \{0\} \subset \sigma_{e5}(A + B) \backslash \{0\}.$$

So, we have proved (4.29).

(v) Suppose that $\lambda \notin \sigma_{e1}(A) \bigcup \sigma_{e1}(B) \bigcup \{0\}$, then $A - \lambda \in \Phi_+(X)$ and $B - \lambda \in \Phi_+(X)$. Using Theorem 2.14.3, we have $(A - \lambda)(B - \lambda) \in \Phi_+(X)$. Since $AB \in \mathcal{PR}(\Phi_+(X))$ and

$$AB(0) \subset BA(0) = B(0) \subset (A + B)(0),$$

then by using (4.31) and Proposition 3.3.2 (i), we have $A + B - \lambda \in \Phi_+(X)$. So, $\lambda \notin \sigma_{e1}(A + B)$. Therefore,

$$\sigma_{e1}(A + B) \backslash \{0\} \subset \sigma_{e1}(A) \bigcup \sigma_{e1}(B) \bigcup \{0\}.$$

Suppose $\lambda \notin \sigma_{e1}(A + B) \bigcup \{0\}$, then $A + B - \lambda \in \Phi_+(X)$. Since AB and BA are in $\mathcal{PR}(\Phi_+(X))$, and by applying Eqs. (4.31), (4.32), we have

$$(A - \lambda)(B - \lambda) \in \Phi_+(X), \ (B - \lambda)(A - \lambda) \in \Phi_+(X).$$

It is clear that $A - \lambda \in \Phi_+(X)$ and $B - \lambda \in \Phi_+(X)$. Hence, $\lambda \notin \sigma_{e1}(A) \bigcup \sigma_{e1}(B)$. Therefore,

$$\left[\sigma_{e1}(A) \bigcup \sigma_{e1}(B)\right] \backslash \{0\} \subset \sigma_{e1}(A + B) \backslash \{0\}.$$

This proves (4.30).

(vi) Now, suppose that $\lambda \notin \sigma_{eap}(A) \bigcup \sigma_{eap}(B) \bigcup \{0\}$, then by Corollary 4.1.1, we have $A - \lambda \in \Phi_+(X)$, $i(A - \lambda) \leq 0$, $B - \lambda \in \Phi_+(X)$ and $i(B - \lambda) \leq 0$. Using Theorem 2.14.3, we have $(A - \lambda)(B - \lambda) \in \Phi_+(X)$ and

$$i\left((A - \lambda)(B - \lambda)\right) = i(A - \lambda) + i(B - \lambda) + \dim\left(\frac{X}{R(B - \lambda) + \mathcal{D}(A - \lambda)}\right) - $$
$$\dim[A(0) \bigcap N(A - \lambda)].$$

Since $R(B - \lambda) + \mathcal{D}(A - \lambda) = X$, then

$$i\left((A - \lambda)(B - \lambda)\right) = i(A - \lambda) + i(B - \lambda) - \dim[A(0) \bigcap N(A - \lambda)] \leq 0.$$

On the other hand, since $AB \in \mathcal{PR}(\Phi_+(X))$, then $A + B - \lambda \in \Phi_+(X)$ and $i((A - \lambda)(B - \lambda)) = i(A + B - \lambda) \le 0$. Again, applying Corollary 4.1.1, it is clear that $\lambda \notin \sigma_{eap}(A + B)$. Hence,

$$\sigma_{eap}(A + B)\backslash\{0\} \subset \sigma_{eap}(A)\bigcup\sigma_{eap}(B)\bigcup\{0\}.$$

The rest of the proof is analogous to the previous case, it use Lemma 2.13.2 and $\mathcal{D}(A) + R(B) = X$.

The proofs of (vi) and (vii) may be achieved by using the same reasoning as (iii). Q.E.D.

Theorem 4.3.3 *Let* $\lambda \in \mathbb{C}$, T, $S \in BCR(X)$, *and* $S(0) \subset T(0)$. *If there exists* H_λ *such that* $H_\lambda(\lambda - T - S) = I - K$ *and* $-\lambda^{-1}TSH_\lambda$ *is demicompact, then* $\lambda \in \sigma_{e1}(T + S)\backslash\{0\}$ *implies that* $\lambda \in [\sigma_{e1}(T)\bigcup\sigma_{e1}(S)]\backslash\{0\}$. ◇

Proof. Since $T \in BCR(X)$, then $(\lambda - T)(\lambda - S) = \lambda(\lambda - T) - (\lambda - T)S$. So,

$$\lambda(\lambda - T) - (\lambda - T)S = \lambda(\lambda - T - S) + TS.$$

Hence, $TSK(0) = TS(0) \subset (\lambda(\lambda - T - S) + TS)(0)$ and $\mathcal{D}(\lambda(\lambda - T - S) + TS) = \mathcal{D}(TSK) = X$. Thus,

$$(\lambda - T)(\lambda - S) = \lambda(\lambda - T - S) + TS + TSK - TSK.$$

Clearly, $\mathcal{D}(TS) = X$, then

$$\begin{aligned} (\lambda - T)(\lambda - S) &= \lambda(\lambda - T - S) + TS(I - K) + TSK \\ &= \lambda(\lambda - T - S) + TSH_\lambda(\lambda - T - S) + TSK. \end{aligned}$$

So, $\lambda(\lambda - T - S) + TSH_\lambda(\lambda - T - S) + TSK = \lambda(I + \lambda^{-1}TSH_\lambda)(\lambda - T - S) + TSK$. We conclude

$$(\lambda - T)(\lambda - S) = \lambda(I + \lambda^{-1}TSH_\lambda)(\lambda - T - S) + TSK.$$

Since $\mathcal{D}(TSK) = \mathcal{D}(\lambda(I + \lambda^{-1}TSH_\lambda)(\lambda - T - S))$ and $TSK(0) \subset \lambda(\lambda - T - S) + TS(I - K)(0) = \lambda(I + \lambda^{-1}TSH_\lambda)(\lambda - T - S)(0)$, then

$$(\lambda - T)(\lambda - S) - TSK = \lambda(I + \lambda^{-1}TSH_\lambda)(\lambda - T - S). \tag{4.39}$$

Let $\lambda \notin [\sigma_{e1}(T)\bigcup\sigma_{e1}(S)]\backslash\{0\}$, then $\lambda - T \in \Phi_+(X)$ and $\lambda - S \in \Phi_+(X)$. Hence, $TSK \in \mathcal{PR}(\Phi_+(X))$. In view of both Lemma 2.14.3 (iii) and Proposition 2.7.2, we get

$$(\lambda - T)(\lambda - S) - TSK \in \Phi_+(X).$$

Again, by using Lemma 2.14.3 (v) and (4.39), we show that $\lambda - T - S \in \Phi_+(X)$. So, we deduce that $\lambda \notin \sigma_{e1}(T + S)\backslash\{0\}$. Q.E.D.

4.4 S-Essential Spectra of Linear Relations

4.4.1   Characterization of S-Essential Spectra of Linear Relations

Theorem 4.4.1 *Let $T \in \mathcal{CR}(X)$ and $S \in \mathcal{BR}(X)$ such that $S(0) \subset T(0)$ and $\|S\| \neq 0$. Then,*

(i) $\sigma_{ei,S}(T)$, $i = 1, 2, 3, 4, 5$ are closed.
(ii) The index is constant in any component of $\rho_{ei,S}(\cdot)$, $i = 1, 2, 3, 4, 5$.   ◇

Proof. *(i)* For $i = 5$. Let $\lambda_0 \in \rho_{e5,S}(T)$. Then, $R(T - \lambda_0 S)$ is closed. Hence, $T - \lambda_0 S$ is open. Then, according to Proposition 2.9.3 (i), we have $\gamma(T - \lambda_0 S) > 0$. Thus, $\frac{\gamma(T - \lambda_0 S)}{\|S\|} > 0$. Consider $\mu \in \mathbb{C}$ such that $|\mu - \lambda_0| < \frac{\gamma(T - \lambda_0 S)}{\|S\|}$. Then, according to Corollary 3.3.1, $\lambda_0 S - \mu S - \lambda_0 S + T \in \Phi(X)$ and $i(\lambda_0 S - \mu S - \lambda_0 S + T) = 0$. Since $\lambda_0 S(0) \subset T(0) = (\mu S - T)(0)$ and $\mathcal{D}(\mu S - T) = \mathcal{D}(T) \subset \mathcal{D}(S)$, then the use Proposition 2.3.4 allows us to conclude that $\lambda_0 S - \mu S - \lambda_0 S + T = \mu S - T$. Hence, $\mu S - T \in \Phi(X)$ and $i(\mu S - T) = 0$. Therefore, $\rho_{e5}(T)$ is open. For the other cases, the same arguments have been used.

(ii) Let λ_1 and λ_2 be any two points in $\rho_{ei,S}(T)$, $i = 1, 2, 3, 4, 5$ which are connected by a smooth curve Γ whose points are all in $\rho_{ei,S}(T)$, $i = 1, 2, 3, 4, 5$. Since $\rho_{ei,S}(T)$, $i = 1, 2, 3, 4, 5$ are an open set, then for each $\lambda \in \Gamma$, then by Corollary 3.3.1, there exists an $\varepsilon > 0$ such that, for all $\mu \in \mathbb{C}, 0 < |\mu - \lambda| < \varepsilon$, $\mu \in \rho_{ei,S}(T)$, $i = 1, 2, 3, 4, 5$ and $i(\mu S - T) = i(\lambda S - T)$. By using the Heine-Borel theorem, there exists a finite number of such sets which cover Γ. Since each of these sets overlaps with, at least, another set and since $i(\mu S - T)$ is constant on each one, we see that $i(\lambda_1 S - T) = i(\lambda_2 S - T)$.   Q.E.D.

4.4.2   Characterization of $\sigma_{e5,S}(\cdot)$

Theorem 4.4.2 *Let $T \in \mathcal{CR}(X)$ and $S \in \mathcal{BR}(X)$ such that $S(0) \subset T(0)$, $S \neq 0$, and $S \neq T$. Then,*

$$\sigma_{e5,S}(T) = \bigcap_{K \in \mathcal{K}_T(X)} \sigma_S(T + K).$$ ◇

Proof.   Let $\lambda \notin \sigma_{e5,S}(T)$. Then, $T - \lambda S \in \Phi(X)$ and $i(T - \lambda S) = 0$. Hence, $\alpha(T - \lambda S) = \beta(T - \lambda S) = n < \infty$. So, there exists an everywhere defined single valued K with $\dim R(K) \leq \alpha(T - \lambda S)$ such that $T - \lambda S - K$ is bijective, where K is defined by

$$Kx := \sum_{i=1}^{n} x_i'(x) y_i, \quad x \in X,$$

$\{x_1, \ldots, x_n\}$ is a basis of $N(T - \lambda S)$, and $x_1', \ldots, x_n'$ are linear functionals such that $x_i'(x_j) = \delta_{ij}$. Choose $y_1, \ldots, y_n \in X$ such that $[y_1], \ldots, [y_n] \in X/R(T - \lambda S)$ are linearly independent ($n = \alpha(T - \lambda S) = \beta(T - \lambda S) < \infty$). Hence, it is clear that K is a bounded finite rank operator. So, $K \in \mathcal{KR}(X)$ and $K(0) = 0 \subset (T - \lambda S)(0) = T(0) - S(0) = T(0)$. It is clear that $K \in \mathcal{K}_T(X)$ and also $(\lambda S - T) - K = \lambda S - (T + K)$ is bijective. Therefore, $\lambda \in \rho_S(T + K)$. This shows that $\lambda \notin \bigcap_{K \in \mathcal{K}_T(X)} \sigma_S(T + K)$. So,

$$\bigcap_{K \in \mathcal{K}_T(X)} \sigma_S(T + K) \subset \sigma_{e5,S}(T). \tag{4.40}$$

To prove the inverse inclusion of (4.40). Suppose $\lambda \notin \bigcap_{K \in \mathcal{K}_T(X)} \sigma_S(T + K)$, then there exists $K \in \mathcal{K}_T(X)$ such that $\lambda \in \rho_S(T + K)$. Hence, $T - \lambda S + K \in \Phi(X)$ and $i(T - \lambda S + K) = 0$. Since $K \in \mathcal{K}_T(X)$, it follows from Lemma 2.16.1 that $T - \lambda S + K - K \in \Phi(X)$ and $i(T - \lambda S + K - K) = i(T - \lambda S + K) = 0$. So, by using Lemma 2.16.1, we infer that $T - \lambda S \in \Phi(X)$ and $i(T - \lambda S) = 0$. We conclude $\lambda \notin \sigma_{e5,S}(T)$. Hence, $\sigma_{e5,S}(T) \subset \bigcap_{K \in \mathcal{K}_T(X)} \sigma_S(T + K)$.   Q.E.D.

Corollary 4.4.1 *Let* $T \in \mathcal{CR}(X)$ *and* $S \in \mathcal{BR}(X)$ *such that* $S(0) \subset T(0)$, $S \neq 0$, *and* $S \neq T$. *Then*,

$$\sigma_{e5,S}(T) = \bigcap_{P \in \mathcal{PR}(\Phi(X))} \sigma_S(T + P). \qquad \diamond$$

Proof. Let $\mathcal{O} := \bigcap_{P \in \mathcal{PR}(\Phi(X))} \sigma_S(T + P)$. Since $\mathcal{K}_T(X) \subset \mathcal{PR}(\Phi(X))$, we infer that $\mathcal{O} \subset \sigma_{e5,S}(T)$. Conversely, let $\lambda \notin \mathcal{O}$, then there exists $P \in \mathcal{PR}(\Phi(X))$ such that $\lambda \notin \sigma_S(T + P)$. Therefore, $\lambda \in \rho_S(T + P)$. Hence, $\lambda S - T + P \in \Phi(X)$ and $i(\lambda S - T + P) = 0$. Now, the use of Proposition 3.3.2 makes us conclude that $\lambda S - T + P - P \in \Phi(X)$ and $i(\lambda S - T + P - P) = 0$. Finally, Corollary 2.16.1, shows that $\lambda S - T \in \Phi(X)$ and $i(\lambda S - T) = 0$. So, $\lambda \notin \sigma_{e5,S}(T)$.   Q.E.D.

4.4.3 Relationship Between $\sigma_{e4,S}(\cdot)$ and $\sigma_{e5,S}(\cdot)$

Theorem 4.4.3 *Let* $T \in CR(X)$ *and* $S \in BR(X)$ *such that* $S(0) \subset T(0)$, $S \neq 0$, *and* $S \neq T$. *Then,*

(*i*) $\sigma_{e5,S}(T) = \sigma_{e4,S}(T) \bigcup \{\lambda \in \mathbb{C} \text{ such that } i(T - \lambda S) \neq 0\}$.
(*ii*) *If* $\rho_{e4,S}(T)$ *is connected and* $\rho_S(T) \neq \emptyset$, *then*

$$\sigma_{e4,S}(T) = \sigma_{e5,S}(T). \qquad \diamondsuit$$

Proof. (*i*) Let $\lambda \notin \sigma_{e4,S}(T) \bigcup \{\lambda \in \mathbb{C} \text{ such that } i(T - \lambda S) \neq 0\}$. Then, $T - \lambda S \in \Phi(X)$ and $i(T - \lambda S) = 0$. Hence, $\lambda \notin \sigma_{e5,S}(T)$. So,

$$\sigma_{e5,S}(T) \subset \sigma_{e4,S}(T) \bigcup \{\lambda \in \mathbb{C} \text{ such that } i(T - \lambda S) \neq 0\}. \qquad (4.41)$$

To prove the inverse inclusion of (4.41). Suppose $\lambda \notin \sigma_{e5,S}(T)$, then by using Corollary 4.4.1, there exists $K \in \mathcal{PR}(\Phi(X))$ such that $T - \lambda S - K$ is bijective. Moreover, by virtue of Proposition 2.3.4, we infer that $T - \lambda S = T - \lambda S + K - K$. So, by using Proposition 3.3.2 (*iii*), $T - \lambda S \in \Phi(X)$ and $i(T - \lambda S) = 0$. Thus, $\lambda \notin \sigma_{e4,S}(T)$ and $\lambda \notin \sigma_{e4,S}(T) \bigcup \{\lambda \in \mathbb{C} \text{ such that } i(T - \lambda S) \neq 0\}$. Hence, $\sigma_{e4,S}(T) \bigcup \{\lambda \in \mathbb{C} \text{ such that } i(T - \lambda S) \neq 0\} \subset \sigma_{e5,S}(T)$.

(*ii*) Clearly $\sigma_{e4,S}(T) \subset \sigma_{e5,S}(T)$. The opposite inclusion can be obtained by showing that

$$\rho_{e4,S}(T) \bigcap \sigma_{e5,S}(T) = \emptyset.$$

Assume that there exists

$$\lambda_0 \in \rho_{e4,S}(T) \bigcap \sigma_{e5,S}(T). \qquad (4.42)$$

Let $\lambda_1 \in \rho_S(T)$, then $\lambda_1 S - T \in \Phi_+(X) \bigcap \Phi_-(X)$ and $i(\lambda_1 S - T) = 0$. Since $\rho_{e4,S}(T)$ is connected, it follows from Theorem 4.4.1 (*ii*) that $i(\lambda S - T)$ is constant on any component of $\rho_{e4,S}(T)$. Therefore, $i(\lambda_1 S - T) = i(\lambda_0 S - T) = 0$. Hence, $\lambda_0 \notin \sigma_{e5,S}(T)$ which contradicts the (4.42). Q.E.D.

4.5 Racočević and Schmoeger S-Essential Spectra of a Linear Relations

The main purpose of this section is to prove that the properties of the Racočević and Schmoeger S-essential spectra of a linear relation for closed densely defined operators in Banach spaces obtained in Refs. [2, 65] are valid for closed linear relations.

Theorem 4.5.1 *Let $T \in \mathcal{CR}(X)$ and $S \in \mathcal{BR}(X)$ such that $S(0) \subset T(0)$ and $S \neq 0$. Then,*

(i) $\sigma_{eap,S}(T) = \sigma_{e1,S}(T) \bigcup \{\lambda \in \mathbb{C}$ such that $i(T - \lambda S) > 0\}$.
(ii) $\sigma_{e\delta,S}(T) = \sigma_{e2,S}(T) \bigcup \{\lambda \in \mathbb{C}$ such that $i(T - \lambda S) < 0\}$.   ◇

Proof.   For all $\lambda \in \mathbb{C}$, we have $\lambda S(0) \subset T(0)$ and $\overline{\mathcal{D}(T)} \subset \mathcal{D}(\lambda S) = X$. So, by using Proposition 2.7.2, we infer that $\lambda S - T$ is closed.

(i) Let $\lambda \notin \sigma_{eap,S}(T)$. Then, there exist $K \in \mathcal{K}_T(X)$ such that

$$
\begin{aligned}
\lambda S - T - K &= (\lambda S - T) - K \\
&= A - K
\end{aligned}
$$

is bounded below. Since $A_\lambda := \lambda S - T$ is closed, $K(0) \subset T(0) = A_\lambda(0)$ and $\mathcal{D}(A_\lambda) = \mathcal{D}(T) \subset \mathcal{D}(K)$, then by using Proposition 2.7.2, we have $\lambda S - (T + K)$ is closed. Since $\lambda S - (T + K)$ is injective with closed range and open, then $\lambda S - (T + K) \in \Phi_+(X)$ and

$$
\begin{aligned}
i((\lambda S - T) - K) &= \alpha(\lambda S - T - K) - \beta(\lambda S - T - K) \\
&= 0 - \beta(\lambda S - T - K) \leq 0.
\end{aligned}
$$

Therefore, $\lambda S - (T + K) \in \Phi_+(X)$ and $i(\lambda S - (T + K)) \leq 0$. By Lemma 2.16.1 (i), we have $\lambda S - (T + K) + K = (\lambda S - T) + K - K \in \Phi_+(X)$ and $i(\lambda S - (T + K) + K) = i(\lambda S - (T + K)) \leq 0$. Since $K(0) \subset T(0) = (\lambda S - T)(0)$ and $\mathcal{D}(\lambda S - T) = \mathcal{D}(T) \subset \mathcal{D}(K)$, and by using Proposition 2.3.4, we have $\lambda S - T \in \Phi_+(X)$ and $i(\lambda S - T) \leq 0$. Therefore, $\lambda \notin \sigma_{e1,S}(T) \bigcup \{\lambda \in \mathbb{C}$ such that $i(\lambda S - T) > 0\}$. Conversely, let $\lambda \notin \sigma_{e1,S}(T) \bigcup \{\lambda \in \mathbb{C}$ such that $i(\lambda S - T) > 0\}$, then $\lambda S - T \in \Phi_+(X)$ and $i(\lambda S - T) \leq 0$. We define the operator K by

$$
K : X \longrightarrow Kx := \sum_{i=1}^{n} x'_i(x) y_i, \quad x \in X,
$$

where $\{x_1, \ldots, x_n\}$ is a basis of $N(\lambda S - T)$ and $x'_1, \ldots, x'_n$ are a linear functionals such that $x'_i(x_j) = \delta_{ij}$. Choose $y_1, \ldots, y_n \in X$ such that $[y_1], \ldots, [y_n] \in X/R(\lambda S - T)$ are linearly independent (such elements exist since $n \leq \beta(\lambda S - T)$). It is clear that K is continuous, and so that K is a bounded finite rank operator. So, $K \in \mathcal{K}_T(X)$ and also $(\lambda S - T) - K = \lambda S - (T + K)$ is injective. Since $\lambda S - T \in \Phi_+(X)$, we have $\lambda S - (T + K) \in \Phi_+(X)$

with the same index that $\lambda S - T$. In this situation, we have $K \in \mathcal{K}_T(X)$, $\lambda S - (T + K)$ is injective, closed with closed range. Hence, $\lambda S - (T + K)$ is open. So, $\lambda \in \rho_{ap,S}(T + K)$. Therefore, $\lambda \notin \sigma_{eap,S}(T)$.

The proofs of (ii) is similar to the previous one.   Q.E.D.

Corollary 4.5.1 *Let* $T \in CR(X)$ *and* $S \in BR(X)$ *such that* $S(0) \subset T(0)$ *and* $S \neq 0$. *Then,*

(i) $\lambda \notin \sigma_{eap,S}(T)$ *if and only if,* $\lambda S - T \in \Phi_+(X)$ *and* $i(T - \lambda S) \leq 0$.
(iii) $\lambda \notin \sigma_{e\delta,S}(T)$ *if and only if,* $\lambda S - T \in \Phi_-(X)$ *and* $i(T - \lambda S) \geq 0$.   ◇

Proposition 4.5.1 *Let* $T \in CR(X)$ *and* $S \in BR(X)$ *such that* $S(0) \subset T(0)$, $S \neq 0$ *and* $\rho_S(T) \neq \emptyset$. *If* $\Phi_{T,S}$ *is connected, then*

$$\sigma_{e1,S}(T) = \sigma_{eap,S}(T) \ \text{and} \ \sigma_{e2,S}(T) = \sigma_{e\delta,S}(T).$$ ◇

Proof. It is easy to check that $\sigma_{e1,S}(T) \subset \sigma_{eap,S}(T)$. For the second inclusion, we take $\mu \in \rho_{e1,S}(T)$, then

$$\mu \in \Phi_{+T,S} = \Phi_{T,S} \bigcup (\Phi_{+T,S} \setminus \Phi_{T,S}).$$

Hence, we will discuss these two cases:

<u>First case</u> : If $\mu \in \Phi_{T,S}$, then $i(T - \mu S) = 0$. Indeed, let $\mu_0 \in \rho_S(T)$, then $\mu_0 \in \Phi_{T,S}$ and $i(T - \mu_0 S) = 0$. It follows from Theorem 2.12.2 that $i(T - \mu S)$ is constant on any component of $\Phi_{T,S}$. Therefore, $\rho_S(T) \subset \Phi_{T,S}$, then $i(T - \mu S) = 0$ for all $\mu \in \Phi_{T,S}$. This shows that $\mu \in \rho_{eap,S}(T)$.

<u>Second case</u> : If $\mu \in (\Phi_{+T,S} \setminus \Phi_{T,S})$, then $\alpha(T - \mu S) < \infty$ and $\beta(T - \mu S) = +\infty$. So, $i(T - \mu S) = -\infty < 0$. Hence, we obtain the second inclusion from the above two cases. A same reasoning as before leads to the second equality, which completes the proof.   Q.E.D.

4.5.1   Stability of Racočević and Schmoeger S-Essential Spectra of a Linear Relations

We start our investigation with the following result.

Theorem 4.5.2 *Let* X *be complete,* $T \in CR(X)$, A, $S \in LR(X)$ *satisfy* $A(0) \subset S(0) \subset T(0)$ *and* $\dim \mathcal{D}(A) = \infty$. *Let* S *be* T-*bounded with* T-*bound* δ_1

and A is T-precompact with T-bound δ_2. Let $\lambda \in \mathbb{C}$ such that $\delta_2 + |\lambda|\delta_1 < 1$, then

(i) $\lambda \in \sigma_{eap,S}(T + A)$ if and only if, $\lambda \in \sigma_{eap,S}(T)$,
(ii) $\lambda \in \sigma_{e\delta,S}(T + A)$ if and only if, $\lambda \in \sigma_{e\delta,S}(T)$. ◇

Proof. Let A be T-precompact. Then, AG_T is precompact, and X and X_T are complete. By using Proposition 2.11.3, we get AG_T is compact. As a matter of fact, A is T-bounded with T-bound δ_2. Using the fact that S is T-bounded with T-bound δ_1, $\delta_2 + |\lambda|\delta_1 < 1$ and, by applying Lemma 3.1.1 (i), we obtain $\lambda S - (T + A)$ is closed.

(i) Suppose that $\lambda \notin \sigma_{eap,S}(T)$. Then, by using Corollary 4.5.1, we have $\lambda S - T \in \Phi_+(X)$. In view of Proposition 2.14.3 (ii), we get $(\lambda S - T)G_{\lambda S-T} \in \Phi_+(X_T, X)$, which provides by referring to Theorem 2.6.1 that $(\lambda S - T)G_T \in \Phi_+(X_T, X)$. Since AG_T is compact and $\dim \mathcal{D}(A) = \dim \mathcal{D}(AG_T) = \infty$, then by using Corollary 2.16.1 (iv) and Lemma 2.16.2, it follows that $(\lambda S - (T + A))G_T \in \Phi_+(X_T, X)$. The use of Theorem 2.6.1, we obtain $(\lambda S - (T + A))G_{\lambda S-(T+A)} \in \Phi_+(X_T, X)$. Thus, Proposition 2.11.3 yields $\lambda S - (T+A) \in \Phi_+(X)$ and in view of Proposition 2.11.3, we have $i(\lambda S - T) = i(\lambda S - (T + A))$. Using Corollary 4.5.1, we have $\lambda \notin \sigma_{eap,S}(T + A)$ which yields $\sigma_{eap,S}(T + A) \subset \sigma_{eap,S}(T)$. Conversely, let $\lambda \notin \sigma_{eap,S}(T + A)$. Then, by using Corollary 4.5.1, we have $\lambda S - (T + A) \in \Phi_+(X)$. From Proposition 2.14.3 (ii), we get $(\lambda S - (T + A))G_{\lambda S-(T+A)} \in \Phi_+(X_T, X)$, which entails by referring to Theorem 2.6.1, $(\lambda S - (T + A))G_T \in \Phi_+(X_T, X)$. Since AG_T is compact, then by using both Corollary 2.16.1 (iv) and Lemma 2.16.2, it follows that $(\lambda S - T)G_T \in \Phi_+(X_T, X)$. From Theorem 2.6.1, we obtain $(\lambda S - T)G_{\lambda S-T} \in \Phi_+(X_T, X)$. Using Proposition 2.14.3 (ii), we get $\lambda S - T \in \Phi_+(X)$. From Proposition 2.11.3, we have $i(\lambda S - T) = i(\lambda S - (T+A))$. So, by Corollary 4.5.1, $\lambda \notin \sigma_{eap,S}(T)$. Thus, $\sigma_{eap,S}(T + A) = \sigma_{eap,S}(T)$.

(ii) Now, suppose that $\lambda \notin \sigma_{e\delta,S}(T)$, then by Corollary 4.5.1, we have $\lambda S - T \in \Phi_-(X)$. Applying Proposition 2.11.3, we obtain $(\lambda S - T)G_{\lambda S-T} \in \Phi_-(X_T, X)$. Using Theorem 2.6.1, we get $(\lambda S - T)G_T \in \Phi_-(X_T, X)$. Since AG_T is precompact, then in view of Theorem 2.16.3 (i), we obtain $(\lambda S - (T + A))G_T \in \Phi_-(X_T, X)$. Resorting to Theorem 2.6.1, we get $(\lambda S - (T + A))G_{\lambda S-(T+B)} \in \Phi_-(X_T, X)$. As a matter of fact, applying Proposition 2.11.3, we get $\lambda S - (T+A) \in \Phi_-(X)$. From Proposition 2.11.3, we have $i(\lambda S - T) = i(\lambda S - (T+A))$. Hence, by Corollary 4.5.1, $\lambda \notin \sigma_{e\delta,S}(T+A)$. Then, $\sigma_{e\delta,S}(T + A) \subset \sigma_{e\delta,S}(T)$. Conversely, let $\lambda \notin \sigma_{e\delta,S}(T + A)$, then by

Corollary 4.5.1, we obtain $\lambda S - (T + A) \in \Phi_-(X)$. Using Proposition 2.14.3 (i), we get $(\lambda S - (T+A))G_{\lambda S-(T+A)} \in \Phi_-(X_T, X)$. Applying Theorem 2.6.1, we get $(\lambda S - (T + A))G_T \in \Phi_-(X_T, X)$. The latter holds if and only if, by Proposition 2.11.3, $((\lambda S - (T + A))G_T)^* \in \Phi_+(X^*, X_T^*)$. Subsequently, using Propositions 2.8.1 (iii) and 2.14.3 (i), we get $((\lambda S - T)G_T)^* + (AG_T)^* \in \Phi_+(X^*, X_T^*)$. Since AG_T is precompact, then by Proposition 2.14.3 (ii), we have $(AG_T)^*$ is precompact. Applying Theorem 2.16.3 (i), we get $((\lambda S - T)G_T)^* \in \Phi_+(X_T, X^*)$. Besides, using Theorem 2.14.4, we get $(\lambda S - T)G_T \in \Phi_-(X_T, X)$. Hence, by Proposition 2.14.3 (i), $\lambda S - T \in \Phi_-(X)$. Applying Proposition 2.11.3, we have $i(\lambda S - T) = i(\lambda S - (T+A))$. That is by Corollary 4.5.1, $\lambda \notin \sigma_{e\delta,S}(T)$. We conclude that $\sigma_{e\delta,S}(T + A) = \sigma_{e\delta,S}(T)$.   Q.E.D.

Theorem 4.5.3 *Let* $T \in CR(X)$, A, $S \in LR(X)$. *Suppose that* S *is* T-*bounded with* T-*bound* δ_1 *and* A *is* T-*bounded with* T-*bound* δ_2. *Let* $\lambda \in \mathbb{C}$ *such that* $\delta_2 + |\lambda|\delta_1 < 1$. *If* $G(A) \subset G(\lambda S) \subset G(T)$ *and* $\dim \mathcal{D}(A) = \infty$, *then*

(i) $\lambda \in \sigma_{eap,S}(T + A)$ *implies* $\lambda \in \sigma_{eap,S}(T)$.
(ii) *If* $\dim A(0) < \infty$, *then* $\lambda \in \sigma_{eap,S}(T + A)$ *if and only if,* $\lambda \in \sigma_{eap,S}(T)$. $\diamond$

Proof. Since S is T-bounded with T-bound δ_1 and A is T-bounded with T-bound δ_2 such that $\delta_2 + |\lambda|\delta_1 < 1$, then applying Lemma 3.1.1, we obtain $\lambda S - (T + A)$ is closed.

(i) Suppose that $\lambda \notin \sigma_{eap,S}(T)$, then by Corollary 4.5.1, we get $\lambda S - T \in \Phi_+(X)$ and $i(\lambda S - T) \leq 0$. Since $G(A) \subset G(\lambda S) \subset G(T)$, then applying Proposition 2.3.3, we get $G(\lambda S) \subset G(\lambda S - T)$, and $G(A) \subset G(\lambda S - T)$. On the one hand, we have

$$G(\lambda S - (T + A)) = \{(x, y) \in X \times X \text{ such that } (x, y_1) \in G(\lambda S - T) \text{ and } \\ (x, y_2) \in G(A) \subset G(\lambda S - T), \text{ where } y = y_1 + y_2\} \\ \subset G(\lambda S - T).$$

On the other hand,

$$\dim \mathcal{D}(\lambda S - (T + A)) = \dim(\mathcal{D}(\lambda S - T) \bigcap \mathcal{D}(A)) = \dim \mathcal{D}(A) = \infty.$$

Then, by Proposition 2.14.4, we obtain $\lambda S - (T+A) \in \Phi_+(X)$. Hence, $G(\lambda S - (T + A)) \subset G(\lambda S - T)$. By using Lemma 2.4.1, we get $i(\lambda S - (T + A)) \leq i(\lambda S - T) \leq 0$. So, by Corollary 4.5.1, we obtain $\lambda \notin \sigma_{eap,S}(T + A)$. Thus, $\sigma_{eap,S}(T + A) \subset \sigma_{eap,S}(T)$.

(*ii*) Since $G(A) \subset G(\lambda S) \subset G(T)$, applying Proposition 2.3.3, we get $G(\lambda S) \subset G(\lambda S - T)$ and $G(A) \subset G(\lambda S - T)$. By using Proposition 2.3.3, we obtain $G(A) \subset G(\lambda S - (T + A))$. On the one hand,

$$
\begin{aligned}
G(\lambda S - T - A + A) = &\{(x, y + z) \in X \times Y \ such \ that \\
&(x, y) \in G(\lambda S - (T + A)) \ and \ (x, z) \in G(A)\} \\
\subset &\{(x, y + z) \in X \times Y \ such \ that \ (x, y) \in G(\lambda S - (T + A)) \\
&and \ (x, z) \in G(S) \subset G(\lambda S - (T + A))\} \\
= &G(\lambda S - (T + A)).
\end{aligned}
$$

On the other hand,

$$
\dim \mathcal{D}(\lambda S - T - A + A) = \dim(\mathcal{D}(\lambda S) \bigcap \mathcal{D}(T) \bigcap \mathcal{D}(A)) = \dim \mathcal{D}(A) = \infty.
$$

Let $\lambda \notin \sigma_{eap,S}(T + A)$. Then, by Corollary 4.5.1, we have $\lambda S - (T + A) \in \Phi_+(X)$. Since $\lambda S - T$ is closed and $\dim A(0) < \infty$, then $\lambda S - T - A + A$ is closed and $G(\lambda S - T - A + A) \subset G(\lambda S - (T + A))$. Since $\dim \mathcal{D}(\lambda S - T - A + A) = \infty$, then by Proposition 2.14.4, we get $\lambda S - T - A + A \in \Phi_+(X)$. Thus, by Proposition 2.14.3 (*iv*), we obtain $\lambda S - T \in \Phi_+(X)$. Using Lemma 2.4.1, we get $i(\lambda S - T) \leq i(\lambda S - T - A + A) \leq i(\lambda S - (T + A)) \leq 0$. Hence, by Corollary 4.5.1, we obtain $\lambda \notin \sigma_{eap,S}(T)$. Therefore, $\sigma_{eap,S}(T) \subset \sigma_{eap,S}(T + A)$ which implies that $\sigma_{eap,S}(T + A) = \sigma_{eap,S}(T)$. $\hspace{2cm}$ Q.E.D.

4.6   *S*-Essential Spectra of the Sum of Two Linear Relations

In this section, we investigate the S-essential spectra of the sum of two closed linear relations defined on a Banach space by means of S-essential spectra of these two linear relations where their products are Fredholm or semi-Fredholm perturbations.

Theorem 4.6.1 *Let X be a Banach space and let A, $S \in \mathcal{L}(X)$ and $B \in BC\mathcal{R}(X)$ such that $AB \subset BA \subset SB \subset BS$. Then,*

(*i*) *If $AB \in \mathcal{PR}(\Phi(X))$, then*

$$
\sigma_{e4,S^2}(AS + BS) \backslash \{0\} \subset \left[\sigma_{e4,S}(A) \bigcup \sigma_{e4,S}(B) \right] \backslash \{0\}.
$$

If, further, $BA \in \mathcal{PR}(\Phi(X))$, $SA = AS$ and $BS = SB$, then

$$\sigma_{e4,S^2}(AS + BS)\backslash\{0\} = \left[\sigma_{e4,S}(A)\bigcup\sigma_{e4,S}(B)\right]\backslash\{0\}.$$

(ii) If $BA \in \mathcal{PR}(\Phi(X))$, then

$$\sigma_{e5,S^2}(SA + BS) \subset \left[\sigma_{e5,S}(A)\bigcup\sigma_{e5,S}(B)\right]\backslash\{0\}.$$

Moreover, if $AB \in \mathcal{PR}(\Phi(X))$, $SA = AS$, $BS = SB$ and $\Phi_{A,S}$ is connected, then

$$\sigma_{e5,S}(SA + BS)\backslash\{0\} = \left[\sigma_{e5,S}(A)\bigcup\sigma_{e5,S}(B)\right]\backslash\{0\}. \tag{4.43}$$

(iii) If $AB \in \mathcal{PR}(\Phi_+(X))$, then

$$\sigma_{e1,S^2}(SA + SB)\backslash\{0\} \subset \left[\sigma_{e1,S}(A)\bigcup\sigma_{e1,S}(B)\right]\backslash\{0\}.$$

If, further, $BA \in \mathcal{PR}(\Phi_+(X))$, $SA = AS$ and $BS = SB$, then

$$\sigma_{e1,S^2}(AS + SB)\backslash\{0\} = \left[\sigma_{e1,S}(A)\bigcup\sigma_{e1,S}(B)\right]\backslash\{0\}. \tag{4.44}$$

(iv) If the hypotheses of (iii) are satisfied, then

$$\sigma_{eap,S^2}(AS + SB)\backslash\{0\} \subset \left[\sigma_{eap,S}(A)\bigcup\sigma_{eap,S}(B)\right]\backslash\{0\}.$$

If, further, $\Phi_{S,A}$ is connected, $BA \in \mathcal{PR}(\Phi_+(X))$, $SA = AS$ and $BS = SB$, then

$$\sigma_{eap,S^2}(AS + SB)\backslash\{0\} = \left[\sigma_{eap,S}(A)\bigcup\sigma_{eap,S}(B)\right]\backslash\{0\}.$$

(v) If the hypotheses of (iv) are satisfied, then

$$\sigma_{e\delta}(A + B)\backslash\{0\} \subset \left[\sigma_{e\delta}(A)\bigcup\sigma_{e\delta}(B)\right]\backslash\{0\}.$$

If, further, $\Phi_{S,A}$ is connected, $BA \in \mathcal{PR}(\Phi_-(X))$, $SA = AS$ and $BS = SB$, then

$$\sigma_{e\delta,S^2}(AS + SB)\backslash\{0\} = \left[\sigma_{e\delta,S}(A)\bigcup\sigma_{e\delta,S}(B)\right]\backslash\{0\}.$$

(vi) If $AB \in \mathcal{PR}(\Phi_+(X))\bigcap\mathcal{PR}(\Phi_-(X))$, then

$\sigma_{e3,S^2}(AS + SB)\backslash\{0\} \subset$
$([\sigma_{e3,S}(A)\bigcup\sigma_{e3,S}(B)]\bigcup[\sigma_{e1,S}(A)\bigcap\sigma_{e2,S}(B)]\bigcup[\sigma_{e2,S}(A)\bigcap\sigma_{e1,S}(B)])\backslash\{0\}.$

Moreover, if $BA \in \mathcal{PR}(\Phi_+(X))\bigcap\mathcal{PR}(\Phi_-(X))$, $SA = AS$ and $BS = SB$, then

$\sigma_{e3}(SA + SB)\backslash\{0\} =$
$([\sigma_{e3,S}(A)\bigcup\sigma_{e3,S}(B)]\bigcup[\sigma_{e1,S}(A)\bigcap\sigma_{e2,S}(B)]\bigcup[\sigma_{e2,S}(A)\bigcap\sigma_{e1,S}(B)])\backslash\{0\}.\diamond$

Proof. We first prove, for $\lambda \in \mathbb{C}$

$$(\lambda S - A)(\lambda S - B) = AB + \lambda(\lambda S^2 - AS - SB) \qquad (4.45)$$

and

$$(\lambda S - B)(\lambda S - A) = BA + \lambda(\lambda S^2 - SA - BS). \qquad (4.46)$$

Indeed, since $\mathcal{D}(\lambda S - A) = \mathcal{D}(\lambda S) \cap \mathcal{D}(A) = X$ and, by using Proposition 2.3.6, we have $(\lambda S - A)(\lambda S - B) = (\lambda S - A)\lambda S - (\lambda S - A)B$. Further, by using Proposition 2.3.6, we have $(\lambda S - A)B \subset \lambda SB - AB$. Then, $(\lambda SB - AB)(0) = \lambda SB(0) - AB(0) \subset \lambda SB(0) - BA(0) = B(0) \subset (\lambda S - A)B(0)$. Hence, $(\lambda S - A)B(0) = (\lambda SB - AB)(0)$. Since

$$
\begin{aligned}
\mathcal{D}((\lambda S - A)B) &= \left\{ x \in \mathcal{D}(B) = X : Bx \cap \mathcal{D}(\lambda S - A) \neq \emptyset \right\} \\
&= \left\{ x \in \mathcal{D}(B) = X : Bx \cap X \neq \emptyset \right\} \\
&= \{ x \in \mathcal{D}(B) = X : Bx \neq \emptyset \} \\
&= \mathcal{D}(B) \\
&= X
\end{aligned}
$$

and $\mathcal{D}(\lambda SB - AB) = \mathcal{D}(\lambda SB) \cap \mathcal{D}(AB) = X$, we infer from Proposition 2.3.6 that

$$(\lambda S - A)B = \lambda SB - AB.$$

Hence,

$$(\lambda S - A)(\lambda S - B) = AB + \lambda(\lambda S^2 - AS - SB).$$

So, (4.45) holds. We note that

$$
\begin{aligned}
(\lambda S - B)(\lambda S - A) &= \lambda S(\lambda S - A) - B(\lambda S - A) \\
&= \lambda^2 S^2 - \lambda SA - \lambda BS + BA.
\end{aligned}
$$

Hence, (4.46) holds.

(i) Let $\lambda \notin \sigma_{e4,S}(A) \bigcup \sigma_{e4,S}(B) \bigcup \{0\}$. Then, $\lambda S - A \in \Phi(X)$ and $\lambda S - B \in \Phi(X)$. So, Theorem 2.14.3 gives $(\lambda S - A)(\lambda S - B) \in \Phi(X)$. Hence, by using (4.45), we have

$$AB + \lambda(\lambda S^2 - AS - SB) \in \Phi(X).$$

Since $AB \in \mathcal{PR}(\Phi(X))$, then

$$AB - (\lambda S - A)(\lambda S - B) \in \Phi(X).$$

This implies that

$$AB - AB - \lambda(\lambda S^2 - AS - SB) \in \Phi(X).$$

Since

$$AB(0) \subset \lambda(\lambda S^2 - AS - SB)(0) = (SB)(0),$$

then by using Lemma 2.16.1, we infer that $\lambda S^2 - AS - SB \in \Phi(X)$. Hence, $\lambda \notin \sigma_{e4,S^2}(AS + SB)$. Therefore,

$$\sigma_{e4,S^2}(AS + BS)\backslash\{0\} \subset \left[\sigma_{e4,S}(A)\bigcup\sigma_{e4,S}(B)\right]\backslash\{0\}. \tag{4.47}$$

To prove the inverse inclusion of (4.47). Suppose $\lambda \notin \sigma_{e4,S^2}(AS + SB)\bigcup\{0\}$, then $\lambda S^2 - AS - SB \in \Phi(X)$. Since $AB \in \mathcal{PR}(\Phi(X))$, $BA \in \mathcal{PR}(\Phi(X))$, and

$$AB(0) \subset (\lambda S^2 - AS - SB)(0) = (SB)(0),$$

then by using (4.45) and (4.46), we have

$$(A - \lambda S)(B - \lambda S) \in \Phi(X) \text{ and } (B - \lambda S)(A - \lambda S) \in \Phi(X). \tag{4.48}$$

Applying Theorem 2.14.3, it is clear that $\lambda S - A \in \Phi(X)$ and $\lambda S - B \in \Phi(X)$. Therefore, $\lambda \notin \sigma_{e4,S}(A)\bigcup\sigma_{e4,S}(B)$. This proves that

$$\left[\sigma_{e4,S}(A)\bigcup\sigma_{e4,S}(B)\right]\backslash\{0\} \subset \sigma_{e4,S^2}(SA + BS)\backslash\{0\}. \tag{4.49}$$

(ii) Let $\lambda \notin [\sigma_{e5,S}(A)\bigcup\sigma_{e5,S}(B)]\backslash\{0\}$. Then, $A - \lambda S \in \Phi(X)$, $i(A - \lambda S) = 0$, $B - \lambda S \in \Phi(X)$ and $i(B - \lambda S) = 0$. Using Theorem 2.14.3, we infer that $(B - \lambda)(A - \lambda S) \in \Phi(X)$. By using Theorem 2.14.3, we have

$$i\left((B - \lambda S)(A - \lambda S)\right) = i(B - \lambda S) + i(A - \lambda S) +$$
$$\dim X/[R(A - \lambda S) + \mathcal{D}(B - \lambda S)] - \dim[A(0)\bigcap N(B - \lambda S)]$$

and clearly,

$$R(S - \lambda S) + \mathcal{D}(T - \lambda S) = X \text{ and } A(0)\bigcap N(B - \lambda S) = \{0\}\bigcap N(B - \lambda S) = \{0\}.$$

Then,

$$i\left((B - \lambda S)(A - \lambda S)\right) = i(B - \lambda S) + i(A - \lambda S) = 0 \tag{4.50}$$

implies that $i\left(BA + \lambda(\lambda S^2 - SA - BS)\right) = 0$. Moreover, $BA \in \mathcal{PR}(\Phi(X))$ and

$$BA(0) = \lambda(\lambda S^2 - SA - BS)(0) = BS(0) = B(0).$$

Hence, by using Theorem 2.14.3, we infer that $\lambda S^2 - SA - BS \in \Phi(X)$ and $i\left(\lambda S^2 - SA - BS\right) = 0$. Therefore, $\lambda \notin \sigma_{e5,S^2}(SA + BS)$, whence

$$\sigma_{e5,S^2}(SA + BS) \subset \left[\sigma_{e5,S}(A) \bigcup \sigma_{e5,S}(B)\right] \setminus \{0\}. \tag{4.51}$$

To prove the inverse inclusion of (4.51). Let $\lambda \notin \sigma_{e5,S^2}(SA + BS)$, then $\lambda S^2 - SA - BS \in \Phi(X)$ and $i(\lambda S^2 - SA - BS) = 0$. Since $AB \in \mathcal{PR}(\Phi(X))$, $BA \in \mathcal{PR}(\Phi(X))$ and S commute with A and B, it is easy to show that $A - \lambda S \in \Phi(X)$ and $B - \lambda S \in \Phi(X)$. On the other hand, applying (4.45), (4.48), (4.50) and Theorem 2.14.3, we have

$$\begin{aligned} i[(B - \lambda S)(A - \lambda S)] &= i(A - \lambda S) + i(B - \lambda S) \\ &= i(\lambda S^2 - SA - BS) \\ &= 0. \end{aligned} \tag{4.52}$$

Since A is bounded single valued, we get $\rho_S(A) \neq \emptyset$. Besides, $\Phi_{S,A}$ is connected, this together with Theorem 4.4.3 (ii), allow us to deduce that

$$\sigma_{e4,S}(A) = \sigma_{e5,S}(A).$$

Using the last equality and the fact that $A - \lambda S \in \Phi(X)$, we deduce that $i(A - \lambda S) = 0$. It follows from (4.52) that $i(B - \lambda S) = 0$. We conclude $\lambda \notin \sigma_{e5,S}(A) \bigcup \sigma_{e5,S}(B)$. Hence,

$$\left[\sigma_{e5,S}(A) \bigcup \sigma_{e5,S}(B)\right] \setminus \{0\} \subset \sigma_{e5,S^2}(SA + BS) \setminus \{0\}.$$

So, we prove (4.43).

(iii) Suppose that $\lambda \notin \sigma_{e1,S}(A) \bigcup \sigma_{e1,S}(B) \bigcup \{0\}$, then $A - \lambda S \in \Phi_+(X)$ and $B - \lambda S \in \Phi_+(X)$. Using Theorem 2.14.3, we have $(A - \lambda S)(B - \lambda S) \in \Phi_+(X)$. Since $AB \in \mathcal{PR}(\Phi_+(X))$, and

$$AB(0) \subset BA(0) = B(0) \subset \lambda(\lambda S^2 - AS - SB)(0) = (SB)(0),$$

we can apply (4.45) and Theorem 2.14.3, we have $\lambda S^2 - AS - SB \in \Phi_+(X)$. So, $\lambda \notin \sigma_{e1,S^2}(AS + SB)$. Therefore,

$$\sigma_{e1,S^2}(AS + SB) \subset \sigma_{e1,S}(A) \bigcup \sigma_{e1}(B, S) \bigcup \{0\}.$$

Suppose $\lambda \notin \sigma_{e1}(A + B) \bigcup \{0\}$, then $A + B - \lambda \in \Phi_+(X)$. Since $AB \in \mathcal{PR}(\Phi_+(X))$ and $BA \in \mathcal{PR}(\Phi_+(X))$, and applying Eqs. (4.45), (4.46), we have

$$(A - \lambda S)(B - \lambda S) \in \Phi_+(X), \ (B - \lambda S)(A - \lambda S) \in \Phi_+(X).$$

By using Theorem 2.14.3, it is clear that $A - \lambda S \in \Phi_+(X)$ and $B - \lambda S \in \Phi_+(X)$. Hence, $\lambda \notin \sigma_{e1,S}(A) \bigcup \sigma_{e1,S}(B)$. Therefore,

$$\left[\sigma_{e1,S}(A) \bigcup \sigma_{e1,S}(B) \right] \backslash \{0\} \subset \sigma_{e1,S^2}(AS + SB) \backslash \{0\}.$$

This proves (4.44). The rest of proof is analogous to the previous case. This completes the proof. $\hspace{8cm}$ Q.E.D.

4.7 Pseudospectra and ε-Pseudospectra of Linear Relations

4.7.1 Some Properties of Pseudospectra and ε-Pseudospectra of Linear Relations

In this subsection, we define the pseudospectra of linear relation and study some properties.

Proposition 4.7.1 *Let X be a normed vector space and $T \in LR(X)$. Then,*

$$\bigcap_{\varepsilon > 0} \sigma_\varepsilon(T) = \sigma(T) = \bigcap_{\varepsilon > 0} \Sigma_\varepsilon(T). \hspace{3cm} \diamondsuit$$

Proof. It is clear that $\sigma(T) \subset \sigma_\varepsilon(T) \subset \Sigma_\varepsilon(T)$ for all $\varepsilon > 0$, then

$$\sigma(T) \subset \bigcap_{\varepsilon > 0} \sigma_\varepsilon(T) \subset \bigcap_{\varepsilon > 0} \Sigma_\varepsilon(T). \hspace{2.5cm} (4.53)$$

Conversely, if $\lambda \notin \sigma(T)$, then $\lambda \in \rho(T)$. Hence, $(\lambda - \widetilde{T})^{-1}$ is a bounded linear operator, where $\widetilde{T}$ is given in (2.4). So, there exists $\varepsilon > 0$ such that $\|(\lambda - \widetilde{T})^{-1}\| < \frac{1}{\varepsilon}$. Thus, $\lambda \notin \Sigma_\varepsilon(T)$. Hence, $\lambda \notin \bigcap_{\varepsilon > 0} \Sigma_\varepsilon(T)$. Then,

$$\bigcap_{\varepsilon > 0} \Sigma_\varepsilon(T) \subset \sigma(T). \hspace{3cm} (4.54)$$

Thus, the use of (4.53) and (4.54) makes us conclude that

$$\bigcap_{\varepsilon > 0} \sigma_\varepsilon(T) = \sigma(T) = \bigcap_{\varepsilon > 0} \Sigma_\varepsilon(T). \hspace{2cm} \text{Q.E.D.}$$

In the sequel of this section, X will denote Banach space over the complex field $\mathbb{C}$.

Proposition 4.7.2 *Let $\varepsilon > 0$ and $T \in CR(X)$ be injective with dense range. Then,*

$$\sigma_\varepsilon(T) \subset \{\lambda \in \mathbb{C} \text{ such that } |\lambda| > \gamma(T) - \varepsilon\}. \qquad \diamond$$

Proof. We will discuss three cases:

<u>First case</u> : If $\gamma(T) < \varepsilon$, then the result is obviously.

<u>Second case</u> : If $\gamma(T) = \varepsilon$. Then, T is open. Hence, $0 \in \rho(T)$. Furthermore, $\|T^{-1}\| = \frac{1}{\varepsilon}$. Then, $0 \notin \sigma_\varepsilon(T)$. Therefore, $\sigma_\varepsilon(T) \subset \{\lambda \in \mathbb{C} \text{ such that } |\lambda| > 0\}$.

<u>Third case</u> : If $\gamma(T) > \varepsilon$, then T is open and injective with dense range. On the other hand, for $\lambda \in \mathbb{C}$ such that $|\lambda| \leq \gamma(T) - \varepsilon$, then $|\lambda| < \gamma(T)$. By using Lemma 2.9.3, the relation $\lambda - T$ is open, injective with dense range, i.e., $\lambda \in \rho(T)$. For $x \in \mathcal{D}(T)$, we have

$$\begin{aligned} \|Tx\| &= \|Tx - \lambda x + \lambda x\| \\ &\leq \|Tx - \lambda x\| + \|\lambda x\|. \end{aligned}$$

Consequently, for $x \in \mathcal{D}(T)$, we obtain

$$\|(T - \lambda)x\| \geq \|Tx\| - |\lambda|\|x\| \geq (\gamma(T) - |\lambda|)\|x\|.$$

Therefore,

$$\gamma(\lambda - T) \geq \gamma(T) - |\lambda| \geq \gamma(T) - \gamma(T) + \varepsilon = \varepsilon.$$

So,

$$\|(\lambda - T)^{-1}\| \leq \frac{1}{\varepsilon}.$$

Then, $\lambda \in \rho_\varepsilon(T)$. Hence, $\sigma_\varepsilon(T) \subset \{\lambda \in \mathbb{C} \text{ such that } |\lambda| > \gamma(T) - \varepsilon\}$. Q.E.D.

Theorem 4.7.1 *Let $T \in CR(X)$ and $\varepsilon > 0$. Then,*

(i) $\sigma_\varepsilon(T) = \sigma_\varepsilon(T^*)$.
(ii) $\Sigma_\varepsilon(T) = \Sigma_\varepsilon(T^*)$. $\qquad \diamond$

Proof. *(i)* Let $\lambda \in \rho_\varepsilon(T^*)$. Then, $\lambda \in \rho(T^*)$ and $\|(\lambda - T^*)^{-1}\| \leq \frac{1}{\varepsilon}$. By using Remark 2.18.1, we have $\lambda \in \rho(T)$. So, $(\lambda - T)^{-1}$ is continuous. From Proposition 2.8.1 *(vi)*, we have

$$\|(\lambda - T)^{-1}\| = \|(\lambda - T^*)^{-1}\| \leq \frac{1}{\varepsilon}.$$

So, $\lambda \in \rho_\varepsilon(T)$. Conversely, if $\lambda \in \rho_\varepsilon(T)$, then $\|(\lambda - T)^{-1}\| \leq \frac{1}{\varepsilon}$. By using Proposition 2.8.1 (vi), we have

$$\|(\lambda - T)^{-1}\| = \|(\lambda - T^*)^{-1}\| \leq \frac{1}{\varepsilon}.$$

Furthermore, $\lambda \in \rho(T^*)$ by Remark 2.18.1 (iii). Then, $\lambda \in \rho_\varepsilon(T^*)$.

(ii) Let us assume that $\lambda \in \Sigma_\varepsilon(T)$. Then, using Remark 2.20.2 (i_1), we obtain

$$\lambda \in \sigma_\varepsilon(T) \bigcup \left\{ \lambda \in \mathbb{C} : \|(\lambda - T)^{-1}\| = \frac{1}{\varepsilon} \right\}.$$

There are two possible cases:

First case : Let $\lambda \in \sigma_\varepsilon(T)$. Then, by using (i) and Remark 2.20.2 (i), we deduce that $\lambda \in \Sigma_\varepsilon(T^*)$.

Second case : Let us assume that $\lambda \notin \sigma_\varepsilon(T)$ and $\|(\lambda - T)^{-1}\| = \frac{1}{\varepsilon}$. Then, $\lambda \in \rho(T)$, $\|(\lambda - T)^{-1}\| \leq \frac{1}{\varepsilon}$ and $\|(\lambda - T)^{-1}\| = \frac{1}{\varepsilon}$. This implies that $(\lambda - T)^{-1}$ is a bounded linear operator and $\|(\lambda - T)^{-1}\| = \frac{1}{\varepsilon}$. Hence, by referring to Proposition 2.8.1 (vi), we infer that $\|(\lambda - T^*)^{-1}\| = \|(\lambda - T)^{-1}\| = \frac{1}{\varepsilon}$. Consequently, $\lambda \in \Sigma_\varepsilon(T^*)$. This shows that

$$\Sigma_\varepsilon(T) \subset \Sigma_\varepsilon(T^*). \tag{4.55}$$

In order to prove the inverse inclusion, it is sufficient to proceed by the same reasoning as (4.55).   Q.E.D.

Lemma 4.7.1 *Let $T \in CR(X)$ and $\varepsilon > 0$. If $\lambda \notin \sigma(T)$, then $\lambda \in \sigma_\varepsilon(T)$ if and only if there exists $x \in D(T)$ such that $\|(\lambda - T)x\| < \varepsilon\|x\|$.*   ◇

Proof. If $\lambda \in \sigma_\varepsilon(T)$ and $\lambda \notin \sigma(T)$, then $\|(\lambda - T)^{-1}\| > \frac{1}{\varepsilon}$. So, there exists a non-zero vector $y \in X$ such that

$$\|(\lambda - T)^{-1}y\| > \frac{1}{\varepsilon}\|y\|. \tag{4.56}$$

Putting $x := (\lambda - T)^{-1}y$, then $x \in D(T)$ and

$$\begin{aligned}(\lambda - T)x &= (\lambda - T)(\lambda - T)^{-1}y \\ &= y + (\lambda - T)(0).\end{aligned}$$

By using Lemma 2.3.1, we have $y \in (\lambda - T)x$. On the other hand, $(\lambda - T)(0) = T(0)$. By Lemma 2.3.1, $0 \in T(0)$ and, from Lemma 2.5.7, we have

$$\begin{aligned}
\|(\lambda - T)x\| &= \text{dist}(y, (\lambda - T)(0)) \\
&= \text{dist}(y, T(0)) \\
&\leq \text{dist}(y, 0) \\
&\leq \|y\|.
\end{aligned} \tag{4.57}$$

So, from (4.56) and (4.57), we have

$$\|x\| > \frac{1}{\varepsilon}\|y\| \geq \frac{1}{\varepsilon}\|(\lambda - T)x\|.$$

Hence, $\|(\lambda - T)x\| < \varepsilon\|x\|$. Conversely, assume that there exists $x \in \mathcal{D}(T)$ such that $\|(\lambda - T)x\| < \varepsilon\|x\|$. Since $\lambda \in \rho(T)$, then $\lambda - T$ is injective, open and, we have

$$\gamma(\lambda - T)\|x\| \leq \|(\lambda - T)x\| < \varepsilon\|x\|.$$

Hence, $0 < \gamma(\lambda - T) < \varepsilon$. Using Lemma 2.5.7, we have $\gamma(\lambda - T) = \|(\lambda - T)^{-1}\|^{-1}$. So, $\|(\lambda - T)^{-1}\| > \frac{1}{\varepsilon}$. \hfill Q.E.D.

Proposition 4.7.3 *Let X be a Banach space, $T \in \mathcal{CR}(X)$ and $\varepsilon > 0$. Then,*

$$\Sigma_\varepsilon(T) = \sigma(T) \bigcup \left\{ \lambda \in \mathbb{C} : \exists \ (x_n) \subset \mathcal{D}(T), \|x_n\| = 1 \ and \ \lim_{n \to \infty} \|(\lambda - T)x_n\| \leq \varepsilon \right\}.$$

Proof. Let $\lambda \in \left\{ \lambda \in \mathbb{C} : \exists \ (x_n) \subset \mathcal{D}(T), \|x_n\| = 1 \ and \ \lim_{n \to \infty} \|(\lambda - T)x_n\| \leq \varepsilon \right\}$ and $\lambda \notin \sigma(T)$. Since $\lambda \in \rho(T)$, then $(\lambda - T)^{-1}$ is a bounded operator. This implies from Lemma 2.9.1 that $\gamma(\lambda - T) = \|(\lambda - T)^{-1}\|^{-1} < \infty$. By referring to Lemma 2.5.7 (i), we have

$$\begin{aligned}
\|(\lambda - T)x_n\| &= \text{dist}\left((\lambda - T)x_n, 0\right) \\
&\geq \gamma(\lambda - T) \ \text{dist}\left(x_n, (\lambda - T)^{-1}(0)\right).
\end{aligned}$$

The fact that $(\lambda - T)^{-1}$ is an operator implies that $(\lambda - T)^{-1}(0) = 0$. Hence,

$$\begin{aligned}
\|(\lambda - T)x_n\| &\geq \gamma(\lambda - T) \ \text{dist}(x_n, 0) \\
&\geq \gamma(\lambda - T) \ \|x_n\| \\
&\geq \gamma(\lambda - T) \ \text{as} \ \|x_n\| = 1.
\end{aligned}$$

This implies from Lemma 2.9.1 that

$$\|(\lambda - T)x_n\| \geq \|(\lambda - T)^{-1}\|^{-1}.$$

Therefore, $\|(\lambda - T)^{-1}\| \geq \frac{1}{\varepsilon}$. As a result, $\lambda \in \Sigma_\varepsilon(T)$, as desired. Conversely, let us assume that $\lambda \in \rho(T)$ and $\|(\lambda - T)^{-1}\| \geq \frac{1}{\varepsilon}$. In view of Lemma 2.5.7

(ii) implies that for every $n \in \mathbb{N}\setminus\{0\}$, there exists $(y_n)_n$ such that $\|y_n\| \leq 1$ and

$$\|(\lambda - T)^{-1}\| - \frac{1}{n} \leq \|(\lambda - T)^{-1}y_n\| \leq \|(\lambda - T)^{-1}\|.$$

Consequently,

$$
\begin{aligned}
\lim_{n\to+\infty} \|(\lambda - T)^{-1}y_n\| &= \lim_{n\to+\infty} \|(\lambda - T)^{-1}\| \\
&= \|(\lambda - T)^{-1}\| \\
&\geq \frac{1}{\varepsilon}.
\end{aligned}
\tag{4.58}
$$

Since $(\lambda - T)^{-1}$ is an operator, then we can take $x_n = \|(\lambda - T)^{-1}y_n\|^{-1}(\lambda - T)^{-1}y_n$. Hence, $(x_n)_n \subset \mathcal{D}(T), \|x_n\| = 1$ and $y_n \in \|(\lambda - T)^{-1}y_n\| (\lambda - T)x_n$. Therefore,

$$
\begin{aligned}
\|\|(\lambda - T)^{-1}y_n\| (\lambda - T)x_n\| &= \text{dist}\left(y_n, \|(\lambda - T)^{-1}y_n\| (\lambda - T)(0)\right) \\
&= \text{dist}\left(y_n, (\lambda - T)(0)\right).
\end{aligned}
$$

Consequently,

$$
\begin{aligned}
\|(\lambda - T)x_n\| &\leq \|(\lambda - T)^{-1}y_n\|^{-1}\|y_n\| \\
&\leq \|(\lambda - T)^{-1}y_n\|^{-1}.
\end{aligned}
$$

Hence, by applying (4.58), we obtain $\lim_{n\to+\infty} \|(\lambda - T)x_n\| \leq \varepsilon$. This enables us to conclude that

$$\Sigma_\varepsilon(T) = \sigma(T) \bigcup \left\{\lambda \in \mathbb{C} : \exists\ (x_n) \subset \mathcal{D}(T), \|x_n\| = 1 \text{ and } \lim_{n\to\infty} \|(\lambda - T)x_n\| \leq \varepsilon\right\}.$$

This completes the proof. Q.E.D.

4.7.2 Characterization of Pseudospectra of Linear Relations

Theorem 4.7.2 *Let $T \in \mathcal{CR}(X)$ and $\varepsilon > 0$. Then, the following three conditions are equivalent.*

(i) $\lambda \in \sigma_\varepsilon(T)$.

(ii) *There exists a continuous linear relation $S \in \mathcal{LR}(X)$ satisfying $\mathcal{D}(S) \supset \mathcal{D}(T)$, $S(0) \subset T(0)$, $\|S\| < \varepsilon$ such that $\lambda \in \sigma(T + S)$.*

(iii) *Either $\lambda \in \sigma(T)$ or $\|(\lambda - T)^{-1}\| > \dfrac{1}{\varepsilon}$.* $\diamondsuit$

Proof. $(i) \implies (ii)$ Assume that $\lambda \in \sigma_\varepsilon(T)$. We will discuss these two cases:

<u>First case</u> : If $\lambda \in \sigma(T)$, then we may put $S = 0$.

<u>Second case</u> : if $\lambda \notin \sigma(T)$, then by using Lemma 4.7.1, there exists $x_0 \in \mathcal{D}(T)$, $\|x_0\| = 1$ such that

$$\|(\lambda - T)x_0\| < \varepsilon.$$

By using Theorem 2.8.1, there exists $x' \in X^*$ such that $\|x'\| = 1$ and $x'(x_0) = \|x_0\|$. We define the relation $S : X \longrightarrow X$ by $S(x) := x'(x)(\lambda - T)x_0$. It is clear that $\mathcal{D}(S) = X$ and $S(0) = 0$ which implies that S is everywhere defined and single value. Moreover, for $x \in X$, we have

$$
\begin{aligned}
\|Sx\| &\leq \|x'(x)(\lambda - T)x_0)\| \\
&\leq \|x'\|\|x\|\|(\lambda - T)x_0\| \\
&\leq \varepsilon\|x\|.
\end{aligned}
$$

This implies that $S \in BR(X)$ and $\|S\| \leq \varepsilon$. In addition, we have

$$
\begin{aligned}
(\lambda - (T + S))x_0 &= (\lambda - T - S)x_0 \\
&= (\lambda - T)x_0 - Sx_0 \\
&= (\lambda - T)x_0 - x'(x_0)(\lambda - T)x_0 \\
&= (\lambda - T)x_0 - (\lambda - T)x_0 \\
&= (\lambda - T)(0) \\
&= (\lambda - T)(0) - S(0) \\
&= (\lambda - (T + S))(0),
\end{aligned}
$$

then $0 \neq x_0 \in N(\lambda - (T + S))$. Hence, $\lambda - (T + S)$ is not injective. So, $\lambda \in \sigma(T + S)$.

$(ii) \implies (iii)$ We derive a contradiction from the assumption that $\lambda \notin \sigma(T)$ and $\|(\lambda - T)^{-1}\| \leq \frac{1}{\varepsilon}$. By using Lemma 2.5.7, we have $\lambda \in \rho(T)$ and $\gamma(\lambda - T) \geq \varepsilon$. So, from Remark 2.18.1 (i), $\lambda - T$ is injective, open with dense range. Furthermore, $S(0) \subset T(0) = (\lambda - T)(0)$, $\mathcal{D}(S) \supset \mathcal{D}(T) = \mathcal{D}(\lambda - T)$ and $\|S\| < \varepsilon \leq \gamma(\lambda - T)$. Then, by Lemma 2.9.3, $\lambda - T - S$ is injective, open with dense range. So, $\lambda \in \rho(T + S)$. This is a contradiction.

$(iii) \implies (i)$ Trivial. Q.E.D.

Remark 4.7.1 *It follows, immediately, from Theorem 4.7.2, that for $T \in CR(X)$ and $\varepsilon > 0$,*

$$\sigma_\varepsilon(T) = \bigcup_{\substack{\|S\| < \varepsilon \\ S(0) \subset T(0) \\ \mathcal{D}(S) \supset \mathcal{D}(T)}} \sigma(T + S). \qquad \diamond$$

4.7.3   Stability of Pseudospectra of Linear Relations

Proposition 4.7.4 *Let* $T \in LR(X)$ *and* $\varepsilon > 0$. *Then, for any* $\alpha, \beta \in \mathbb{C}$ *with* $\beta \neq 0$, *we have*

$$\sigma_\varepsilon(\alpha + \beta T) = \alpha + \beta \, \sigma_{\varepsilon/|\beta|}(T).$$ ◇

Proof. For $\alpha, \lambda \in \mathbb{C}$ and $\beta \in \mathbb{C}\backslash\{0\}$, we have

$$\lambda - \alpha - \beta \widetilde{T} = \beta \left(\frac{\lambda - \alpha}{\beta} - \widetilde{T} \right),   \tag{4.59}$$

where $\widetilde{T}$ is given in (2.4). Let $\lambda \in \sigma_\varepsilon(\alpha + \beta T)$. There are two possible cases:

<u>First case</u> : If $\lambda \in \sigma(\alpha I + \beta T)$, then by using (4.59), we infer that $\frac{\lambda - \alpha}{\beta} \in \sigma(\widetilde{T}) = \sigma(T) \subset \sigma_{\varepsilon/|\beta|}(T)$. This leads to $\lambda \in \alpha + \beta \, \sigma_{\varepsilon/|\beta|}(T)$.

<u>Second case</u> : If $\lambda \in \sigma_\varepsilon(\alpha I + \beta T)\backslash\sigma(\alpha I + \beta T)$, then by applying (4.59), we obtain

$$\left\| \left(\frac{\lambda - \alpha}{\beta} - \widetilde{T} \right)^{-1} \right\| = \left\| \left(\beta^{-1}(\lambda - \alpha - \beta\widetilde{T}) \right)^{-1} \right\|$$

$$= |\beta| \left\| \left(\lambda - \alpha - \beta\widetilde{T} \right)^{-1} \right\|$$

$$> \frac{|\beta|}{\varepsilon}.$$

Hence, we deduce that $\frac{\lambda - \alpha}{\beta} \in \sigma_{\varepsilon/|\beta|}(T)$. Thus, $\lambda \in \alpha + \beta\sigma_{\varepsilon/|\beta|}(T)$. As a result, $\sigma_\varepsilon(\alpha I + \beta T) \subset \alpha + \beta \, \sigma_{\varepsilon/|\beta|}(T)$, as desired. Conversely, a same reasoning as before leads to the result.   Q.E.D.

Proposition 4.7.5 *Let* X *be a Banach space,* $T \in CR(X)$ *and* $\varepsilon > 0$. *Then, for* $\delta > 0$, *we have*

$$\sigma_\varepsilon(T) \subset \mathbb{D}(0, \delta) + \sigma_\varepsilon(T) \subset \sigma_{\varepsilon+\delta}(T).$$ ◇

Proof. Let $\lambda \in \mathbb{D}(0, \delta) + \sigma_\varepsilon(T)$. Then, there exists $\lambda_1 \in \mathbb{D}(0, \delta)$ and $\lambda_2 \in \sigma_\varepsilon(T)$ such that $\lambda = \lambda_1 + \lambda_2$. Assume that $\lambda \notin \sigma_{\varepsilon+\delta}(T)$, then $\lambda_1 + \lambda_2 \in \rho(T)$ and $\|(\lambda_1 + \lambda_2 - T)^{-1}\| \leq \frac{1}{\varepsilon+\delta}$. This implies that $\lambda_1 + \lambda_2 - T$ is injective, open with dense range and $\gamma(\lambda_1 + \lambda_2 - T) \geq \varepsilon + \delta$. Since $\lambda_1 \in \mathbb{D}(0, \delta)$, then $|\lambda_1| \leq \delta < \varepsilon + \delta \leq \gamma(\lambda_1 + \lambda_2 - T)$. By using Lemma 2.9.3, we infer that $\lambda_1 + \lambda_2 - T - \lambda_1 = \lambda_2 - T$ is injective and open with dense range. Therefore,

$\lambda_2 \in \rho(T)$. Now, by using Lemma 2.9.2, we have

$$
\begin{aligned}
\gamma(\lambda_2 - T) &= \gamma(\lambda_2 - \lambda_1 + \lambda_1 - T) \\
&\geq \gamma(\lambda_1 + \lambda_2 - T) - |\lambda_1| \\
&\geq \varepsilon + \delta - \delta \\
&\geq \varepsilon.
\end{aligned}
$$

Hence, $\|(\lambda_2 - T)^{-1}\| \leq \frac{1}{\varepsilon}$. We conclude that $\lambda_2 \notin \sigma_\varepsilon(T)$. This is a contradiction. $\hspace{2cm}$ Q.E.D.

Proposition 4.7.6 *Let X be a Banach space and $T \in CR(X)$, then for any $\varepsilon > 0$ and $S \in LR(X)$ such that $S(0) \subset T(0)$, $\mathcal{D}(S) \supset \mathcal{D}(T)$ and $\|S\| < \varepsilon$, we have*

$$
\sigma_{\varepsilon - \|S\|}(T) \subset \sigma_\varepsilon(T + S) \subset \sigma_{\varepsilon + \|S\|}(T). \hspace{1.5cm} \diamond
$$

Proof. Let $\lambda \notin \sigma_{\varepsilon + \|S\|}(T)$. Then, $\lambda \in \rho(T)$ and $\|(\lambda - T)^{-1}\| \leq \frac{1}{\varepsilon + \|S\|}$. Hence, $\lambda - T$ is injective, open with dense range and $\gamma(\lambda - T) \geq \varepsilon + \|S\|$. The fact that $\|S\| < \varepsilon + \|S\| \leq \gamma(\lambda - T)$ implies from Lemma 2.9.3 that $\lambda - T - S$ is injective and open with dense range. Then, $\lambda \in \rho(T + S)$. On the other hand, by Lemma 2.9.2, we have

$$
\begin{aligned}
\gamma(\lambda - T - S) &\geq \gamma(\lambda - T) - \|S\| \\
&\geq \varepsilon + \|S\| - \|S\| \\
&\geq \varepsilon.
\end{aligned}
$$

Then, $\|(\lambda - T - S)^{-1}\| \leq \frac{1}{\varepsilon}$. Hence, $\lambda \notin \sigma_\varepsilon(T + S)$. So, $\sigma_\varepsilon(T + S) \subset \sigma_{\varepsilon + \|S\|}(T)$. Let $\lambda \notin \sigma_\varepsilon(T + S)$. Then, $\lambda \in \rho(T + S)$ and $\|(\lambda - T - S)^{-1}\| \leq \frac{1}{\varepsilon}$. Hence, $\lambda - T - S$ is injective, open with dense range and $\gamma(\lambda - T - S) \geq \varepsilon$. Using the fact $\|S\| < \gamma(\lambda - T - S)$, we infer from Lemma 2.9.3 that $\lambda - T - S + S = \lambda - T$ is injective, open with dense range. Then, $\lambda \in \rho(T)$. On the other hand, by Lemma 2.9.2, we have

$$
\begin{aligned}
\gamma(\lambda - T) &= \gamma(\lambda - T - S + S) \\
&\geq \gamma(\lambda - T - S) - \|S\| \\
&\geq \varepsilon - \|S\|.
\end{aligned}
$$

Then, $\|(\lambda - T)^{-1}\| \leq \frac{1}{\varepsilon - \|S\|}$. Hence, $\lambda \notin \sigma_{\varepsilon - \|S\|}(T)$. So, $\sigma_{\varepsilon + \|S\|}(T) \subset \sigma_\varepsilon(T + S)$. This completes the proof. $\hspace{2cm}$ Q.E.D.

Theorem 4.7.3 *Let X be a Banach space and $T \in CR(X)$. Assume that $V \in L(X)$ such that $0 \in \rho(V)$. Let $k = \|V\|\|V^{-1}\|$ and $S = VTV^{-1}$. Then,*

$$\sigma(S) = \sigma(T)$$

and, for $\varepsilon > 0$, we have

$$\sigma_{\varepsilon/k}(T) \subset \sigma_\varepsilon(S) \subset \sigma_{k\varepsilon}(T). \qquad \Diamond$$

Proof. We first show that S is closed. Since $0 \in \rho(V)$ and V is closed, then V has a closed range ($R(V) = X$) and, we have $\alpha(V) = 0 < \infty$ and $\gamma(V) > 0$ (as V is injective and open). By using Proposition 2.9.4, we have VT is closed. Moreover, since V^{-1} is single valued and bounded, then VTV^{-1} is closed. Hence, S is closed. Using the fact V and V^{-1} are bounded single valued, then

$$\begin{aligned} \lambda - S &= \lambda - VTV^{-1} \\ &= V(\lambda\, V^{-1} - TV^{-1}) \\ &= V(\lambda - T)V^{-1}. \end{aligned} \qquad (4.60)$$

Now, let $\lambda \in \rho(T)$. Then, $(\lambda - T)^{-1}$ is a bounded single valued. This implies that $\lambda - T$ is a bounded below and surjective. Since S is closed, then by using Proposition 2.9.4 and (4.60), we have $\lambda - S$ is also closed, bounded below, surjective. Hence, $\lambda \in \rho(S)$. Conversely, if $\lambda \in \rho(S)$, then $(\lambda - S)^{-1}$ is a bounded single valued. This leads to $\lambda - S$ is a bounded below and surjective. Since V and V^{-1} are bounded single valued, then by (4.60), we have

$$\lambda - T = V^{-1}(\lambda - S)V. \qquad (4.61)$$

The fact that S is closed implies from Propositions 2.7.2 and 2.9.4, and (4.61) that $\lambda - T$ is also closed, bounded below, surjective. Hence, $\lambda \in \rho(T)$, which implies the first result. Now, it follows from (4.60) and (4.61) that

$$(\lambda - T)^{-1} = V^{-1}(\lambda - S)^{-1}V \text{ and } (\lambda - S)^{-1} = V(\lambda - T)^{-1}V^{-1}.$$

Thus,

$$\begin{aligned} \|(\lambda - S)^{-1}\| &= \|V(\lambda - T)^{-1}V^{-1}\| \\ &\leq \|V(\lambda - T)^{-1}\|\,\|V^{-1}\| \\ &\leq \|V\|\,\|(\lambda - T)^{-1}\|\,\|V^{-1}\| \\ &\leq k\,\|(\lambda - T)^{-1}\|. \end{aligned}$$

In the same way,

$$\|(\lambda - T)^{-1}\| \leq k\|(\lambda - S)^{-1}\|.$$

For $\lambda \in \sigma_{\varepsilon/k}(T)$, we have

$$\lambda \in \sigma(T) \text{ or } \|(\lambda - T)^{-1}\| > \frac{k}{\varepsilon}.$$

Then,

$$\lambda \in \sigma(S) \text{ or } \|(\lambda - S)^{-1}\| \geq \frac{1}{k} \|(\lambda - T)^{-1}\| > \frac{1}{\varepsilon}.$$

Hence,

$$\lambda \in \sigma_\varepsilon(S).$$

Therefore,

$$\sigma_{\varepsilon/k}(T) \subset \sigma_\varepsilon(S).$$

On the other hand, for $\lambda \in \sigma_\varepsilon(S)$, we have

$$\lambda \in \sigma(S) \text{ or } \|(\lambda - S)^{-1}\| > \frac{1}{\varepsilon}.$$

Hence,

$$\lambda \in \sigma(T) \text{ or } \|(\lambda - T)^{-1}\| \geq \frac{1}{k} \|(\lambda - S)^{-1}\| > \frac{1}{k\varepsilon}.$$

So,

$$\lambda \in \sigma_{k\varepsilon}(T).$$

Therefore, $\sigma_\varepsilon(S) \subset \sigma_{k\varepsilon}(T)$.   Q.E.D.

4.8   Localization of Pseudospectra of Linear Relations

We give some further results on the location of the pseudospectra. We start with the following general result. Although the result is well known, we include the proof.

Theorem 4.8.1 *Let X be a Banach space, $T \in CR(X)$, and $\varepsilon > 0$. Then,*

(i) If $\lambda \notin \sigma(T)$, then

$$\|(\lambda - T)^{-1}\| \geq \frac{1}{\text{dist}(\lambda, \sigma(T))}.$$

(ii) If $\lambda \notin \sigma_\varepsilon(T)$, then

$$\|(\lambda - T)^{-1}\| \geq \frac{1}{\text{dist}(\lambda, \sigma_\varepsilon(T)) + \varepsilon}. \qquad \diamond$$

Proof. (*i*) Let $\lambda \in \rho(T)$. Since $\mathrm{dist}(\lambda, \sigma(T)) = \inf\{|z - \lambda| \text{ such that } z \in \sigma(T)\}$, then for all $\eta > 0$, there exists $z_\eta \in \sigma(T)$ such that

$$|\lambda - z_\eta| < \mathrm{dist}(\lambda, \sigma(T)) + \eta.$$

Suppose $|\lambda - z_\eta| < \gamma(\lambda - T)$. Since $\lambda - T$ is injective open with dense range, $(z_\eta - \lambda)I(0) = 0 \subset (\lambda - T)(0)$, $\mathcal{D}((z_\eta - \lambda)I) = \mathcal{D}(I) = X \supset \mathcal{D}((\lambda - T)) = \mathcal{D}(T)$ and $|\lambda - z_\eta| < \gamma(\lambda - T)$, then by using Lemma 2.9.3, we have $\lambda - T + z_\eta - \lambda = z_\eta - T$ is injective open with dense range. Hence, $z_\eta \in \rho(T)$. This is a contradiction. Therefore, $|\lambda - z_\eta| \geq \gamma(\lambda - T)$ for all $\eta > 0$. Thus,

$$\begin{aligned} \gamma(\lambda - T) &\leq |\lambda - z_\eta| \\ &< \mathrm{dist}(\lambda, \sigma(T)) + \eta \quad \text{for all } \eta > 0. \end{aligned}$$

So,

$$\gamma(\lambda - T) \leq \mathrm{dist}(\lambda, \sigma(T)).$$

Hence,

$$\|(\lambda - T)^{-1}\| \geq \frac{1}{\mathrm{dist}(\lambda, \sigma(T))}.$$

(*ii*) Let $\lambda \in \rho_\varepsilon(T)$. Assume that $\|(\lambda - T)^{-1}\| < \frac{1}{\mathrm{dist}(\lambda, \sigma_\varepsilon(T)) + \varepsilon}$, i.e., $\gamma(\lambda - T) > \mathrm{dist}(\lambda, \sigma_\varepsilon(T)) + \varepsilon$. Since $\mathrm{dist}(\lambda, \sigma_\varepsilon(T)) = \inf\{|z - \lambda| \text{ such that } z \in \sigma_\varepsilon(T)\}$, for $\eta, 0 < \eta \leq \varepsilon$, there exists $z_\eta \in \sigma_\varepsilon(T)$ such that

$$|\lambda - z_\eta| < \mathrm{dist}(\lambda, \sigma_\varepsilon(T)) + \eta. \tag{4.62}$$

Therefore,

$$\begin{aligned} |\lambda - z_\eta| &< \mathrm{dist}(\lambda, \sigma_\varepsilon(T)) + \varepsilon \\ &< \gamma(\lambda - T). \end{aligned}$$

But $(z_\eta - \lambda)I(0) = 0 \subset (\lambda - T)(0)$, $\mathcal{D}((z_\eta - \lambda)I) = \mathcal{D}(I) = X \supset \mathcal{D}((\lambda - T)) = \mathcal{D}(T)$ and $\lambda - T$ is injective open with dense range (as $\lambda \in \rho(T)$), then by using Lemma 2.9.3, we have $\lambda - T + z_\eta - \lambda = z_\eta - T$ is injective open with dense range. Therefore, $z_\eta \in \rho(T)$. But $z_\eta \in \sigma_\varepsilon(T)$, then $\|(z_\eta - T)^{-1}\| > \frac{1}{\varepsilon}$, i.e., $\gamma(z_\eta - T) < \varepsilon$. On the other hand, by Lemma 2.9.2,

$$\begin{aligned} \gamma(z_\eta - T) &= \gamma(\lambda - T + z_\eta - \lambda) \\ &\geq \gamma(\lambda - T) - |z_\eta - \lambda|. \end{aligned}$$

Therefore, $\gamma(\lambda - T) \leq \gamma(z_\eta - T) + |z_\eta - \lambda|$. This implies from (4.62) that for all $0 < \eta \leq \varepsilon$

$$\gamma(\lambda - T) < \varepsilon + \mathrm{dist}(\lambda, \sigma_\varepsilon(T)) + \eta.$$

Thus, $\gamma(\lambda - T) \leq \varepsilon + \text{dist}(\lambda, \sigma_\varepsilon(T))$. This is a contradiction. Hence,

$$\|(\lambda - T)^{-1}\| \geq \frac{1}{\text{dist}(\lambda, \sigma_\varepsilon(T)) + \varepsilon}. \qquad \text{Q.E.D.}$$

Corollary 4.8.1 *Let X be a Banach space, $T \in CR(X)$, and $\varepsilon > 0$. Then,*

$$\{\lambda \in \mathbb{C} \;\; \text{such that } \text{dist}(\lambda, \sigma(T)) < \varepsilon\} \subset \sigma_\varepsilon(T). \qquad \diamond$$

Proof. Let $\lambda \notin \sigma_\varepsilon(T)$. Then, by using Theorem 4.8.1, we have

$$\frac{1}{\text{dist}(\lambda, \sigma(T))} \leq \|(\lambda - T)^{-1}\| \leq \frac{1}{\varepsilon}.$$

Therefore, $\text{dist}(\lambda, \sigma(T)) \geq \varepsilon$. $\qquad \text{Q.E.D.}$

4.9 Characterization of ε-Pseudospectra of Linear Relations

In this section, we investigate the characterization of ε-pseudospectra and discuss the pseudospectra of a sequence of closed linear relations in a Banach space.

Theorem 4.9.1 *Let X be a Banach space, $\varepsilon > 0$ and let $T \in CR(X)$. Then,*

$$\bigcup_{\substack{\|S\| \leq \varepsilon \\ S(0) \subset T(0) \\ \mathcal{D}(S) \supset \mathcal{D}(T)}} \sigma(T + S) \subset \Sigma_\varepsilon(T). \qquad \diamond$$

Proof. We derive a contradiction from the assumption that $\lambda \notin \Sigma_\varepsilon(T)$. Then,

$$\lambda \in \rho(T) \quad \text{and} \quad \|(\lambda - T)^{-1}\| < \frac{1}{\varepsilon}.$$

This leads to $\lambda - T$ is injective, open with dense range and

$$\gamma(\lambda - T) > \varepsilon. \qquad (4.63)$$

Let $S \in LR(X)$ satisfy $\|S\| \leq \varepsilon$, $S(0) \subset T(0)$ and $\mathcal{D}(S) \supset \mathcal{D}(T)$. Then, by using (4.63), we infer that $\|S\| < \gamma(\lambda - T)$. Thus, by applying Lemma 2.9.3, we deduce that $\lambda - T - S$ is injective and open with dense range. This is equivalent to saying that $\lambda \in \rho(T + S)$. $\qquad \text{Q.E.D.}$

Remark 4.9.1 (*i*) *We should notice that if X is a Banach space and $T \in CR(X)$, then*

$$\bigcup_{\substack{\|S\| \leq \varepsilon \\ S(0) \subset T(0) \\ \mathcal{D}(S) \supset \mathcal{D}(T)}} \sigma(T + S) \subset \Sigma_\varepsilon(T).$$

In fact, it suffices to consider the following example: let $X = l_p(\mathbb{N})$ be the space of sequences $x : \mathbb{N} \longrightarrow \mathbb{C}$ summable with a power $p \in [1, \infty)$ with the standard norm and $\varepsilon > 0$. Consider linear relation T defined by

$$G(T) = \{(x, y) \in X \times X : x(n) = \varepsilon \, y(n - 1), n \geq 2\}.$$

It is clear that $T = \varepsilon L^{-1}$, where L is the left shift single valued operator defined by

$$L(x(n)) = x(n + 1), \quad n \geq 1 \quad and \quad x \in X.$$

We know that L is bounded operator with $\|L\| = 1$. Therefore, $\mathcal{D}\left(\varepsilon^{-1}L\right) = X$ and

$$\left\|\varepsilon^{-1}L\right\| = \varepsilon^{-1} \, \|L\| = \varepsilon^{-1} < \infty.$$

This implies that $\varepsilon^{-1}L$ is bounded operator. Hence, $\varepsilon^{-1}L$ is closed which yields T is closed. Moreover, $\|T^{-1}\| = \frac{1}{\varepsilon}$. Thus, we conclude that

$$0 \in \Sigma_\varepsilon(T).$$

Now, we show that

$$0 \notin \bigcup_{\substack{\|S\| \leq \varepsilon \\ S(0) \subset T(0) \\ \mathcal{D}(S) \supset \mathcal{D}(T)}} \sigma(T + S). \tag{4.64}$$

Since $N(T) = (\varepsilon^{-1}L)(0) = \varepsilon^{-1}L(0) = \{0\}$, $\gamma(T) = \|T^{-1}\|^{-1} = \varepsilon > 0$ and $R(T) = X$, then T is injective and open with dense range. Let $S \in LR(X)$ satisfy that $\|S\| \leq \varepsilon$, $S(0) \subset T(0)$ and $\mathcal{D}(S) \supset \mathcal{D}(T)$. In view of Lemma 2.9.3 implies that $T + S$ is injective and open with dense range. This is equivalent to say that $0 \in \rho(T + S)$. As a result, (4.64) holds, as desired.

(*ii*) *Note that*

$$\overline{\sigma_\varepsilon(T)} \subset \Sigma_\varepsilon(T). \tag{4.65}$$

Indeed, we have $\sigma_\varepsilon(T) \subset \Sigma_\varepsilon(T)$. Using the fact that $\Sigma_\varepsilon(T)$ is closed, then we obtain $\overline{\sigma_\varepsilon(T)} \subset \Sigma_\varepsilon(T)$. $\diamond$

Now, the goal is achieved the equality in (4.65). It is well known that if the resolvent norm of the closed linear operator T acting in a Banach space cannot be constant on an open set, then

$$\overline{\sigma_\varepsilon(T)} = \Sigma_\varepsilon(T) \text{ (see [46])}.$$

Note that, even if X is Banach space, $T \in C\mathcal{R}(X)$ and $\lambda \in \rho(T)$, we have $(\lambda - T)^{-1}$ is bounded operator in a Banach space X.

Theorem 4.9.2 *Let X be a Banach space, $\varepsilon > 0$, and $T \in C\mathcal{R}(X)$ such that the resolvent norm of T cannot be constant on an open set. Then,*

$$\Sigma_\varepsilon(T) \subset \overline{\sigma_\varepsilon(T)}. \qquad \diamond$$

Proof. It is clear that

$$\Sigma_\varepsilon(T) \setminus \left\{ \lambda \in \mathbb{C} : \|(\lambda - T)^{-1}\| = \frac{1}{\varepsilon} \right\} \subset \overline{\sigma_\varepsilon(T)}.$$

Then, it is sufficient to show that

$$\left\{ \lambda \in \mathbb{C} : \|(\lambda - T)^{-1}\| = \frac{1}{\varepsilon} \right\} \subset \overline{\sigma_\varepsilon(T)}.$$

Let us assume that $\|(\lambda - T)^{-1}\| = \frac{1}{\varepsilon}$. Then, there exists μ belongs to the neighborhood of λ and $\mu \in \rho(T)$ such that $\|(\mu - T)^{-1}\| > \|(\lambda - T)^{-1}\| = \frac{1}{\varepsilon}$. This leads to $\mu \in \sigma_\varepsilon(T)$. This enables us to conclude that $\lambda \in \overline{\sigma_\varepsilon(T)}$. Q.E.D.

As an immediate consequence of Remark 4.9.1 and Theorem 4.9.2, we have:

Corollary 4.9.1 *Let X be a Banach space, $\varepsilon > 0$, and $T \in C\mathcal{R}(X)$ such that the resolvent norm of T cannot be constant on an open set. Then,*

$$\Sigma_\varepsilon(T) = \overline{\sigma_\varepsilon(T)}. \qquad \diamond$$

The following theorem is devoted to give some equivalent definitions of ε-pseudospectra.

Theorem 4.9.3 *Let X be a Banach space, $\varepsilon > 0$, and $T \in C\mathcal{R}(X)$ such that the resolvent norm of T cannot be constant on an open set. Then, the following propositions are all equivalent:*

(i) $\lambda \in \displaystyle\bigcup_{\substack{\|S\| \leq \varepsilon \\ S(0) \subset T(0) \\ \mathcal{D}(S) \supset \mathcal{D}(T)}} \sigma(T + S).$

(ii) $\lambda \in \sigma(T) \bigcup \left\{ \lambda \in \mathbb{C} : \|(\lambda - T)^{-1}\| \geq \frac{1}{\varepsilon} \right\}.$

(iii) $\lambda \in \sigma(T) \bigcup \overline{\{\lambda \in \mathbb{C} : \exists\, x \in \mathcal{D}(T), \|x\| = 1 \text{ and } \|(\lambda - T)x\| < \varepsilon\}}.$ $\qquad \diamond$

Proof. $(i) \Longrightarrow (ii)$ Since $\Sigma_\varepsilon(T)$ is closed, then the use of Theorem 4.9.1 gives the wanted inclusion and achieves the proof of $(i) \Longrightarrow (ii)$.

$(ii) \Longrightarrow (iii)$ Let us assume that $\lambda \in \sigma(T) \cup \left\{ \lambda \in \mathbb{C} : \|(\lambda - T)^{-1}\| \geq \dfrac{1}{\varepsilon} \right\}$. Then, we will discuss the following two cases:

<u>First case</u> : If $\lambda \in \sigma(T)$, then the result is trivial.

<u>Second case</u> : If $\lambda \in \Sigma_\varepsilon(T) \backslash \sigma(T)$, then $\lambda \in \rho(T)$ and $\|(\lambda - T)^{-1}\| \geq \frac{1}{\varepsilon}$. We infer that there exists μ belongs to the neighborhood of λ such that $\|(\mu - T)^{-1}\| > \|(\lambda - T)^{-1}\|$ which yields

$$\|(\mu - T)^{-1}\| > \frac{1}{\varepsilon}.$$

This implies that there exists $y \in X$ such that $\|y\| = 1$ and $\|(\mu - T)^{-1}y\| > \dfrac{1}{\varepsilon}$. Since $\mu \in \rho(T)$, then $(\mu - T)^{-1}$ is an operator. Putting

$$x = \|(\mu - T)^{-1}y\|^{-1}(\mu - T)^{-1}y,$$

then $x \in \mathcal{D}(T)$, $\|x\| = 1$ and $y \in \|(\mu - T)^{-1}y\| \, (\mu - T)x$. We infer that

$$\| \|(\mu - T)^{-1}y\| \, (\mu - T)x\| = \mathrm{dist}\,(y, (\mu - T)(0)) \,.$$

Consequently,

$$
\begin{aligned}
\|(\mu - T)x\| &= \|(\mu - T)^{-1}y\|^{-1} \, \mathrm{dist}\,(y, (\mu - T)(0)) \\
&\leq \|(\mu - T)^{-1}y\|^{-1}\|y\| \\
&< \varepsilon.
\end{aligned}
$$

This enables us to conclude that

$$\lambda \in \overline{\{\lambda \in \mathbb{C} : \exists \, x \in \mathcal{D}(T), \|x\| = 1 \text{ and } \|(\lambda - T)x\| < \varepsilon\}}.$$

$(iii) \Longrightarrow (i)$ Let $\lambda \in \sigma(T) \cup \{\lambda \in \mathbb{C} : \exists \, x \in \mathcal{D}(T), \|x\| = 1 \text{ and } \|(\lambda - T)x\| < \varepsilon\}$. We notice the existence of two cases.

<u>First case</u> : Let $\lambda \in \sigma(T)$. Then, we may put $S = 0$.

<u>Second case</u> : Let $\lambda \in \{\lambda \in \mathbb{C} : \exists \, x \in \mathcal{D}(T), \|x\| = 1 \text{ and } \|(\lambda - T)x\| < \varepsilon\}$ and $\lambda \in \rho(T)$. Then, there exists $x' \in X^*$ such that $\|x'\| = 1$ and $x'(x) = 1$. Now, we consider the following linear relation $S(y) = x'(y) \, (\lambda - T)x$. Then, $S(0) = 0$ and $\mathcal{D}(S) = X$. Let $y \in X$, we have

$$
\begin{aligned}
\|S(y)\| &\leq \|x'\| \, \|y\| \, \|(\lambda - T)x\| \\
&\leq \varepsilon \|y\|,
\end{aligned}
$$

which implies that $\|S\| \leq \varepsilon$. Moreover,

$$
\begin{aligned}
(\lambda - T - S)x &= (\lambda - T)(0) \\
&= (\lambda - T)(0) - S(0) \\
&= (\lambda - T - S)(0),
\end{aligned}
$$

and by applying Lemma 4.7.2, we infer that $0 \in (\lambda - T - S)x$. This leads to $0 \neq x \in N(\lambda - T - S)$. So, $\lambda - T - S$ is not injective. Hence, $\lambda \in \sigma(T + S)$. This completes the proof. $\qquad$ Q.E.D.

As a direct consequence of Theorem 4.9.3, we have the following result.

Corollary 4.9.2 *Let X be a Banach space, $\varepsilon > 0$, and $T \in CR(X)$ such that the resolvent norm of T cannot be constant on an open set. Then,*

$$
\overline{\bigcup_{\substack{\|S\| \leq \varepsilon \\ S(0) \subset T(0) \\ \mathcal{D}(S) \supset \mathcal{D}(T)}} \sigma(T + S)} = \Sigma_\varepsilon(T). \qquad \Diamond
$$

The following result shows the relation between the pseudospectra of a closed linear relation on a Banach space and the ε-pseudospectra of all perturbed linear relations.

Proposition 4.9.1 *Let X be a Banach space, $T \in CR(X)$ and δ, $\varepsilon > 0$. For any continuous $S \in LR(X)$ such that $S(0) \subset T(0)$ and $\mathcal{D}(S) \supset \overline{\mathcal{D}(T)}$. Then,*

$$
\sigma_\varepsilon(T) \subset \Sigma_{\varepsilon + \|S\|}(T + S) \subset \sigma_{\varepsilon + \delta + 2\|S\|}(T). \qquad \Diamond
$$

Proof. It follows from Proposition 2.7.2 and Remark 2.20.2 (i) that

$$
\sigma_\varepsilon(T) \subset \Sigma_{\varepsilon + \|S\|}(T + S).
$$

Now, we have to prove $\Sigma_{\varepsilon + \|S\|}(T + S) \subset \sigma_{\varepsilon + \delta + 2\|S\|}(T)$. Let $\lambda \notin \sigma_{\varepsilon + \delta + 2\|S\|}(T)$. Then, $\lambda \in \rho(T)$ and $\|(\lambda - T)^{-1}\| \leq \frac{1}{\varepsilon + \delta + 2\|S\|}$, which implies that $\lambda - T$ is open, injective with dense range and

$$
\gamma(\lambda - T) \geq \varepsilon + \delta + 2\|S\| > \|S\|.
$$

The fact that $S(0) \subset T(0) = \overline{T(0)}$ and $\mathcal{D}(S) \supset \overline{\mathcal{D}(T)} \supset \mathcal{D}(T)$ implies from (ii) and (iii) of Lemma 2.9.3 that $\lambda - T - S$ is open, injective with dense range and

$$
\begin{aligned}
\gamma(\lambda - T - S) &\geq \gamma(\lambda - T) - \|S\| \\
&\geq \varepsilon + \delta + 2\|S\| - \|S\| \\
&> \varepsilon + \|S\|.
\end{aligned}
$$

This leads to $\lambda \in \rho(S+T)$ and $\|(\lambda - T - S)^{-1}\| < \frac{1}{\varepsilon + \|S\|}$. Moreover, $T + S$ is closed. Indeed, since $S(0) \subset T(0)$ and $\mathcal{D}(S) \supset \overline{\mathcal{D}(T)}$, then by (i) of Lemma 2.9.3, we deduce that $T + S$ is closed. This enables us to conclude that

$$\Sigma_{\varepsilon + \|S\|}(T + S) \subset \sigma_{\varepsilon + \delta + 2\|S\|}(T).\qquad\qquad \text{Q.E.D.}$$

Corollary 4.9.3 *Let X be a Banach space, $T \in C\mathcal{R}(X)$ and $\varepsilon > 0$. For any continuous $S \in L\mathcal{R}(X)$ such that $S(0) \subset T(0)$ and $\mathcal{D}(S) \supset \overline{\mathcal{D}(T)}$. Then,*

$$\Sigma_{\varepsilon}(T) \subset \Sigma_{\varepsilon + \|S\|}(T + S) \subset \Sigma_{\varepsilon + 2\|S\|}(T).\qquad\qquad \Diamond$$

Proof. Since $S(0) \subset T(0)$ and $\mathcal{D}(S) \supset \overline{\mathcal{D}(T)}$, then by Lemma 2.9.3, we get $T + S$ is closed. Let us assume that $\lambda \notin \Sigma_{\varepsilon + 2\|S\|}(T)$. Then, $\lambda \in \rho(T)$ and $\|(\lambda - T)^{-1}\| < \frac{1}{\varepsilon + 2\|S\|}$. This leads to $\lambda - T$ is open, injective with dense range and

$$\gamma(\lambda - T) > \varepsilon + 2\|S\|. \qquad\qquad (4.66)$$

The fact that $\|S\| < \gamma(\lambda - T)$ implies from (4.66) that $\lambda - T - S$ is open, injective with dense range. Moreover, it follows from Lemma 2.9.3 (iii) that

$$\begin{aligned} \gamma(\lambda - T - S) &> \gamma(\lambda - T) - \|S\| \\ &> \varepsilon + 2\|S\| - \|S\| \\ &> \varepsilon + \|S\|. \end{aligned}$$

Hence, $\lambda \in \rho(T + S)$ and $\|(\lambda - T - S)^{-1}\| < \frac{1}{\varepsilon + \|S\|}$. This is equivalent to saying that

$$\lambda \notin \Sigma_{\varepsilon + \|S\|}(T + S).$$

Now, we prove $\Sigma_{\varepsilon}(T) \subset \Sigma_{\varepsilon + \|S\|}(T+S)$. Let us assume that $\lambda \notin \Sigma_{\varepsilon + \|S\|}(T+S)$, then $\lambda \in \rho(T + S)$ and $\|(\lambda - T - S)^{-1}\| < \frac{1}{\varepsilon + \|S\|}$. This implies that $\lambda - T - S$ is open, injective with dense range and

$$\begin{aligned} \gamma(\lambda - T - S) &> \varepsilon + \|S\| \\ &> \|S\|. \end{aligned}$$

Combining the fact that $S(0) \subset T(0) = \overline{T(0)}$, $\mathcal{D}(S) \supset \mathcal{D}(T)$ and (ii) of Lemma 2.9.3, we obtain $\lambda - T - S + S$ is open and injective with dense range. Using (iv) of Lemma 2.9.3, we can write $\lambda - T = \lambda - T - S + S$. Finally, the use of Lemma 2.9.3 (iii) gives

$$\begin{aligned} \gamma(\lambda - T) &\geq \gamma(\lambda - T - S) - \|S\| \\ &> \varepsilon. \end{aligned}$$

Consequently, $\lambda \notin \Sigma_{\varepsilon}(T)$. This shows that $\Sigma_{\varepsilon}(T) \subset \Sigma_{\varepsilon + \|S\|}(T + S)$. Q.E.D.

4.9.1 The Pseudospectra and the ε-Pseudospectra of a Sequence of Closed Linear Relations

The purpose of this subsection is to establish the relation between the pseudospectra (respectively, ε-pseudospectra) of the sequence T_n and its limit.

Theorem 4.9.4 *Let X be a Banach space, $(T_n)_n$ be a sequence of closed linear relations from X into X, $T \in C\mathcal{R}(X)$ such that $\|T_n - T\| \to 0$ when $n \to \infty$, $\mathcal{D}(T_n) = \mathcal{D}(T)$ and there exists $n_1 \in \mathbb{N}$ such that $T(0) = T_n(0)$, for all $n \geq n_1$. Then,*

(i) For every triplet $(\varepsilon_1, \varepsilon_2, \varepsilon_3)$ of real numbers with $0 < \varepsilon_1 < \varepsilon_2 < \varepsilon_3$, there exists a positive integers $n_2 \geq n_1$ such that

$$\sigma_{\varepsilon_1}(T) \subset \sigma_{\varepsilon_2}(T_n) \subset \sigma_{\varepsilon_3}(T), \quad \text{for all } n \geq n_2.$$

(ii) For every real number $\varepsilon > 0$, there exists a positive integers $n_2 \geq n_1$ such that

$$\Sigma_\varepsilon(T) = \Sigma_\varepsilon(T_n), \quad \text{for all } n \geq n_2. \qquad \diamond$$

Proof. (i) First, we prove that $\sigma_{\varepsilon_1}(T) \subset \sigma_{\varepsilon_2}(T_n)$. Let us assume that $\lambda \notin \sigma_{\varepsilon_2}(T_n)$, for all $n \geq 0$. Then, $\lambda \in \rho(T_n)$ and $\|(\lambda - T_n)^{-1}\| \leq \frac{1}{\varepsilon_2}$. This leads to $\lambda - T_n$ is open, injective with dense range and $\gamma(\lambda - T_n) \geq \varepsilon_2$. Using the fact that $0 < \varepsilon_1 < \varepsilon_2$, we obtain $\gamma(\lambda - T_n) > \varepsilon_1$. Since $\|T_n - T\| \to 0$ when $n \to \infty$. Let $n_0 \in \mathbb{N}$ such that

$$\|T_n - T\| < \gamma(\lambda - T_n) - \varepsilon_1 < \gamma(\lambda - T_n), \quad \text{for all } n \geq n_0.$$

Using the fact $\mathcal{D}(\lambda - T_n) = \mathcal{D}(T_n) = \mathcal{D}(T_n) \bigcap \mathcal{D}(T) = \mathcal{D}(T_n - T)$ and $(T - T_n)(0) = T_n(0) = (\lambda - T_n)(0) = \overline{(\lambda - T_n)(0)}$, for all $n \geq n_1$, then by Lemma 2.9.3 (ii), we deduce that $\lambda - T_n + T_n - T$ is open and injective with dense range. Now, using Proposition 2.3.4, we can write

$$\lambda - T = \lambda - T - T_n + T_n$$

for all $n \geq n_1$. Now, we propose to show that $\gamma(\lambda - T) \geq \varepsilon_1$. Let $n_2 = \max\{n_0, n_1\}$. The use of Lemma 2.9.2 makes us to conclude, for all $n \geq n_2$, that

$$\begin{aligned} \gamma(\lambda - T) &\geq \gamma(\lambda - T_n) - \|T_n - T\| \\ &\geq \gamma(\lambda - T_n) - \gamma(\lambda - T_n) + \varepsilon_1 \\ &\geq \varepsilon_1. \end{aligned}$$

Thus, $\lambda \in \rho(T)$ and $\gamma(\lambda - T) \geq \varepsilon_1$ which yields $\lambda \notin \sigma_{\varepsilon_1}(T)$. This shows that

$$\sigma_{\varepsilon_1}(T) \subset \sigma_{\varepsilon_2}(T_n), \text{ for all } n \geq n_2.$$

A similar reasoning allows us to reach $\sigma_{\varepsilon_2}(T_n) \subset \sigma_{\varepsilon_3}(T)$, for all $n \geq n_2$.

(ii) Let $\lambda \notin \Sigma_\varepsilon(T_n)$, for all $n \geq 0$. Then, $\lambda \in \rho(T_n)$ and $\|(\lambda - T_n)^{-1}\| < \frac{1}{\varepsilon}$. This implies that $\lambda - T_n$ is open, injective with dense range and $\gamma(\lambda - T_n) > \varepsilon$. Since $\|T_n - T\| \to 0$ as $n \to 0$, then there is $n_0 \in \mathbb{N}$ such that

$$\|T_n - T\| < \gamma(\lambda - T_n) - \varepsilon < \gamma(\lambda - T_n), \text{ for all } n \geq n_0.$$

Since $\mathcal{D}(\lambda - T_n) = \mathcal{D}(T_n - T)$ and $(T - T_n)(0) = \overline{(\lambda - T_n)(0)}$, for all $n \geq n_1$, then by using Proposition 2.3.4, we infer that

$$\lambda - T = \lambda - T_n + T_n - T$$

is open and injective with dense range. Let $n_2 = \max\{n_0, n_1\}$. Then, using Lemma 2.9.2, for all $n \geq n_2$, we conclude that

$$
\begin{aligned}
\gamma(\lambda - T) &\geq \gamma(\lambda - T_n) - \|T_n - T\| \\
&\geq \gamma(\lambda - T_n) - \gamma(\lambda - T_n) + \varepsilon \\
&> \varepsilon.
\end{aligned}
$$

Hence, we obtain $\lambda \notin \Sigma_\varepsilon(T)$. This shows that $\Sigma_\varepsilon(T) \subset \Sigma_\varepsilon(T_n)$, for all $n \geq n_2$. The opposite inclusion is analogous.   Q.E.D.

The following result can be directly derived from Proposition 4.7.6, Corollary 4.9.3 and Theorem 4.9.4.

Corollary 4.9.4 *Let X be a Banach space, ε, δ_1, $\delta_2 > 0$, $(T_n)_n$ be a sequence of closed linear relations from X into X and $T \in C\mathcal{R}(X)$ such that $\|T_n - T\| \to 0$ when $n \to \infty$. We suppose that there exists $n_1 \in \mathbb{N}$ such that $\mathcal{D}(T_n) = \mathcal{D}(T)$ and $T(0) = T_n(0)$, for all $n \geq n_1$. Then, for any continuous $S \in L\mathcal{R}(X)$ such that $S(0) \subset T(0)$ and $\mathcal{D}(S) \supset \overline{\mathcal{D}(T)}$, there exists a positive integers $n_2 \geq n_1$ such that*

$$\sigma_\varepsilon(T) \subset \sigma_{\varepsilon + \delta_1 + \|S\|}(T_n + S) \subset \sigma_{\varepsilon + \delta_1 + \delta_2 + 2\|S\|}(T), \text{ for all } n \geq n_2,$$

and

$$\Sigma_\varepsilon(T) \subset \Sigma_{\varepsilon + \|S\|}(T_n + S) \subset \Sigma_{\varepsilon + 2\|S\|}(T), \text{ for all } n \geq n_2. \qquad \diamondsuit$$

4.10   Essential Pseudospectra of Linear Relations

In this section, we study some properties of essential pseudospectra of a linear relation and establish some results of perturbation on the context of linear relations. Throughout this section, $X, Y, \ldots$ will denote Banach spaces over the complex field $\mathbb{C}$.

Theorem 4.10.1 *Let $T \in C\mathcal{R}(X)$ and $\varepsilon > 0$. Then, $\sigma_{ei,\varepsilon}(T)$ is a closed set with $i \in \{1, 2, 3, 4\}$.* $\diamond$

Proof. The fact that $T \in C\mathcal{R}(X)$ and $S \in \mathcal{U}_T(X)$ implies that $\lambda - T - S \in C\mathcal{R}(X)$ for all $\lambda \in \mathbb{C}$.

<u>For $i = 1$</u>: let $\lambda \notin \sigma_{e1,\varepsilon}(T)$, then for all $S \in \mathcal{U}_T(X)$, we have $\lambda - T - S \in \Phi_+(X)$. Since $\lambda - T - S$ is closed and $R(\lambda - T - S)$ is closed, then applying Proposition 2.9.3 (i), we obtain $\gamma(\lambda - T - S) > 0$. Let $r > 0$ such that $r < \gamma(\lambda - T - S)$, and let $\mu \in \mathbb{D}(\lambda, r)$, then $|\lambda - \mu| < r < \gamma(\lambda - T - S)$. Hence,

$$\mu - T - S = \lambda - T - S - (\lambda - \mu) \in \Phi_+(X).$$

Consequently, $\mu \notin \sigma_{e1,\varepsilon}(T)$ which implies that $\mu \in \mathbb{C} \backslash \sigma_{e1,\varepsilon}(T)$. This leads to $\mathbb{D}(\lambda, r) \subset \mathbb{C} \backslash \sigma_{e1,\varepsilon}(T)$. Therefore, $\sigma_{e1,\varepsilon}(T)$ is a closed. By using the same reasoning as above, we get $\sigma_{ei,\varepsilon}(T)$ is a closed set with $i \in \{2, 3, 4\}$. Q.E.D.

Proposition 4.10.1 *Let $T \in C\mathcal{R}(X)$. Then,*

(i) If $0 < \varepsilon_1 < \varepsilon_2$, then $\sigma_{ei}(T) \subset \sigma_{ei,\varepsilon_1}(T) \subset \sigma_{ei,\varepsilon_2}(T)$, with $i \in \{1, 2, 3, 4\}$.
(ii) Let $\varepsilon > 0$, then $\sigma_{ei,\varepsilon}(T) \subset \sigma_\varepsilon(T)$, with $i \in \{1, 2, 3, 4\}$.
(iii) $\bigcap_{\varepsilon > 0} \sigma_{ei,\varepsilon}(T) = \sigma_{ei}(T)$, with $i \in \{1, 2, 3, 4\}$. $\diamond$

Proof. The fact that $T \in C\mathcal{R}(X)$ and $S \in \mathcal{U}_T(X)$ implies that $\lambda - T - S \in C\mathcal{R}(X)$ for all $\lambda \in \mathbb{C}$.

(i) <u>For $i = 1$</u>: let $\lambda \notin \sigma_{e1,\varepsilon_2}(T)$, then for all $S \in \mathcal{U}_T(X)$, we have $\lambda - T - S \in \Phi_+(X)$. Since $\varepsilon_1 < \varepsilon_2$, then for all continuous $S \in L\mathcal{R}(X)$ such that $S(0) \subset T(0), \mathcal{D}(S) \supset \overline{\mathcal{D}(T)}$ and $\|S\| < \varepsilon_1$, we infer that $\lambda - T - S \in \Phi_+(X)$. Hence, we conclude that $\lambda \notin \sigma_{e1,\varepsilon_1}(T)$. Therefore, $\sigma_{e1,\varepsilon_1}(T) \subset \sigma_{e1,\varepsilon_2}(T)$. Now, we claim that $\sigma_{e1}(T) \subset \sigma_{e1,\varepsilon_1}(T)$. Suppose that $\lambda \notin \sigma_{e1,\varepsilon_2}(T)$ which is equivalent to $\lambda - T - S \in \Phi_+(X)$ for all $S \in \mathcal{U}_T(X)$. In particular, for $S = 0$, then $\lambda - T - S \in \Phi_+(X)$. Hence, we infer that $\lambda \notin \sigma_{e1}(T)$.

<u>For $i = 2$</u>: let $\lambda \notin \sigma_{e2,\varepsilon_2}(T)$, then for all $S \in \mathcal{U}_T(X)$, we have $\lambda - T - S \in \Phi_-(X)$. This implies from Theorem 2.14.4 (ii) that $(\lambda - T - S)^* \in \Phi_+(X^*)$. Since S is continuous, then by Proposition 2.8.1, we get $(\lambda - T - S)^* = \lambda - T^* - S^*$. It follows from Proposition 2.8.1 that $\|S\| = \|S^*\|$. Consequently, $\lambda - T^* - S^* \in \Phi_+(X^*)$. Since $\mathcal{D}(T) \subset \mathcal{D}(S)$ and $S(0) \subset T(0)$, then

$$S^*(0) = \mathcal{D}(S)^\perp \subset \mathcal{D}(T)^\perp = T^*(0),$$

and

$$\mathcal{D}(T^*) = T(0)^\perp \subset S(0)^\perp = \mathcal{D}(S^*).$$

Hence, $\lambda - T^* - S^* \in \Phi_+(X^*)$, for all continuous linear relation S^* such that $S^*(0) \subset T^*(0), \mathcal{D}(S^*) \supset \overline{\mathcal{D}(T^*)}$ and $\|S^*\| \leq \varepsilon_2$. This implies that $\lambda \notin \sigma_{e1,\varepsilon_2}(T^*)$. By referring to the case that $i = 1$, we conclude that $\lambda \notin \sigma_{e1,\varepsilon_1}(T^*)$. This lead to $\lambda - T^* - S^* \in \Phi_+(X^*)$, for all continuous linear relation S^* such that $\|S^*\| \leq \varepsilon_1$. Therefore, $\lambda - T - S \in \Phi_-(X)$, for all $S \in \mathcal{U}_T(X)$. Thus, $\lambda \notin \sigma_{e2,\varepsilon_1}(T)$. Now, let us assume that $\lambda \notin \sigma_{e2,\varepsilon_1}(T)$, then $\lambda - T - S \in \Phi_-(X)$, for all $S \in \mathcal{U}_T(X)$. The fact that $0 \in \mathcal{U}_T(X)$ implies that $\lambda \notin \sigma_{e2}(T)$.

<u>For $i = 3$</u>: since $\sigma_{e3}(T) = \sigma_{e1}(T) \bigcap \sigma_{e2}(T)$, $\sigma_{e3,\varepsilon_1}(T) = \sigma_{e1,\varepsilon_1}(T) \bigcap \sigma_{e2,\varepsilon_1}(T)$ and $\sigma_{e3,\varepsilon_2}(T) = \sigma_{e1,\varepsilon_2}(T) \bigcap \sigma_{e2,\varepsilon_2}(T)$, then $\sigma_{e3}(T) \subset \sigma_{e3,\varepsilon_1}(T) \subset \sigma_{e3,\varepsilon_2}(T)$.

<u>For $i = 4$</u>: this assertion is immediately deduced from (i) and (ii).

(ii) Let $\lambda \notin \sigma_\varepsilon(T)$. Then, $\lambda \in \rho(T)$ and $\|(\lambda - T)^{-1}\| \leq \frac{1}{\varepsilon}$. This implies that $\lambda - T$ is injective, open with dense range and $\gamma(\lambda - T) \geq \varepsilon$. Now, if we take $S \in \mathcal{U}_T(X)$, then $\|S\| \leq \varepsilon \leq \gamma(\lambda - T)$. Thus implies that $\lambda - T - S$ is injective, open with dense range. This implies that $\alpha(\lambda - T - S) = 0$, $\beta(\lambda - T - S) = 0$ and $R(\lambda - T - S)$ is closed. Hence, $\lambda - T - S \in \Phi(X)$. Therefore, $\lambda \in \sigma_{ei,\varepsilon}(T)$, for all $i \in \{1, 2, 3, 4\}$.

(iii) <u>For $i = 1$</u>: from (i) and (ii), we have $\sigma_{e1}(T) \subset \sigma_\varepsilon(T)$ for all $\varepsilon > 0$, then $\sigma_{e1}(T) \subset \bigcap_{\varepsilon > 0} \sigma_\varepsilon(T)$. Conversely, let $\lambda \notin \sigma_{e1}(T)$, then $\lambda - T \in \Phi_+(X)$. Therefore, $R(\lambda - T)$ is closed. Hence, $\gamma(\lambda - T) > 0$. Let ε such that $0 < \varepsilon \leq \gamma(\lambda - T)$ and let $S \in \mathcal{U}_T(X)$. Since $\gamma(\lambda - T) \geq \varepsilon > \|S\|$, then $\alpha(\lambda - T - S) \leq \alpha(\lambda - T) < \infty$. Hence, $\lambda - T - S$ is open, and $R(\lambda - T - S)$ is closed. So, $\lambda \notin \sigma_{e1,\varepsilon}(T)$. Thus, we deduce that $\lambda \notin \bigcap_{\varepsilon > 0} \sigma_{e1,\varepsilon}(T)$.

By using the same reasoning as above, we get the next inclusion. Q.E.D.

Theorem 4.10.2 *Let $T \in CR(X)$ and $\varepsilon > 0$. Then, the following properties are equivalent:*

(i) $\lambda \notin \sigma_{e5,\varepsilon}(T)$.
(ii) For all continuous linear relations $S \in LR(X)$ such that $\mathcal{D}(S) \supset \mathcal{D}(T)$, $S(0) \subset T(0)$ and $\|S\| < \varepsilon$, we have $T + S - \lambda \in \Phi(X)$ and $i(T + S - \lambda) = 0$.
(iii) For all $D \in \mathcal{L}(X)$ such that $\|D\| < \varepsilon$, we have $T + D - \lambda \in \Phi(X)$ and $i(T + D - \lambda) = 0$. $\diamond$

Proof. $(i) \Longrightarrow (ii)$ Let $\lambda \notin \sigma_{e5,\varepsilon}(T)$. Then, there exists $K \in \mathcal{K}_T(X)$ such that $\lambda \notin \sigma_{\varepsilon}(T + K)$. Using Theorem 4.7.2, for all continuous linear relations $S \in LR(X)$ such that $\mathcal{D}(S) \supset \mathcal{D}(T + K) = \mathcal{D}(T) \bigcap \mathcal{D}(K) = \mathcal{D}(T)$ as $\mathcal{D}(K) \supset \mathcal{D}(T)$, $S(0) \subset (T + K)(0) = T(0)$ as $K(0) \subset T(0)$ and $\|S\| < \varepsilon$, we have $\lambda \in \rho(T + S + K)$. Then, $T + S + K - \lambda$ is open, injective with dense range. Since K is compact, then K is continuous. This implies that $S + K - \lambda$ is continuous. Furthermore, $(S + K - \lambda)(0) \subset T(0)$ and $\mathcal{D}(S + K - \lambda) = \mathcal{D}(S) \bigcap \mathcal{D}(K)$. Since T is closed, then by using Proposition 2.7.2, $T + S + K - \lambda$ is closed. Hence, $R(T + S + K - \lambda)$ is closed. So, $R(T + S + K - \lambda) = X$. Therefore, $T + S + K - \lambda \in \Phi(X)$ and $i(T + S + K - \lambda) = 0$, for all continuous linear relations $S \in LR(X)$ such that $\mathcal{D}(S) \supset \mathcal{D}(T)$, $S(0) \subset T(0)$ and $\|S\| < \varepsilon$. It comes from Proposition 2.3.4 for all continuous linear relations $S \in LR(X)$ such that $\mathcal{D}(S) \supset \mathcal{D}(T)$, $S(0) \subset T(0)$ and $\|S\| < \varepsilon$, we have $T + S - \lambda \in \Phi(X)$ and $i(T + S - \lambda) = 0$.
$(ii) \Longrightarrow (iii)$ is trivial.
$(iii) \Longrightarrow (i)$ We suppose that for all $D \in \mathcal{L}(X)$ such that $\|D\| < \varepsilon$, we have $T + D - \lambda \in \Phi(X)$ and $i(T + D - \lambda) = 0$. By Proposition 2.8.1, $N((T + D - \lambda)^*) = R(T + D - \lambda)^{\perp}$. Let $n = \alpha(T + D - \lambda) = \beta(T + D - \lambda)$, $\{x_1, \ldots, x_n\}$ be basis for $N(T + D - \lambda)$ and $\{y'_1, \ldots, y'_n\}$ be basis for the $N((T + D - \lambda)^*)$. Then, there are functionals $x'_1, \ldots, x'_n \in X^*$ (the adjoint space of X) and elements $y_1, \ldots, y_n$ such that

$$x'_j(x_k) = \delta_{jk} \quad \text{and} \quad y'_j(y_k) = \delta_{jk}, \quad 1 \leq j, k \leq n,$$

where $\delta_{jk} = 0$ if $j \neq k$ and $\delta_{jk} = 1$ if $j = k$. The single valued operator K is defined by

$$Kx = \sum_{k=1}^{n} x'_k(x) y_k, \quad x \in X.$$

K is bounded, since $\mathcal{D}(K) = X$ and

$$\|Kx\| \leq \|x\| \left(\sum_{k=1}^{n} \|x'_k\| \|y_k\| \right).$$

Moreover, the range of K is contained in a finite subspace of X. Then, K is a finite rank relation in X. Hence, K is compact. We prove that

$$N(T+D-\lambda)\bigcap N(K) = \{0\} \quad \text{and} \quad R(T+D-\lambda)\bigcap R(K) = \{0\}, \quad (4.67)$$

for all $D \in \mathcal{L}(X)$ such that $\|D\| < \varepsilon$. Let $x \in N(T+D-\lambda)$, then

$$x = \sum_{k=1}^{n} \alpha_k x_k.$$

Therefore, $x'_j(x) = \alpha_j$, $1 \le j \le n$. On the other hand, if $x \in N(K)$, then $x'_j(x) = 0$, $1 \le j \le n$. This proves the first relation in (4.67). The second inclusion is similar. In fact, if $y \in R(K)$, then

$$y = \sum_{k=1}^{n} \alpha_k y_k.$$

Hence, $y'_j(y) = \alpha_j$, $1 \le j \le n$. But, if $y \in R(T+D-\lambda)$, then $y'_j(y) = 0$, $1 \le j \le n$. This gives the second relation in (4.67). On the other hand $K \in \mathcal{K}_T(X) = \mathcal{K}_{T+D-\lambda}(X)$ since $K(0) \subset T(0) = (T+D-\lambda)(0)$ and $\mathcal{D}(K) \supset \mathcal{D}(T) = \mathcal{D}(T+D-\lambda)$. We deduce from Corollary 2.16.1 (vii) that $T + D + K - \lambda \in \Phi(X)$ and $i(T+D+K-\lambda) = 0$. If $x \in N(T+D+K-\lambda)$, then $0 \in Tx+Dx+Kx-\lambda x$. Hence, $-Kx \in (T+D-\lambda)x$. So, $Kx \in R(K)\bigcap R(T+D-\lambda) = \{0\}$. Therefore, $Kx = 0$ and $0 \in (T+D-\lambda)x$. This implies that $x \in N(T+D-\lambda)\bigcap N(K)$. Hence, $x = 0$. Thus, $\alpha(T+D+K-\lambda) = 0$. In the some way, one proves that $R(T+D+K-\lambda) = X$. Since $T+D+K-\lambda$ is closed and from Theorem 2.5.4 (ii), $T + D + K - \lambda$ is open. Furthermore, it is injective with dense range. Hence, $\lambda \in \rho(T+D+K)$ for all continuous single valued operators $D \in LR(X)$ such that $\mathcal{D}(D) \supset \mathcal{D}(T)$ and $\|D\| < \varepsilon$. This implies that $\lambda \in \rho(T+D+K)$ for all $D \in \mathcal{L}(X)$ such that $\|D\| < \varepsilon$. This is equivalent to saying that

$$\lambda \notin \bigcup_{\substack{\|D\|<\varepsilon \\ \mathcal{D}(D)\supset\mathcal{D}(T)}} \sigma_\varepsilon(T+K+D).$$

It follow from Theorem 4.7.2 that $\lambda \notin \sigma_\varepsilon(T+K)$. This implies that

$$\lambda \notin \bigcap_{K\in\mathcal{K}_T(X)} \sigma_\varepsilon(T+K).$$

So, $\lambda \notin \sigma_{e5,\varepsilon}(T)$. Q.E.D.

Remark 4.10.1 *It follow immediately, from Theorem 4.10.2 that for $T \in CR(X)$ and $\varepsilon > 0$, we have*

$$\sigma_{e5,\varepsilon}(T) = \bigcup_{\substack{\|D\| < \varepsilon \\ \mathcal{D}(D) \supset \mathcal{D}(T)}} \sigma_{e5}(T + D) = \bigcup_{\substack{\|S\| < \varepsilon \\ S(0) \subset T(0) \\ \mathcal{D}(S) \supset \mathcal{D}(T)}} \sigma_{e5}(T + S). \qquad \diamond$$

Theorem 4.10.3 *Let $T \in CR(X)$ and $\varepsilon > 0$. Then,*

$$\sigma_{e5,\varepsilon}(T) = \bigcap_{P \in \mathcal{PR}(\Phi(X))} \sigma_{\varepsilon}(T + P). \qquad \diamond$$

Proof. Let $\mathcal{O} := \bigcap_{P \in \mathcal{PR}(\Phi(X))} \sigma_{\varepsilon}(T+P)$. Since $\mathcal{K}_T(X) \subset \mathcal{PR}(\Phi(X))$, we infer that $\mathcal{O} \subset \sigma_{e5,\varepsilon}(T)$. Conversely, let $\lambda \notin \mathcal{O}$, then there exist $P \in \mathcal{PR}(\Phi(X))$ such that $\lambda \notin \sigma_{\varepsilon}(T + P)$. Since P is continuous, then by using Proposition 2.7.2, we have $T + P$ is closed. Thus, by using Theorem 4.7.2, we see that $\lambda \in \rho(T + S + P)$ for all continuous linear relations $S \in LR(X)$ such that $S(0) \subset (T + P)(0) = T(0)$, $\mathcal{D}(S) \supset \mathcal{D}(T + P) = \mathcal{D}(T) \bigcap \mathcal{D}(P) = \mathcal{D}(T)$ and $\|S\| < \varepsilon$. Using the fact that $(S+P-\lambda)(0) \subset T(0)$, $\mathcal{D}(S+P-\lambda) = \mathcal{D}(S) \bigcap \mathcal{D}(P) \supset \mathcal{D}(T)$ and T is closed, we infer Proposition 2.7.2 that $T + S + P - \lambda$ is closed, and $T + S + P - \lambda$ is open as $\lambda \in \rho(T + P + S)$. So, from Theorem 2.5.4 (ii), $R(T+S+P-\lambda)$ is closed. Hence, $T+S+P-\lambda$ is injective and surjective. Then, $T + S + P - \lambda \in \Phi(X)$ and $i(T + S + P - \lambda) = 0$. Since $-P \in \mathcal{PR}(\Phi(X))$, then by using Theorem 3.3.2, we obtain $T + S + P - P - \lambda \in \Phi(X)$ and $i(T + S - P + P - \lambda) = 0$. The fact that $-P(0) = P(0) \subset (T + S + P - \lambda)(0)$ and $\mathcal{D}(-P) = \mathcal{D}(P) \supset \mathcal{D}(T + S + P - \lambda) = \mathcal{D}(T) \bigcap \mathcal{D}(S) \bigcap \mathcal{D}(P)$ implies from Proposition 2.3.4 that $T + S - P + P - \lambda = T + S - \lambda$. Hence, for all continuous linear relations $S \in LR(X)$ such that $S(0) \subset T(0)$, $\mathcal{D}(S) \supset \mathcal{D}(T)$ and $\|S\| < \varepsilon$, we have $T + S - \lambda \in \Phi(X)$ and $i(T + S - \lambda) = 0$. Finally, the use of Theorem 4.10.2 allows us to conclude that $\lambda \notin \sigma_{e5,\varepsilon}(T)$.   Q.E.D.

Corollary 4.10.1 *Let $T \in CR(X)$, $\varepsilon > 0$ and $\mathcal{J}(X)$ be a subset of $LR(X)$. If $\mathcal{K}_T(X) \subset \mathcal{J}(X) \subset \mathcal{PR}(\Phi(X))$, then*

$$\sigma_{e5,\varepsilon}(T) = \bigcap_{J \in \mathcal{J}(X)} \sigma_{\varepsilon}(T + J). \qquad \diamond$$

Proof. It is clear to see that

$$\sigma_{e5,\varepsilon}(T) = \bigcap_{P \in \mathcal{PR}(\Phi(X))} \sigma_{\varepsilon}(T + P) \subset \bigcap_{J \in \mathcal{J}(X)} \sigma_{\varepsilon}(T + J) \subset \bigcap_{K \in \mathcal{K}_T(X)} \sigma_{\varepsilon}(T + K).$$

This completes the proof.   Q.E.D.

Remark 4.10.2 *It follows from the definition of essential pseudospectra, and both Theorem 4.10.3 and Corollary 4.10.1 that for $T \in CR(X)$ and $\varepsilon > 0$, we have*

(i) $\sigma_{e5,\varepsilon}(T + K) = \sigma_{e5,\varepsilon}(T)$ *for all* $K \in \mathcal{K}_T(X)$.
(ii) $\sigma_{e5,\varepsilon}(T + P) = \sigma_{e5,\varepsilon}(T)$ *for all* $P \in \mathcal{PR}(\Phi(X))$.
(iii) *If* $\mathcal{K}_T(X) \subset \mathcal{J}(X) \subset \mathcal{PR}(\Phi(X))$ *and if for all* $J_1, J_2 \in \mathcal{J}(X)$, $J_1 \pm J_2 \in \mathcal{J}(X)$, *then* $\sigma_{e5,\varepsilon}(T + J) = \sigma_{e5,\varepsilon}(T)$, *for all* $J \in \mathcal{J}(X)$. ◇

Proposition 4.10.2 *Let* $T \in CR(X)$. *Then,*

(i) *If* $0 < \varepsilon_1 < \varepsilon_2$, *then* $\sigma_{e5}(T) \subset \sigma_{e5,\varepsilon_1}(T) \subset \sigma_{e5,\varepsilon_2}(T)$.
(ii) *For* $\varepsilon > 0$, $\sigma_{e5,\varepsilon}(T) \subset \sigma_\varepsilon(T)$.
(iii) $\bigcap_{\varepsilon > 0} \sigma_{e5,\varepsilon(T)} = \sigma_{e5}(T)$. ◇

Proof. (i) If $\lambda \notin \sigma_{e5,\varepsilon_2}(T)$, then by using Theorem 4.10.2 and for all continuous linear relations $S \in LR(X)$ such that $\mathcal{D}(S) \supset \mathcal{D}(T)$, $S(0) \subset T(0)$ and $\|S\| < \varepsilon_2$, we have $T + S - \lambda \in \Phi(X)$ and $i(T + S - \lambda) = 0$. Hence, for all continuous linear relations $S \in LR(X)$ such that $\mathcal{D}(S) \supset \mathcal{D}(T)$, $S(0) \subset T(0)$ and $\|S\| < \varepsilon_1$, we have $T + S - \lambda \in \Phi(X)$ and $i(T + S - \lambda) = 0$. Then, $\lambda \notin \sigma_{e5,\varepsilon_1}(T)$. On the other hand, if $\lambda \notin \sigma_{e5,\varepsilon_1}(T)$, then for all continuous linear relations $S \in LR(X)$ such that $\mathcal{D}(S) \supset \mathcal{D}(T)$, $S(0) \subset T(0)$ and $\|S\| < \varepsilon_1$, we have $T + S - \lambda \in \Phi(X)$ and $i(T + S - \lambda) = 0$. In particular for $S = 0$, we have $\lambda \notin \sigma_{e5}(T)$.

(ii) $\sigma_{e5,\varepsilon}(T) = \bigcap_{K \in \mathcal{K}_T(X)} \sigma_\varepsilon(T + K) \subset \sigma_\varepsilon(T + K)$ for all $K \in \mathcal{K}_T(X)$. In particular, $K = 0$.

(iii) From (i), $\sigma_{e5}(T) \subset \sigma_{e5,\varepsilon}(T)$ for all $\varepsilon > 0$. Then, $\sigma_{e5}(T) \subset \bigcap_{\varepsilon > 0} \sigma_{e5,\varepsilon}(T)$. Conversely, If $\lambda \notin \sigma_{e5}(T)$, then $T - \lambda \in \Phi(X)$ and $i(T - \lambda) = 0$. Therefore, $R(T - \lambda)$ is closed and by using Theorem 2.5.4 (ii), we have $T - \lambda$ is open. So, $\gamma(T - \lambda) > 0$. Let ε such that $0 < \varepsilon \le \gamma(T - \lambda)$ and let D be a single valued linear relation in $LR(X)$ such that $\mathcal{D}(D) \supset \mathcal{D}(T)$ and $\|D\| < \varepsilon \le \gamma(T - \lambda)$. By using Proposition 2.9.3 (iii), $\gamma(T - \lambda) = \gamma((T - \lambda)^*)$, then $\|D\| < \varepsilon \le \gamma((T - \lambda)^*)$. So, from Corollary 2.16.1 (iii), $T + D - \lambda \in \Phi_-(X)$. Now, by Theorem 2.9.3, $\alpha(T + D - \lambda) \le \alpha(T - \lambda) < \infty$. Hence, $T + D - \lambda \in \Phi_+(X)$. So, $T + D - \lambda \in \Phi(X)$. Furthermore, $i(T + D - \lambda) = i(T - \lambda) = 0$. Hence, by using Theorem 4.10.2, $\lambda \notin \sigma_{e5,\varepsilon}(T)$. So, $\lambda \notin \bigcap_{\varepsilon > 0} \sigma_{e5,\varepsilon}(T)$. It follows that $\bigcap_{\varepsilon > 0} \sigma_{e5,\varepsilon}(T) \subset \sigma_{e5}(T)$. This completes the proof. Q.E.D.

Theorem 4.10.4 *Let $T \in CR(X)$ and let $\varepsilon > 0$. Then, $\sigma_{e5,\varepsilon}(T)$ is a closed set.* $\diamond$

Proof. Let $\lambda \in \rho_{e5,\varepsilon}(T)$, then $\lambda \notin \sigma_{e5,\varepsilon}(T)$. Let D be a single valued continuous linear relation such that $\mathcal{D}(D) \supset \mathcal{D}(T)$ and $\|D\| < \varepsilon$. Hence, by using Theorem 4.10.2, $T + D - \lambda \in \Phi(X)$ and $i(T + D - \lambda) = 0$. So, $R(T+D-\lambda)$ is closed. Hence, $T+D-\lambda$ is open. Thus, $\gamma(T+D-\lambda) > 0$. Then, by using Proposition 2.8.1, $\gamma(T + D - \lambda) = \gamma((T + D - \lambda)^*)$. Let $r > 0$ such that $r < \gamma(T+D-\lambda)$, and let $\mu \in \mathbb{D}(\lambda, r)$. Then, $|\lambda - \mu| < r < \gamma(T+D-\lambda) = \gamma((T + D - \lambda)^*)$. Therefore, $T + D - \lambda + \lambda - \mu = T + D - \mu \in \Phi_-(X)$. Also, $\alpha(T + D - \mu) \leq \gamma(T + D - \lambda) < \infty$ as $T + D - \lambda \in \Phi(X)$. Then, $T + D - \mu \in \Phi_+(X)$. Hence, $T + D - \mu \in \Phi(X)$. On the other hand,

$$i(T + D - \mu) = i(T + D - \lambda + \lambda - \mu) = i(T + D - \lambda) = 0.$$

Then, for all continuous single valued operators $D \in LR(X)$ such that $\mathcal{D}(D) \supset \mathcal{D}(T)$ and $\|D\| < \varepsilon$, we have $T + D - \mu \in \Phi(X)$ and $i(T + D - \mu) = 0$. By using Theorem 4.10.2, $\mu \in \rho_{e5,\varepsilon}(T)$. Thus, there exists $r > 0$ such that $\mathbb{D}(\lambda, r) \subset \rho_{e5,\varepsilon}(T)$, Hence, $\rho_{e5,\varepsilon}(T)$ is an open set. Q.E.D.

Theorem 4.10.5 *Let $T \in CR(X)$ and $\varepsilon > 0$. Then, $\sigma_{e5,\varepsilon}(T) = \sigma_{e5,\varepsilon}(T^*)$.* $\diamond$

Proof. Let $K \in \mathcal{K}_T(X)$, then K is compact. Therefore, it is continuous by Lemma 2.12.1 and, by using Proposition 2.11.2, we have K^* is compact. Furthermore, $\mathcal{D}(K) \supset \mathcal{D}(T)$. Hence, by using Proposition 2.8.1, $K^*(0) = \mathcal{D}(K)^\perp \subset T^*(0) = \mathcal{D}(T)^\perp$, and $K(0) \subset T(0)$ since K is continuous. So, $\mathcal{D}(K^*) = K(0)^\perp \supset T(0)^\perp = \overline{\mathcal{D}(T^*)} \supset \mathcal{D}(T^*)$. Hence,

$$\mathcal{K}_T(X) \subset \{K \in LR(X) : K \text{ is continuous and } K^* \in \mathcal{K}_{T^*}(X^*)\}.$$

Now, let $K \in \{K \in LR(X)$ such that K is continuous and $K^* \in \mathcal{K}_{T^*}(X^*)\}$, then by using Proposition 2.11.2, we have K is compact. Since, $K^*(0) = \mathcal{D}(K)^\perp \subset T^*(0) = \mathcal{D}(T)^\perp$, then $\mathcal{D}(K) \supset \mathcal{D}(T)$, also $K(0) \subset \overline{K(0)} = \mathcal{D}(K^*)^\top \subset \mathcal{D}(T^*)^\top = T(0)$ as T is closed. Then, $K \in \mathcal{K}_T(X)$. Hence,

$$\mathcal{K}_T(X) = \{K \in LR(X) : K \text{ is continuous and } K^* \in \mathcal{K}_{T^*}(X^*)\}. \quad (4.68)$$

On the other hand for $K \in \mathcal{K}_T(X)$, $T + K$ is closed. Using Theorem 4.7.1, we deduce that $\sigma_\varepsilon(T + K) = \sigma_\varepsilon((T + K)^*)$. But $\mathcal{D}(K) \supset \mathcal{D}(T)$ and K is continuous, then by using Proposition 2.8.1, we have $(T + K)^* = T^* + K^*$. Hence, $\sigma_\varepsilon(T + K) = \sigma_\varepsilon(T^* + K^*)$ for all $K \in \mathcal{K}_T(X)$. Therefore, using Eq (4.68), we have

$$\sigma_{e5,\varepsilon}(T) = \bigcap_{K \in \mathcal{K}_T(X)} \sigma_\varepsilon(T + K)$$

$$= \bigcap_{\substack{K^* \in \mathcal{K}_{T^*}(X^*) \\ K \in LR(X) \\ K \ continuous}} \sigma_\varepsilon(T^* + K^*)$$

$$\supset \bigcap_{\substack{K^* \in \mathcal{K}_{T^*}(X^*) \\ K \in LR(X)}} \sigma_\varepsilon(T^* + K^*)$$

$$= \sigma_{e5,\varepsilon}(T^*).$$

Let

$$\mathcal{O} := \bigcap_{\substack{K^* \in \mathcal{K}_{T^*}(X^*) \\ K \in LR(X) \\ K \ continuous}} \sigma_\varepsilon(T^* + K^*),$$

then

$$\mathcal{O} = \sigma_{e5,\varepsilon}(T) \supset \sigma_{e5,\varepsilon}(T^*). \tag{4.69}$$

On the other hand, let $\lambda \in \mathcal{O}$, then for all $K \in LR(X)$, K is continuous such that $K^* \in \mathcal{K}_{T^*}(X^*)$, we have $\lambda \in \sigma_\varepsilon(T^* + K^*)$. Let $K \in LR(X)$ such that $K^* \in \mathcal{K}_{T^*}(X)$, then $\tilde{K}$ is compact, where $\tilde{K}$ is given in (2.4). Since $G(\tilde{K})$ is the completion of $G(K)$, then $G(\tilde{K}) = \overline{G(K)} = G(\overline{K})$. Hence, $\tilde{K} = \overline{K}$ is compact. Hence, $\overline{K}$ is continuous. Furthermore, $G(\overline{K}^*) = G(-\overline{K}^{-1})^\perp = \overline{G(-K^{-1})}^\perp = G(-K^{-1})^\perp = G(K^*)$. Hence, $\overline{K}^* = K^*$. Thus, $\overline{K}$ is in $LR(X)$, $\overline{K}$ is continuous and $\overline{K}^* = K^* \in \mathcal{K}_{T^*}(X^*)$, since $\lambda \in \mathcal{O}$, then $\lambda \in \sigma_\varepsilon(T^* + \overline{K}^*) = \sigma_\varepsilon(T^* + K^*)$. We conclude that if $\lambda \in \mathcal{O}$, for all $K \in LR(X)$ such that $K^* \in \mathcal{K}_{T^*}(X^*)$ and $\lambda \in \sigma_\varepsilon(T^* + K^*)$, then

$$\lambda \in \bigcap_{\substack{K^* \in \mathcal{K}_{T^*}(X^*) \\ K \in LR(X)}} \sigma_\varepsilon(T^* + K^*) = \sigma_{e5,\varepsilon}(T^*).$$

Hence, $\mathcal{O} \subset \sigma_{e5,\varepsilon}(T^*)$. Using (4.69), we have $\mathcal{O} = \sigma_{e5,\varepsilon}(T) = \sigma_{e5,\varepsilon}(T^*)$. This completes the proof. Q.E.D.

4.10.1 Stability of Essential Pseudospectra of Linear Relations

Proposition 4.10.3 *Let $T \in LR(X)$ and $\varepsilon > 0$, then for any $\alpha, \beta \in \mathbb{C}$ with $\beta \neq 0$, we have*

$$\sigma_{e5,\varepsilon}(\alpha + \beta T) = \alpha + \sigma_{w,\varepsilon/|\beta|}(T)\beta. \qquad \diamondsuit$$

Proof. Let $\alpha, \beta \in \mathbb{C}$ with $\beta \neq 0$. Then,

$$
\begin{aligned}
\sigma_{e5,\varepsilon}(\alpha + \beta T) &= \bigcap_{K \in \mathcal{K}_T(X)} \sigma_\varepsilon(\alpha + \beta T + K) \\
&= \bigcap_{K \in \mathcal{K}_T(X)} \sigma_\varepsilon(\beta(\beta^{-1}\alpha + T + \beta^{-1}K)).
\end{aligned}
$$

It is simply to verify that $\{\beta^{-1}K : K \in \mathcal{K}_T(X)\} = \mathcal{K}_T(X)$. In fact, if $K \in \mathcal{K}_T(X)$, then $(\beta^{-1}K)(0) = K(0) \subset T(0)$ and $\mathcal{D}(\beta^{-1}K) = \mathcal{D}(K) \supset \mathcal{D}(T)$. Moreover, since K is compact, then $\overline{Q_{\beta^{-1}K}\beta^{-1}KB_X} = \overline{\beta^{-1}Q_K K B_X} = \beta^{-1}\overline{Q_K K B_X}$ is compact. Hence, $\beta^{-1}K$ is compact. So, $\beta^{-1}K \in \mathcal{K}_T(X)$. Therefore, $\{\beta^{-1}K : K \in \mathcal{K}_T(X)\} \subset \mathcal{K}_T(X)$. Conversely, by the same way, if $K \in \mathcal{K}_T(X)$, $\beta K \in \mathcal{K}_T(X)$, then $\mathcal{K}_T(X) \subset \{\beta^{-1}K : K \in \mathcal{K}_T(X)\}$. Therefore,

$$
\begin{aligned}
\sigma_{e5,\varepsilon}(\alpha I + \beta T) &= \bigcap_{K \in \mathcal{K}_T(X)} \sigma_\varepsilon(\beta(\beta^{-1}\alpha I + T + K)) \\
&= \bigcap_{K \in \mathcal{K}_T(X)} \sigma_\varepsilon(\alpha I + \beta(T + K)).
\end{aligned}
$$

Using Proposition 4.7.4, we have

$$
\begin{aligned}
\sigma_{e5,\varepsilon}(\alpha I + \beta T) &= \bigcap_{K \in \mathcal{K}_T(X)} (\alpha + \sigma_{\varepsilon/|\beta|}(T + K)\beta) \\
&= \alpha + \left(\bigcap_{K \in \mathcal{K}_T(X)} \sigma_{\varepsilon/|\beta|}(T + K) \right) \beta. \\
&= \alpha + \sigma_{w,\frac{\varepsilon}{|\beta|}}(T)\beta.
\end{aligned}
$$

This completes the proof. Q.E.D.

Proposition 4.10.4 *Let X be a Banach space, $T \in C\mathcal{R}(X)$, and $\varepsilon > 0$. Then, for $\delta > 0$, we have*

$$\sigma_{e5,\varepsilon}(T) \subset \mathbb{D}(0, \delta) + \sigma_{e5,\varepsilon}(T) \subset \sigma_{e5,\varepsilon+\delta}(T). \qquad \diamondsuit$$

Proof. Let $K \in \mathcal{K}_T(X)$, then K is compact. Hence, it is continuous (by Lemma 2.12.1). Using Proposition 2.7.2, $T + K$ is closed. By using Proposition 4.7.5,

$$\mathbb{D}(0, \delta) + \sigma_\varepsilon(T + K) \subset \sigma_{\varepsilon+\delta}(T + K).$$

Then,

$$\mathbb{D}(0, \delta) + \bigcap_{K \in \mathcal{K}_T(X)} \sigma_\varepsilon(T + K) \subset \bigcap_{K \in \mathcal{K}_T(X)} \sigma_{\varepsilon+\delta}(T + K).$$

Hence,

$$\mathbb{D}(0,\delta) + \sigma_{e5,\varepsilon}(T) \subset \sigma_{e5,\varepsilon+\delta}(T). \hspace{2cm} \text{Q.E.D.}$$

Proposition 4.10.5 *Let X be a Banach space and $T \in CR(X)$, then for any $\varepsilon > 0$ and $S \in LR(X)$ such that $S(0) \subset T(0)$, $\mathcal{D}(S) \supset \mathcal{D}(T)$ and $\|S\| < \varepsilon$, we have*

$$\sigma_{e5,\varepsilon-\|S\|}(T) \subset \sigma_{e5,\varepsilon}(T+S) \subset \sigma_{e5,\varepsilon+\|S\|}(T). \hspace{1cm} \diamond$$

Proof. Let $K \in \mathcal{K}_T(X)$, then K is compact. Hence, it is continuous (by Lemma 2.12.1). Using Proposition 2.7.2, $T + K$ is closed. Moreover, let $S \in LR(X)$ such that $S(0) \subset T(0) = (T+K)(0)$, $\mathcal{D}(S) \supset \mathcal{D}(T) = \mathcal{D}(T+K)$ and $\|S\| < \varepsilon$. Then, from Proposition 4.7.6,

$$\sigma_{\varepsilon-\|S\|}(T+K) \subset \sigma_{\varepsilon}(T+S+K) \subset \sigma_{\varepsilon+\|S\|}(T+K).$$

Hence,

$$\bigcap_{K \in \mathcal{K}_T(X)} \sigma_{\varepsilon-\|S\|}(T+K) \subset \bigcap_{K \in \mathcal{K}_T(X)} \sigma_{\varepsilon}(T+S+K) \subset \bigcap_{K \in \mathcal{K}_T(X)} \sigma_{\varepsilon+\|S\|}(T+K).$$

But, since $T(0) = (T+S)(0)$ and $\mathcal{D}(T) = \mathcal{D}(T+S)$, then $\mathcal{K}_T(X) = \mathcal{K}_{(T+S)}(X)$. Thus,

$$\bigcap_{K \in \mathcal{K}_T(X)} \sigma_{\varepsilon-\|S\|}(T+K) \subset \bigcap_{K \in \mathcal{K}_{T+S}(X)} \sigma_{\varepsilon}(T+S+K) \subset \bigcap_{K \in \mathcal{K}_T(X)} \sigma_{\varepsilon+\|S\|}(T+K).$$

Therefore,

$$\sigma_{e5,\varepsilon-\|S\|}(T) \subset \sigma_{e5,\varepsilon}(T+S) \subset \sigma_{e5,\varepsilon+\|S\|}(T). \hspace{1cm} \text{Q.E.D.}$$

Theorem 4.10.6 *Let X be a Banach space, $\varepsilon > 0$, $T \in CR(X)$ and $V \in \mathcal{L}(X)$ such that $0 \in \rho(V)$. Let $k = \|V\|\|V^{-1}\|$ and $S = VTV^{-1}$. Then,*

$$\sigma_{e5,\varepsilon/k}(T) \subset \sigma_{e5,\varepsilon}(S) \subset \sigma_{e5,k\varepsilon}(T).$$

In particular

$$\sigma_{e5}(S) = \sigma_{e5}(T). \hspace{2cm} \diamond$$

Proof. Let $K \in \mathcal{K}_T(X)$, then K is compact. Hence, it is continuous (by Lemma 2.12.1). Using Proposition 2.7.2, $T + K$ is closed. Using the fact V and V^{-1} are bounded single valued, together with Proposition 2.3.6, we obtain

$$\begin{aligned} S + VKV^{-1} &= VTV^{-1} + VKV^{-1} \\ &= V(TV^{-1} + KV^{-1}) \\ &= V(T+K)V^{-1}. \end{aligned}$$

From Theorem 4.7.3, we have for any $\varepsilon > 0$,

$$\sigma_{\varepsilon/k}(T+K) \subset \sigma_\varepsilon(S+VKV^{-1}) \subset \sigma_{k\varepsilon}(T+K).$$

Hence,

$$\bigcap_{K \in \mathcal{K}_T(X)} \sigma_{\varepsilon/k}(T+K) \subset \bigcap_{K \in \mathcal{K}_T(X)} \sigma_\varepsilon(S+VKV^{-1}) \subset \bigcap_{K \in \mathcal{K}_T(X)} \sigma_{k\varepsilon}(T+K).$$

But $\{VKV^{-1} : K \in \mathcal{K}_T(X)\} = \mathcal{K}_S(X)$. In fact, if $K \in \mathcal{K}_T(X)$, then $K(0) \subset T(0)$. Hence, $VKV^{-1}(0) = VK(0) \subset VT(0) = VTV^{-1}(0) = S(0)$. Moreover, since V is bounded, $\mathcal{D}(VK) = \{x \in X : Kx \cap \mathcal{D}(V) \neq \emptyset\} = \mathcal{D}(K)$ and $\mathcal{D}(VT) = \{x \in X : Tx \cap \mathcal{D}(V) \neq \emptyset\} = \mathcal{D}(T)$, then $\mathcal{D}(VKV^{-1}) = \{x \in X : V^{-1}x \cap \mathcal{D}(VK) \neq \emptyset\} \supset \{x \in X : V^{-1}x \cap \mathcal{D}(VT) \neq \emptyset\} = \mathcal{D}(VTV^{-1}) = \mathcal{D}(S)$. On the other hand, K is compact, V is continuous, and $V(0) = \{0\} \subset \mathcal{D}(K)$, then VK is precompact. Hence, $\overline{\Gamma}_0(VK) = 0$. Furthermore, since V^{-1} is single valued $(0 \in \rho(V))$, then $\overline{\Gamma}_0(VKV^{-1}) \leq \overline{\Gamma}_0(VK)\overline{\Gamma}_0(V^{-1}) = 0$. Hence, VKV^{-1} is precompact and, we have X is complete then VKV^{-1} is compact. Therefore,

$$\left\{VKV^{-1} : K \in \mathcal{K}_T(X)\right\} \subset \mathcal{K}_S(X).$$

In the similar way, if $K \in \mathcal{K}_S(X)$, then $K(0) \subset S(0)$. Hence, $V^{-1}KV(0) = V^{-1}K(0) \subset V^{-1}S(0) = V^{-1}SV(0) = T(0)$. Moreover, since V^{-1} is bounded as $0 \in \rho(V)$, $\mathcal{D}(V^{-1}K) = \{x \in X : Kx \cap \mathcal{D}(V^{-1}) \neq \emptyset\} = \mathcal{D}(K)$ and $\mathcal{D}(VS) = \{x \in X : Sx \cap \mathcal{D}(V^{-1}) \neq \emptyset\} = \mathcal{D}(S)$, then $\mathcal{D}(V^{-1}KV) = \{x \in X : Vx \cap \mathcal{D}(V^{-1}K) \neq \emptyset\} \supset \{x \in X : Vx \cap \mathcal{D}(V^{-1}S) \neq \emptyset\} = \mathcal{D}(V^{-1}SV) = \mathcal{D}(T)$. Moreover, K is compact, V^{-1} is continuous and $V^{-1}(0) = \{0\} \subset \mathcal{D}(K)$ (as $0 \in \rho(V)$), then by using Theorem 2.12.2, we obtain $V^{-1}K$ is precompact. So, $\overline{\Gamma}_0(V^{-1}K) = 0$. Furthermore, since V is single valued, then $\overline{\Gamma}_0(V^{-1}KV) \leq \overline{\Gamma}_0(V^{-1}K)\overline{\Gamma}_0(V) = 0$. Hence, $V^{-1}KV$ is precompact and, we have X is complete. Thus, $V^{-1}KV$ is compact. Therefore, $V^{-1}KV \in \mathcal{K}_T(X)$. Hence,

$$\mathcal{K}_S(X) \subset \{VKV^{-1} : K \in \mathcal{K}_T(X)\}.$$

So,

$$\bigcap_{K \in \mathcal{K}_T(X)} \sigma_{\varepsilon/k}(T+K) \subset \bigcap_{K \in \mathcal{K}_S(X)} \sigma_\varepsilon(S+K) \subset \bigcap_{K \in \mathcal{K}_T(X)} \sigma_{k\varepsilon}(T+K).$$

Thus, for any $\varepsilon > 0$

$$\sigma_{e5,\varepsilon/k}(T) \subset \sigma_{e5,\varepsilon}(S) \subset \sigma_{e5,k\varepsilon}(T).$$

Moreover,

$$\bigcap_{\varepsilon>0} \sigma_{e5,\varepsilon/k}(T) \subset \bigcap_{\varepsilon>0} \sigma_{e5,\varepsilon}(S) \subset \bigcap_{\varepsilon>0} \sigma_{e5,k\varepsilon}(T).$$

From the proof of Theorem 4.7.3, S is closed. Then,

$$\sigma_{e5}(T) \subset \sigma_{e5}(S) \subset \sigma_w(T).$$

This completes the proof.   Q.E.D.

Theorem 4.10.7 *Let X be a Banach space, $T \in CR(X)$, and $\varepsilon > 0$. Then,*

(i) If $\lambda \notin \sigma_{e5}(T)$, then

$$\|(\lambda - T)^{-1}\| \geq \frac{1}{\operatorname{dist}(\lambda, \sigma_{e5}(T))}.$$

(ii) If $\lambda \notin \sigma_{e5,\varepsilon}(T)$, then

$$\|(\lambda - T)^{-1}\| > \frac{1}{\operatorname{dist}(\lambda, \sigma_{e5,\varepsilon}(T)) + \varepsilon}. \qquad \diamond$$

Proof. Let $\lambda \in \rho_{e5}(T)$. Then, $\operatorname{dist}(\lambda, \sigma_{e5}(T)) = \inf\{|z - \lambda| \text{ such that } z \in \sigma_{e5}(T)\}$. Hence, for all $\eta > 0$, there exists $z_\eta \in \sigma_{e5}(T)$ such that

$$|\lambda - z_\eta| < \operatorname{dist}(\lambda, \sigma_{e5}(T)) + \eta.$$

Suppose $|\lambda - z_\eta| < \gamma(\lambda - T)$. Since $\lambda - T \in \Phi(X)$ and $i(\lambda - T) = 0$, $(z_\eta - \lambda)I(0) = 0 \subset (\lambda - T)(0)$, $\mathcal{D}((z_\eta - \lambda)I) = \mathcal{D}(I) = X \supset \overline{\mathcal{D}((\lambda - T))} = \overline{\mathcal{D}(T)}$ and $|(z_\eta - \lambda)I| = |\lambda - z_\eta| < \gamma(\lambda - T)$, then $\lambda - T + z_\eta - \lambda = z_\eta - T \in \Phi(X)$ and $i(z_\eta - T) = i(\lambda - T + z_\eta - \lambda) = i(\lambda - T) = 0$. Hence, $z_\eta \in \rho_{e5}(T)$. This is a contradiction. Therefore, $|\lambda - z_\eta| \geq \gamma(\lambda - T)$ for all $\eta > 0$. Thus,

$$\begin{aligned} \gamma(\lambda - T) &\leq |\lambda - z_\eta| \\ &< \operatorname{dist}(\lambda, \sigma_{e5}(T)) + \eta \quad \text{for all } \eta > 0. \end{aligned}$$

So,

$$\gamma(\lambda - T) \leq \operatorname{dist}(\lambda, \sigma_{e5}(T)).$$

Hence,

$$\|(\lambda - T)^{-1}\| \geq \frac{1}{\operatorname{dist}(\lambda, \sigma_{e5,\varepsilon}(T))}.$$

(ii) Let $\lambda \notin \sigma_{e5,\varepsilon}(T)$. Assume that $\|(\lambda - T)^{-1}\| \leq \frac{1}{\operatorname{dist}(\lambda, \sigma_{e5,\varepsilon}(T)) + \varepsilon}$, i.e., $\gamma(\lambda - T) \geq \operatorname{dist}(\lambda, \sigma_{e5,\varepsilon}(T)) + \varepsilon$. Since $\operatorname{dist}(\lambda, \sigma_{e5,\varepsilon}(T)) = \inf\{|z - \lambda| \text{ such that } z \in$

$\sigma_{e5,\varepsilon}(T)\}$, then for η, $0 < \eta \leq \varepsilon$, there exists $z_\eta \in \sigma_{e5,\varepsilon}(T)$ such that $|\lambda - z_\eta| <$ dist$(\lambda, \sigma_{e5,\varepsilon}(T)) + \eta$. Therefore,

$$\begin{aligned} |\lambda - z_\eta| &< \text{dist}(\lambda, \sigma_{e5,\varepsilon}(T)) + \varepsilon \\ &< \gamma(\lambda - T). \end{aligned}$$

Let S be a linear relation such that $S(0) \subset T(0)$, $\mathcal{D}(S) \supset \mathcal{D}(T)$ and $\|S\| < \varepsilon$. If $\lambda \notin \sigma_{e5,\varepsilon}(T)$, then by using Theorem 4.10.2, we have $\lambda - T - S \in \Phi(X)$ and $i(\lambda - T - S) = 0$. Moreover, T is closed, $(\lambda - S)(0) = S(0) \subset T(0)$, $\mathcal{D}((\lambda - S)) = \mathcal{D}(S) \supset \mathcal{D}(T)$, and $\lambda - S$ is continuous, then by using Proposition 2.7.2, we have $\lambda - T - S$ is closed. Now, suppose $|\lambda - z_\eta| < \gamma(\lambda - T - S)$. Since $(z_\eta - \lambda)I(0) = 0 \subset (\lambda - T - S)(0)$, $\mathcal{D}((z_\eta - \lambda)I) = \mathcal{D}(I) = X \supset \overline{\mathcal{D}((\lambda - T - S))} = \overline{\mathcal{D}(T)}$ and $|(z_\eta - \lambda)I| = |\lambda - z_\eta| < \gamma(\lambda - T - S)$, then $\lambda - T - S + z_\eta - \lambda = z_\eta - T - S \in \Phi(X)$ and $i(z_\eta - T - S) = i(\lambda - T - S + z_\eta - \lambda) = i(\lambda - T - S) = 0$. Hence, $z_\eta \in \rho_{e5,\varepsilon}(T)$. This is a contradiction. Therefore, $\gamma(\lambda - T - S) \leq |\lambda - z_\eta|$ for all linear relations S such that $S(0) \subset T(0)$, $\mathcal{D}(S) \supset \mathcal{D}(T)$ and $\|S\| < \varepsilon$. In particular, we put $S = 0$. Then,

$$\begin{aligned} \gamma(\lambda - T) &\leq |\lambda - z_\eta| \\ &< \text{dist}(\lambda, \sigma_{e5,\varepsilon}(T)) + \eta. \end{aligned}$$

Since $\eta \leq \varepsilon$, we have

$$\gamma(\lambda - T) < \text{dist}(\lambda, \sigma_{e5,\varepsilon}(T)) + \varepsilon.$$

This is a contradiction. Hence,

$$\|(\lambda - T)^{-1}\| > \frac{1}{\text{dist}(\lambda, \sigma_{e5,\varepsilon}(T)) + \varepsilon}. \qquad \text{Q.E.D.}$$

4.11   The Essential ε-Pseudospectra of Linear Relations

The aim of this section is to introduce and study the essential ε-pseudospectra of linear relations in a Banach space. Let $T \in CR(X)$ and $\varepsilon > 0$. Let us consider $\mathcal{J}_{T,\varepsilon}(X)$ the set

$$\mathcal{J}_{T,\varepsilon}(X) := \{S \in LR(X) : S(0) \subset T(0), \mathcal{D}(S) \supset \overline{\mathcal{D}(T)} \text{ and } \|S\| \leq \varepsilon\}.$$

Proposition 4.11.1 *Let X be a Banach space, $T \in CR(X)$ and $\varepsilon > 0$. Then,*

(i) *If* $\lambda \notin \Sigma_{e1,\varepsilon}(T)$, *then* $\lambda - T - S \in \Phi_+(X)$, *for all* $S \in \mathcal{J}_{T,\varepsilon}(X)$,

(ii) *If* $\lambda \notin \Sigma_{e2,\varepsilon}(T)$, *then* $\lambda - T - S \in \Phi_-(X)$, *for all* $S \in \mathcal{J}_{T,\varepsilon}(X)$,

(iii) *If* $\lambda \notin \Sigma_{e3,\varepsilon}(T)$, *then* $\lambda - T - S \in \Phi_\pm(X)$, *for all* $S \in \mathcal{J}_{T,\varepsilon}(X)$,

(iv) *If* $\lambda \notin \Sigma_{e4,\varepsilon}(T)$, *then* $\lambda - T - S \in \Phi(X)$, *for all* $S \in \mathcal{J}_{T,\varepsilon}(X)$,

(v) *If* $\lambda \notin \Sigma_{e5,\varepsilon}(T)$, *then* $\lambda - T - S \in \Phi(X)$ *and* $i(\lambda - T - S) = 0$, *for all* $S \in \mathcal{J}_{T,\varepsilon}(X)$. $\diamondsuit$

Proof. The fact that $T \in C\mathcal{R}(X)$ implies from Proposition 2.7.2 that for all $S \in \mathcal{J}_{T,\varepsilon}(X)$ and $\lambda \in \mathbb{C}$, we have $\lambda - T - S \in C\mathcal{R}(X)$.

(i) Let $\lambda \notin \Sigma_{e1,\varepsilon}(T)$. Then, $\alpha(\lambda - T) < \infty$ and $\|(\lambda - T)^{-1}\| < \frac{1}{\varepsilon}$. By virtue of Lemma 2.9.1, we have $\gamma(\lambda - T) > \varepsilon$ which implies that $\lambda - T$ is open. Combining the fact that $T \in C\mathcal{R}(X)$ and Proposition 2.7.2 (i), we infer that $\lambda - T$ is closed. Hence, using Theorem 2.7.2 (ii), we get $R(\lambda - T)$ is closed. Consequently, $\lambda - T \in \Phi_+(X)$. It is clear for any $S \in \mathcal{J}_{T,\varepsilon}(X)$ that $\|S\| < \gamma(\lambda - T)$. Finally, the use of Corollary 2.16.1 (iv) allows us to conclude that

$$\lambda - T - S \in \Phi_+(X).$$

(ii) Let $\lambda \notin \Sigma_{e2,\varepsilon}(T)$. Then, $\beta(\lambda - T) < \infty$ and $\gamma(\lambda - T) > \varepsilon$. In view of Proposition 2.9.3 (i) implies that $\lambda - T$ is closed. Now, using the fact $\lambda - T$ is open and Theorem 2.7.2 (ii), we infer that $R(\lambda - T)$ is closed. Hence, $\lambda - T \in \Phi_-(X)$. Since $\|S\| < \gamma(\lambda - T)$ for any $S \in \mathcal{J}_{T,\varepsilon}(X)$, then the use of Corollary 2.16.1 (ii) allows us to conclude that

$$\lambda - T - S \in \Phi_-(X).$$

(iii) Let us assume that $\lambda \notin \Sigma_{e3,\varepsilon}(T)$. Then, using (iii) of Remark 2.20.4, we obtain $\lambda \notin \Sigma_{e2,\varepsilon}(T) \bigcap \Sigma_{e1,\varepsilon}(T)$. In view of (i) and (ii) implies that

$$\lambda - T - S \in \Phi_-(X) \bigcup \Phi_+(X) = \Phi_\pm(X), \text{ for all } S \in \mathcal{J}_{T,\varepsilon}(X).$$

(iv) Let us assume that $\lambda \notin \Sigma_{e4,\varepsilon}(T)$. Then, using (iv) of Remark 2.20.4, we infer that $\lambda \notin \Sigma_{e2,\varepsilon}(T) \bigcup \Sigma_{e1,\varepsilon}(T)$. The use of (i) and (ii) makes us conclude that

$$\lambda - T - S \in \Phi_-(X) \bigcap \Phi_+(X) = \Phi(X), \text{ for all } S \in \mathcal{J}_{T,\varepsilon}(X).$$

(v) Let us assume that $\lambda \notin \Sigma_{e5,\varepsilon}(T)$. Then, using (v) of Remark 2.20.4, we deduce that $\lambda \notin \Sigma_{e4,\varepsilon}(T)$ and $i(\lambda - T) = 0$. Let $S \in \mathcal{J}_{T,\varepsilon}(X)$. It follows from (iv) that $\lambda - T - S \in \Phi(X)$. The fact that $\|S\| \leq \varepsilon$, then

$$i(\lambda - T - S) = i(\lambda - T) = 0.$$

This completes the proof of (v).   Q.E.D.

The following result gives a characterization of the essential ε-pseudospectra of a closed linear relations by means of Fredholm relations and minimum modulus.

Proposition 4.11.2 *Let X be a Banach space, $T \in C\mathcal{R}(X)$ and $\varepsilon > 0$. Then,*

(i) $\lambda \notin \Sigma_{e1,\varepsilon}(T)$ *if and only if,* $\lambda - T \in \Phi_+(X)$ *and* $\gamma(\lambda - T) > \varepsilon$.
(ii) $\lambda \notin \Sigma_{e2,\varepsilon}(T)$ *if and only if,* $\lambda - T \in \Phi_-(X)$ *and* $\gamma(\lambda - T) > \varepsilon$. $\diamond$

Proof. (i) Let $\lambda \notin \Sigma_{e1,\varepsilon}(T)$. Then, $\alpha(\lambda - T) < \infty$ and $\gamma(\lambda - T) > \varepsilon$. Since $\gamma(\lambda - T) > 0$, then $\lambda - T$ is open. It follows from (i) of Lemma 2.9.3 and (ii) of Theorem 2.7.2 that $R(\lambda - T)$ is closed. Hence, $\lambda - T \in \Phi_+(X)$ and $\gamma(\lambda - T) > \varepsilon$. Conversely, let $\lambda - T \in \Phi_+(X)$ and $\gamma(\lambda - T) > \varepsilon$. Then, $\alpha(\lambda - T) < \infty$ and $\gamma(\lambda - T) > \varepsilon$. This is equivalent to say that $\lambda \notin \Sigma_{e1,\varepsilon}(T)$.
The proof of the assertion (ii) can be checked in the same way as (i). Q.E.D.

From Proposition 4.11.2 and Remark 2.20.4, we can deduce the following corollary.

Corollary 4.11.1 *Let X be a Banach space, $T \in C\mathcal{R}(X)$ and $\varepsilon > 0$. Then,*

(i) $\lambda \notin \Sigma_{e3,\varepsilon}(T)$ *if and only if,* $\lambda - T \in \Phi_\pm(X)$ *and* $\gamma(\lambda - T) > \varepsilon$.
(ii) $\lambda \notin \Sigma_{e4,\varepsilon}(T)$ *if and only if,* $\lambda - T \in \Phi(X)$ *and* $\gamma(\lambda - T) > \varepsilon$.
(iii) $\lambda \notin \Sigma_{e5,\varepsilon}(T)$ *if and only if,* $\lambda - T \in \Phi(X), i(\lambda - T) = 0$ *and* $\gamma(\lambda - T) > \varepsilon$.
                                                            $\diamond$

Lemma 4.11.1 *Let X be a Banach space, $T \in C\mathcal{R}(X)$ and let $\varepsilon > 0$. Then,*

(i) $\Sigma_{e1,\varepsilon}(T) = \Sigma_{e2,\varepsilon}(T^*)$
(ii) $\Sigma_{e2,\varepsilon}(T) = \Sigma_{e1,\varepsilon}(T^*)$.
(iii) $\Sigma_{ei,\varepsilon}(T) = \Sigma_{ei,\varepsilon}(T^*)$, $i = 3, 4, 5$. $\diamond$

Proof. (i) Let us assume that $\lambda \notin \Sigma_{e1,\varepsilon}(T)$. Then, using (i) of Proposition 4.11.2, we have $\lambda - T \in \Phi_+(X)$ and $\gamma(\lambda - T) > \varepsilon$. By virtue of Theorem 2.14.4 (ii) and Proposition 2.8.1 (iii), we get $\lambda - T^* \in \Phi_-(X^*)$. Combining the fact that $\lambda - T$ is open and Propositions 2.9.3 (i), (iii), and 2.8.1 (iii), we infer that $\gamma(\lambda - T^*) > \varepsilon$. Thus, $\lambda \notin \Sigma_{e2,\varepsilon}(T^*)$. This enables us to conclude that $\Sigma_{e1,\varepsilon}(T) \subset \Sigma_{e2,\varepsilon}(T^*)$. A similar reasoning as above gives

$$\Sigma_{e2,\varepsilon}(T^*) \subset \Sigma_{e1,\varepsilon}(T).$$

(ii) Let us assume that $\lambda \notin \Sigma_{e1,\varepsilon}(T^*)$. Then, using ($i$) of Proposition 4.11.2, we have $\lambda - T^* \in \Phi_+(X^*)$ and $\gamma(\lambda - T^*) > \varepsilon$. In view of Theorem 2.14.4 (i) implies that $\lambda - T \in \Phi_-(X)$. It follows from ($iii$) of Theorem 2.7.2 that $\lambda - T$ is open. Then, according to Propositions 2.9.3 (i), (iii), we infer that $\gamma(\lambda - T) > \varepsilon$. Hence, $\lambda \notin \Sigma_{e2,\varepsilon}(T)$. Consequently, $\Sigma_{e2,\varepsilon}(T) \subset \Sigma_{e1,\varepsilon}(T^*)$. Conversely, a same reasoning as before leads to the result.

(iii) For $i = 3$, the use of (i), (ii) and Remark 2.20.4 (iii) allows us to conclude that

$$
\begin{aligned}
\Sigma_{e3,\varepsilon}(T) &= \Sigma_{e1,\varepsilon}(T) \bigcap \Sigma_{e2,\varepsilon}(T) \\
&= \Sigma_{e2,\varepsilon}(T^*) \bigcap \Sigma_{e1,\varepsilon}(T^*) \\
&= \Sigma_{e3,\varepsilon}(T^*).
\end{aligned}
$$

In the same way, one checks easily

$$
\Sigma_{e4,\varepsilon}(T) = \Sigma_{e4,\varepsilon}(T^*). \tag{4.70}
$$

For $i = 5$, let us assume that $\lambda \notin \Sigma_{e5,\varepsilon}(T)$. This implies from ($v$) of Remark 2.20.4 that $\lambda \notin \Sigma_{e4,\varepsilon}(T)$ and $i(\lambda - T) = 0$. By virtue of (4.70), we can get that $\lambda \notin \Sigma_{e4,\varepsilon}(T^*)$. It remains to prove the following

$$
i(\lambda - T^*) = -i(\lambda - T) = 0.
$$

Using the fact that $\lambda \notin \Sigma_{e4,\varepsilon}(T)$, then by Corollary 4.11.1, we infer that $\lambda - T \in \Phi(X)$ and $\gamma(\lambda - T) > \varepsilon$. This implies that $R(\lambda - T)$ is closed and $\lambda - T$ is open. Hence, the use of Theorem 2.14.4 (ii) and Proposition 2.8.1 (iii) makes us conclude that

$$
\alpha(\lambda - T^*) = \beta(\lambda - T) \text{ and } \alpha(\lambda - T) = \beta(\lambda - T^*).
$$

This leads to $i(\lambda - T^*) = -i(\lambda - T) = 0$. Again, by using ($v$) of Remark 2.20.4, we deduce that $\lambda \notin \Sigma_{e5,\varepsilon}(T^*)$. This is equivalent to say that

$$
\Sigma_{e5,\varepsilon}(T^*) \subset \Sigma_{e5,\varepsilon}(T).
$$

Conversely, a same reasoning as before leads to the result.   Q.E.D.

4.11.1   The Essential ε-Pseudospectra of a Sequence of Linear Relations

The aim of this subsection is to discuss the essential ε-pseudospectra of a sequence of linear relations in a Banach space.

Theorem 4.11.1 *Let X be a Banach space, $(T_n)_n$ be a sequence of closed linear relations from X into X and $T \in CR(X)$. We suppose that $\|T_n - T\| \to 0$ when $n \to \infty$. If there exists $n_1 \in \mathbb{N}$ such that $T(0) = T_n(0)$ and $\mathcal{D}(T) = \mathcal{D}(T_n)$, for all $n \geq n_1$. Then, for every $\varepsilon > 0$, there exist $\varepsilon_1, \varepsilon_2 > 0$ and a positive integers n_2 such that*

$$\Sigma_{e1,\varepsilon_1}(T) \subset \Sigma_{e1,\varepsilon_2}(T_n) \subset \Sigma_{e1,\varepsilon}(T), \text{ for all } n \geq n_2. \qquad \Diamond$$

Proof. Let us assume that $\lambda \notin \sigma_{e1,\varepsilon}(T)$. Then, $\alpha(\lambda - T) < \infty$ and $\|(\lambda - T)^{-1}\| < \frac{1}{\varepsilon}$. It follows from Lemma 2.9.1 that $\gamma(\lambda - T) > \varepsilon$. Since $\|T_n - T\| \to 0$ when $n \to \infty$, then there exists $n_0 \in \mathbb{N}$ such that

$$\|T_n - T\| < \|P\|^{-1}\Big(\gamma(\lambda - T) - \varepsilon_3\Big), \text{ for all } n \geq n_0,$$

where P is any continuous projection defined on $\mathcal{D}(\lambda - T)$ with kernel $N(\lambda - T)$. It follows from Remark 2.9.2, for all $n \geq n_0$ that

$$\begin{aligned}
\|T_n - T\| &< \gamma(\lambda - T) - \varepsilon_3 \\
&< \gamma(\lambda - T).
\end{aligned}$$

By using Proposition 2.3.4, we can write

$$\lambda - T_n = \lambda - T + T - T_n, \text{ for all } n \geq n_1. \qquad (4.71)$$

In view of Theorem 2.9.3 (i) and (4.71) implies that $\alpha(\lambda - T_n) \leq \alpha(\lambda - T)$, for all $n \geq n_1$. Consequently, $\alpha(\lambda - T_n) < \infty$, for all $n \geq n_1$. Let $n_2 = \max\{n_0, n_1\}$. Now, the use of Lemma 2.9.6 (i) and (4.71) makes us conclude, for all $n \geq n_2$, that

$$\begin{aligned}
\gamma(\lambda - T_n) &\geq \|P\|^{-1}\gamma(\lambda - T) - \|T_n - T\| \\
&> \varepsilon \|P\|^{-1}.
\end{aligned}$$

Hence, we deduce that $\lambda \notin \sigma_{e1,\varepsilon_2}(T)$, where $\varepsilon_2 = \varepsilon \|P\|^{-1}$. Let us show that

$$\sigma_{e1,\varepsilon_1}(T) \subset \sigma_{e1,\varepsilon_2}(T_n), \text{ for all } n \geq n_2.$$

Assume that $\lambda \notin \sigma_{e1,\varepsilon_2}(T_n)$. Then, $\alpha(\lambda - T_n) < \infty$ and $\|(\lambda - T_n)^{-1}\| < \frac{1}{\varepsilon_2}$. By referring to Lemma 2.9.1, we have $\gamma(\lambda - T_n) > \varepsilon_2$. Since $\|T_n - T\| \to 0$ when $n \to \infty$, then there exists $n_0 \in \mathbb{N}$ such that

$$\|T_n - T\| < \|P_1\|^{-1}\Big(\gamma(\lambda - T_n) - \varepsilon_2\Big), \text{ for all } n \geq n_0,$$

where P_1 is any continuous projection defined on $\mathcal{D}(\lambda - T_n)$ with kernel $N(\lambda - T_n)$. It follows from Remark 2.9.2, for all $n \geq n_0$ that

$$\|T_n - T\| \quad < \quad \gamma(\lambda - T_n) - \varepsilon_2$$
$$< \quad \gamma(\lambda - T_n).$$

By using Proposition 2.3.4, we can write

$$\lambda - T = \lambda - T - T_n + T_n, \text{ for all } n \geq n_1. \tag{4.72}$$

In view of Theorem 2.9.3 (i) and (4.72) implies that $\alpha(\lambda - T) \leq \alpha(\lambda - T_n)$, for all $n \geq n_1$. Consequently, $\alpha(\lambda - T) < \infty$, for all $n \geq n_1$. Let $n_2 = \max\{n_0, n_1\}$. Now, the use of Lemma 2.9.6 (i) and (4.72) makes us conclude, for all $n \geq n_2$, that

$$\gamma(\lambda - T) \quad \geq \quad \|P_1\|^{-1}\gamma(\lambda - T_n) - \|T_n - T\|$$
$$> \quad \varepsilon_2 \|P_1\|^{-1}.$$

Hence, $\lambda \notin \sigma_{e1,\varepsilon_1}(T)$, where $\varepsilon_1 = \varepsilon_2 \|P_1\|^{-1} = \varepsilon \|P\|^{-1} \|P_1\|^{-1}$. \qquad Q.E.D.

Theorem 4.11.2 *Let X be a Banach space, $(T_n)_n$ be a sequence of closed linear relations from X into X and $T \in C\mathcal{R}(X)$. We suppose that $\|T_n - T\| \to 0$ when $n \to \infty$. Then,*

(i) If there exists $n_1 \in \mathbb{N}$ such that $\overline{\mathcal{D}(T)} \subset \mathcal{D}(T_n) \subset \mathcal{D}(T)$ and $T(0) = T_n(0)$, for all $n \geq n_1$, then for every $\varepsilon_2 > 0$, there exist a positive real number ε_1, and a positive integer n_2 such that

$$\Sigma_{e2,\varepsilon_1}(T_n) \subset \Sigma_{e2,\varepsilon_2}(T), \text{ for all } n \geq n_2.$$

(ii) If there exists $n_1 \in \mathbb{N}$ such that $\overline{\mathcal{D}(T_n)} \subset \mathcal{D}(T) \subset \mathcal{D}(T_n)$ and $T(0) = T_n(0)$, for all $n \geq n_1$, then for every $\varepsilon_2 > 0$, there exist a positive real number ε_1, and a positive integer n_2 such that

$$\Sigma_{e2,\varepsilon_1}(T) \subset \Sigma_{e2,\varepsilon_2}(T_n), \text{ for all } n \geq n_2. \qquad \diamond$$

Proof. (i) Let $\lambda \notin \Sigma_{e2,\varepsilon_2}(T)$. Then, by using (ii) of Proposition 4.11.2, we infer that $\lambda - T \in \Phi_-(X)$ and $\gamma(\lambda - T) > \varepsilon_2$. Since $\|T_n - T\| \to 0$ when $n \to \infty$, then there exists $n_0 \in \mathbb{N}$ such that

$$\|T_n - T\| < \|P\|^{-1} (\gamma(\lambda - T) - \varepsilon_2), \text{ for all } n \geq n_0,$$

where P is any continuous projection defined on $\mathcal{D}((\lambda - T)^*)$ with kernel $N((\lambda - T)^*)$. Using Theorem 2.14.4 (i) and the fact that $\lambda - T \in \Phi_-(X)$, we infer that $(\lambda - T)^* \in \Phi_+(X^*)$. This implies that $(\lambda - T)^*$ is open and $\alpha((\lambda - T)^*) < \infty$. It follows from Remark 2.9.2, for all $n \geq n_0$, that

$$\begin{aligned}
\|T_n - T\| &< \gamma(\lambda - T) - \varepsilon_2 \\
&< \gamma(\lambda - T).
\end{aligned} \tag{4.73}$$

Since $\lambda - T$ is open, then by virtue of Proposition 2.9.3 (iii) and (4.73), we get

$$\|T_n - T\| < \gamma((\lambda - T)^*).$$

Let $n_2 = \max\{n_0, n_1\}$. Combining the fact that $(T_n - T)(0) \subset (\lambda - T)(0)$ and $\overline{D(T)} \subset D(T_n - T)$, for all $n \geq n_2$, and Corollary 2.16.1 (ii), we infer that $\lambda - T + T - T_n \in \Phi_-(X)$. Using the fact that for all $n \geq n_1$, $D(\lambda - T_n = D(T_n) \subset D(T)$, $T(0) = T_n(0) = (\lambda - T_n)(0)$, and in view of Lemma 2.9.3 (iv), we can write

$$\lambda - T_n = \lambda - T_n - T + T, \text{ for all } n \geq n_2.$$

Consequently, $\lambda - T_n \in \Phi_-(X)$. Now, by virtue of Proposition 2.9.6, for all $n \geq n_2$, we obtain

$$\begin{aligned}
\gamma(\lambda - T_n) &\geq \|P\|^{-1}\gamma(\lambda - T) - \|T_n - T\| \\
&\geq \|P\|^{-1}\gamma(\lambda - T) - \|P\|^{-1}\left(\gamma(\lambda - T) - \varepsilon_2\right) \\
&> \varepsilon_1,
\end{aligned}$$

where $\varepsilon_1 = \|P\|^{-1}\varepsilon_2$. Finally, the use of Proposition 4.11.2 makes us conclude that $\lambda \notin \Sigma_{e2,\varepsilon_1}(T_n)$, for all $n \geq n_2$.

The proof of (ii) may be achieved by using the same reasoning as (i). Q.E.D.

As an immediate consequence of Theorems 4.11.1 and 4.11.2.

Corollary 4.11.2 *Let X be a Banach space, $\varepsilon > 0$, $(T_n)_n$ be a sequence of closed and bounded linear relations from X into X and $T \in BCR(X)$. We suppose that $\|T_n - T\| \to 0$ when $n \to \infty$. If there exists $n_1 \in \mathbb{N}$ such that $T(0) = T_n(0)$, for all $n \geq n_1$, then for every triplet $(\varepsilon_1, \varepsilon_2, \varepsilon_3)$ such that $0 < \varepsilon_1 < \varepsilon_2 < \varepsilon_3$, there exists a positive integer n_2 such that*

$$\Sigma_{ei,\varepsilon_1}(T) \subset \Sigma_{ei,\varepsilon_2}(T_n) \subset \Sigma_{ei,\varepsilon_3}(T), \text{ for all } n \geq n_2, \text{ for } i = 1, \ldots, 4. \quad \diamond$$

4.12 S-Pseudospectra of Linear Relations

The purpose of this section is to define and characterize the S-pseudospectra of multivalued linear operator and study some properties.

Lemma 4.12.1 *Let X be a Banach space, $\varepsilon > 0$, $T \in CR(X)$ and let $S \in LR(X)$ such that $S(0) \subset T(0)$ and $\mathcal{D}(S) \supset \mathcal{D}(T)$. If $\lambda \notin \sigma_S(T)$, then $\lambda \in \sigma_{\varepsilon,S}(T)$ if and only if there exists $x \in X$ such that $\|(\lambda S - T)x\| < \varepsilon\|x\|$.* $\Diamond$

Proof. Let $\lambda \in \sigma_{\varepsilon,S}(T) \backslash \sigma_S(T)$, then $\|(\lambda S - T)^{-1}\| > \frac{1}{\varepsilon}$. Since $(\lambda S - T)^{-1}$ is bounded operator, there exists a non-zero vector $y \in X$ such that

$$\|(\lambda S - T)^{-1}y\| > \frac{1}{\varepsilon}\|y\|. \tag{4.74}$$

Putting $x := (\lambda S - T)^{-1}y$, then $y \in (\lambda S - T)x$. On the other hand, $(\lambda S - T)(0) = \lambda S(0) - T(0) = T(0)$ (as $S(0) \subset T(0)$). By using Lemma 2.5.7, we have

$$
\begin{aligned}
\|(\lambda S - T)x\| &= \operatorname{dist}(y, (\lambda S - T)(0)) \\
&= \operatorname{dist}(y, T(0)) \\
&\leq \inf_{z \in T(0)} \|y - z\| \\
&= \|y\|. \tag{4.75}
\end{aligned}
$$

From (4.74) and (4.75), we have

$$\|x\| > \frac{1}{\varepsilon}\|y\| \geq \frac{1}{\varepsilon}\|(\lambda S - T)x\|,$$

Hence,

$$\|(\lambda S - T)x\| < \varepsilon\|x\|.$$

Conversely, assume that there exists $x \in X$ such that $\|(\lambda S - T)x\| < \varepsilon\|x\|$. Since $\lambda \in \rho_S(T)$, then $\lambda S - T$ is injective. This implies that

$$\gamma(\lambda S - T)\|x\| \leq \|(\lambda S - T)x\| < \varepsilon\|x\|.$$

So, $0 < \gamma(\lambda S - T) < \varepsilon$. Using Lemma 2.5.7, we have $\gamma(\lambda S - T) = \|(\lambda S - T)^{-1}\|^{-1}$. Therefore, $\lambda \in \sigma_{\varepsilon,S}(T)$. Q.E.D.

Theorem 4.12.1 *Let X be a Banach space, $\varepsilon > 0$ and let $T \in CR(X)$. Then, $\lambda \in \sigma_{\varepsilon,S}(T)$ if and only if there exists a continuous linear relation B satisfying $\mathcal{D}(B) \supset \mathcal{D}(T)$, $B(0) \subset T(0)$, and $\|B\| < \varepsilon$ such that*

$$\lambda \in \sigma_S(T + B). \qquad \qquad \diamondsuit$$

Proof. In the first sense, assume that $\lambda \in \sigma_{\varepsilon,S}(T)$. We will discuss two cases.

<u>First case</u> : If $\lambda \in \sigma_S(T)$, then we may put $B = 0$.

<u>Second case</u> : If $\lambda \notin \sigma_S(T)$, then by Lemma 4.12.1 there exists $x_0 \in X$, $\|x_0\| = 1$ such that

$$\|(\lambda S - T)x_0\| < \varepsilon.$$

By Theorem 2.8.1, there exists $x' \in X^*$ such that $\|x'\| = 1$ and $x'(x_0) = \|x_0\|$. We can define the relation $B : X \longrightarrow X$ by

$$B(x) := x'(x)(\lambda S - T)x_0.$$

It is clear that B is everywhere defined and single valued. Hence,

$$\|Bx\| = \|x'(x)(\lambda S - T)x_0)\| \le \|x'\|\|x\|\|(\lambda S - T)x_0\|.$$

For $x \ne 0$, we have $\frac{\|Bx\|}{\|x\|} \le \|(\lambda S - T)x_0\|$. So, $\|B\| < \varepsilon$. On the other hand, $(\lambda S - (T + B))x_0 = (\lambda S - T)x_0 - Bx_0 = (\lambda S - T)x_0 - x'(x_0)(\lambda S - T)x_0 = (\lambda S - (T + B))(0)$. Therefore,

$$0 \ne x_0 \in N(\lambda S - (T + B)).$$

Hence, $\lambda S - (T + B)$ is not injective. So, $\lambda \in \sigma_S(T + B)$. Conversely, we derive a contradiction from the assumption that $\lambda \notin \sigma_{\varepsilon,S}(T)$, then $\lambda \in \rho_S(T)$ and $\gamma(\lambda S - T) \ge \varepsilon$. Therefore, $\lambda S - T$ is injective, open with dense range. Furthermore, $B(0) \subset T(0) = (\lambda S - T)(0)$, $\mathcal{D}(B) \supset \mathcal{D}(T) = \mathcal{D}(\lambda S - T)$ and $\|B\| < \varepsilon \le \gamma(\lambda S - T)$. Using Lemma 2.9.3, $\lambda S - T - B$ is injective, open with dense range. So, $\lambda \in \rho_S(T + B)$ and this is a contradiction.   Q.E.D.

Remark 4.12.1 *It follows, immediately, from Theorem 4.12.1, that for $T \in CR(X)$ and $\varepsilon > 0$,*

$$\sigma_{\varepsilon,S}(T) = \bigcup_{\substack{\|B\| < \varepsilon \\ B(0) \subset T(0) \\ \mathcal{D}(B) \supset \mathcal{D}(T)}} \sigma_S(T + B).$$

Proposition 4.12.1 *Let X be a Banach space, $\varepsilon > 0$, $T \in CR(X)$ and let $S \in LR(X)$ such that $S(0) \subset T(0)$ and $\mathcal{D}(S) \supset \mathcal{D}(T)$. Then,*

$$\bigcap_{\varepsilon > 0} \sigma_{\varepsilon, S}(T) = \sigma_S(T).$$ ◇

Proof. It is clear that $\sigma_S(T) \subset \sigma_{\varepsilon, S}(T)$ for all $\varepsilon > 0$. Then,

$$\sigma_S(T) \subset \bigcap_{\varepsilon > 0} \sigma_{\varepsilon, S}(T).$$

Conversely, if $\lambda \notin \sigma_S(T)$, then $\lambda \in \rho_S(T)$. Hence, $(\lambda S - T)^{-1}$ is a bounded linear operator. So, there exists $\varepsilon > 0$ such that $\|(\lambda S - T)^{-1}\| \leq \frac{1}{\varepsilon}$. Thus, $\lambda \notin \sigma_{\varepsilon, S}(T)$ and

$$\lambda \notin \bigcap_{\varepsilon > 0} \sigma_{\varepsilon, S}(T).$$ Q.E.D.

Proposition 4.12.2 *Let $T \in CR(X)$ and $\varepsilon > 0$. If T is injective with dense range and $\|S\| \neq 0$, then*

$$\sigma_{\varepsilon, S}(T) \subset \left\{ \lambda \in \mathbb{C} \text{ such that } |\lambda| > \frac{\gamma(T) - \varepsilon}{\|S\|} \right\}.$$ ◇

Proof. If $\gamma(T) < \varepsilon$, then there is nothing to prove. If $\gamma(T) = \varepsilon > 0$, then $T = T - 0S$ is open. Hence, $0 \in \rho_S(T)$. Furthermore, $\|T^{-1}\| = \|(0S - T)^{-1}\| = \frac{1}{\varepsilon}$. So, $0 \notin \sigma_{\varepsilon, S}(T)$. Therefore,

$$\sigma_{\varepsilon, S}(T) \subset \{\lambda \in \mathbb{C} \text{ such that } |\lambda| > 0\}.$$

Now, suppose that $\gamma(T) > \varepsilon$. Then, T open, injective and surjective. On the other hand, for $\lambda \in \mathbb{C}$ such that $|\lambda| \leq \frac{\gamma(T) - \varepsilon}{\|S\|}$, then $\|\lambda S\| = |\lambda| \|S\| \leq \gamma(T) - \varepsilon$. Thus, $\|\lambda S\| < \gamma(T)$. Using Lemma 2.9.3, the relation $\lambda S - T$ is open, injective with dense range, i.e., $\lambda \in \rho_S(T)$. For $x \in \mathcal{D}(T)$, since $S(0) \subset T(0)$, we have

$$\|(\lambda S - T)x\| = \|(T - \lambda S)x\| \geq \|Tx\| - \|\lambda Sx\| \geq (\gamma(T) - |\lambda| \|S\|) \|x\|.$$

Therefore,

$$\gamma(\lambda S - T) \geq \gamma(T) - |\lambda| \|S\| \geq \gamma(T) - \gamma(T) + \varepsilon = \varepsilon.$$

So,

$$\|(\lambda S - T)^{-1}\| \leq \frac{1}{\varepsilon}.$$

Hence,

$$\lambda \in \rho_{\varepsilon, S}(T).$$

Thus,

$$\sigma_{\varepsilon,S}(T) \subset \left\{\lambda \in \mathbb{C} \text{ such that } |\lambda| > \frac{\gamma(T) - \varepsilon}{\|S\|}\right\}. \qquad \text{Q.E.D.}$$

Theorem 4.12.2 *Let $\varepsilon > 0$, $T \in CR(X)$ and let $S \in LR(X)$ such that $S(0) \subset T(0)$ and $\mathcal{D}(S) \supset \mathcal{D}(T)$. Then, $\sigma_{\varepsilon,S}(T) = \sigma_{\varepsilon,S^*}(T^*)$.* ◇

Proof. At first, it is clear from the proof of Theorem 2.19.7 that $(\lambda S - T)^* = \lambda S^* - T^*$. Now, let $\lambda \in \rho_{\varepsilon,S^*}(T^*)$. Then,

$$\lambda \in \rho_{S^*}(T^*) \quad \text{and} \quad \|((\lambda S - T)^*)^{-1}\| \leq \frac{1}{\varepsilon}.$$

By Theorem 2.19.7, $\lambda \in \rho_S(T)$. So, $(\lambda S - T)^{-1}$ is a continuous single valued. From Proposition 2.8.1, we have

$$\|(\lambda S - T)^{-1}\| = \|((\lambda S - T)^{-1})^*\| \leq \frac{1}{\varepsilon}.$$

Thus, $\lambda \in \rho_{\varepsilon,S}(T)$. Conversely, if $\lambda \in \rho_{\varepsilon,S}(T)$, then $\|(\lambda S - T)^{-1}\| \leq \frac{1}{\varepsilon}$. By Proposition 2.8.1,

$$\|(\lambda S - T)^{-1}\| = \|(\lambda S^* - T^*)^{-1}\| \leq \frac{1}{\varepsilon}.$$

Furthermore, $\lambda \in \rho_{S^*}(T^*)$ by Theorem 2.19.7. Then, $\lambda \in \rho_{\varepsilon,S^*}(T^*)$. Q.E.D.

4.12.1 S-Essential Pseudospectra of Linear Relations

In this section, we define the S-essential pseudospectra of a closed linear relation, study some properties and establish some results of perturbation on the context of linear relations.

Definition 4.12.1 *Let T be a linear relation in $CR(X)$ and $\varepsilon > 0$. The S-essential pseudospectra of T is the set*

$$\sigma_{e5,\varepsilon,S}(T) = \bigcap_{K \in \mathcal{K}_T(X)} \sigma_{\varepsilon,S}(T + K).$$

We define the S-essential pseudoresolvent set by

$$\rho_{e5,\varepsilon,S}(T) = \mathbb{C} \backslash \sigma_{e5,\varepsilon,S}(T).$$ ◇

Theorem 4.12.3 *The following properties are equivalent:*

(i) $\lambda \notin \sigma_{e5,\varepsilon,S}(T)$.

(ii) For all continuous linear relations $B \in LR(X)$ such that $\mathcal{D}(B) \supset \mathcal{D}(T)$, $B(0) \subset T(0)$ and $\|B\| < \varepsilon$, we have $T+B-\lambda S \in \Phi(X)$ and $i(T+B-\lambda S) = 0$.
(iii) For all continuous single valued operators $D \in LR(X)$ such that $\mathcal{D}(D) \supset \mathcal{D}(T)$ and $\|D\| < \varepsilon$, we have $T + D - \lambda S \in \Phi(X)$ and $i(T + D - \lambda S) = 0$. ◇

Proof. $(i) \implies (ii)$ Let $\lambda \notin \sigma_{e5,\varepsilon,S}(T)$. Then, there exists $K \in \mathcal{K}_T(X)$ such that $\lambda \notin \sigma_{\varepsilon,S}(T + K)$. Using Theorem 4.12.1, for all continuous linear relations $B \in LR(X)$ such that $\mathcal{D}(B) \supset \mathcal{D}(T + K) = \mathcal{D}(T) \bigcap \mathcal{D}(K) = \mathcal{D}(T)$ as $\mathcal{D}(K) \supset \mathcal{D}(T)$, $B(0) \subset (T + K)(0) = T(0)$ as $K(0) \subset T(0)$ and $\|B\| < \varepsilon$, we have $\lambda \in \rho_S(T + B + K)$. Then, $T + B + K - \lambda S$ is open, injective with dense range. But T is closed, K is compact, then K is continuous. So, $B + K - \lambda S$ is continuous. Furthermore, $(B + K - \lambda S)(0) \subset T(0)$, then by Proposition 2.7.2, $T + B + K - \lambda S$ is closed. Hence, $R(T + B + K - \lambda S)$ is closed. So, $R(T + B + K - \lambda S) = X$. Therefore, $T + B + K - \lambda S \in \Phi(X)$ and $i(T + B + K - \lambda S) = 0$, for all continuous linear relations $B \in LR(X)$ such that $\mathcal{D}(B) \supset \mathcal{D}(T)$, $B(0) \subset T(0)$ and $\|B\| < \varepsilon$. It comes from Proposition 2.3.4 and Theorem 3.5.5 for all continuous linear relations $B \in LR(X)$ such that $\mathcal{D}(B) \supset \mathcal{D}(T)$, $B(0) \subset T(0)$ and $\|B\| < \varepsilon$, we have $T + B - \lambda S \in \Phi(X)$ and $i(T + B - \lambda S) = 0$.

$(ii) \implies (iii)$ is trivial.

$(iii) \implies (i)$ Let D be a continuous single valued operators in $LR(X)$ such that $\mathcal{D}(D) \supset \mathcal{D}(T)$ and $\|D\| < \varepsilon$, then $T + D - \lambda S \in \Phi(X)$ and $i(T + D - \lambda S) = 0$. By using Proposition 2.8.1, we have $N((T+D-\lambda S)^*) = R(T+D-\lambda S)^\perp$. Let $n = \alpha(T+D-\lambda S) = \beta(T+D-\lambda S)$, $\{x_1, \ldots, x_n\}$ be basis for $N(T+D-\lambda S)$ and $\{y'_1, \ldots, y'_n\}$ be basis for $N((T + D - \lambda S)^*)$. Hence, there are functionals $x'_1, \ldots, x'_n \in X^*$ and elements $y_1, \ldots, y_n$ such that

$$x'_j(x_k) = \delta_{jk} \quad \text{and} \quad y'_j(y_k) = \delta_{jk}, \quad 1 \leq j, k \leq n,$$

where $\delta_{jk} = 0$ if $j \neq k$ and $\delta_{jk} = 1$ if $j = k$. The single valued operator K is defined by

$$Kx = \sum_{k=1}^{n} x'_k(x)y_k, \ x \in X.$$

K is bounded, since $\mathcal{D}(K) = X$ and

$$\|Kx\| \leq \|x\| \left(\sum_{k=1}^{n} \|x'_k\| \|y_k\| \right).$$

Moreover, the range of K is contained in a finite subspace of X. Then, K is a finite rank relation in X. Hence, K is compact. Let $x \in N(T+D-\lambda S)$, then

$$x = \sum_{k=1}^{n} \alpha_k x_k.$$

Therefore, $x'_j(x) = \alpha_j$, $1 \leq j \leq n$. On the other hand, if $x \in N(K)$, then $x'_j(x) = 0$, $1 \leq j \leq n$. This proves that $N(T+D-\lambda S) \bigcap N(K) = \{0\}$. Now, if $y \in R(K)$, then

$$y = \sum_{k=1}^{n} \alpha_k y_k.$$

Hence,

$$y'_j(y) = \alpha_j, \ 1 \leq j \leq n.$$

But, if $y \in R(T+D-\lambda S)$, then

$$y'_j(y) = 0, \ 1 \leq j \leq n.$$

So, $N(T+D-\lambda S) \bigcap N(K) = \{0\}$ and $R(T+D-\lambda S) \bigcap R(K) = \{0\}$. On the other hand, $K \in \mathcal{K}_T(X) = \mathcal{K}_{T+D-\lambda S}(X)$ since $K(0) \subset T(0) = (T+D-\lambda S)(0)$ and $\mathcal{D}(K) \supset \mathcal{D}(T) = \mathcal{D}(T+D-\lambda S)$. So, we deduce that $T+D+K-\lambda S \in \Phi(X)$ and $i(T+D+K-\lambda S) = 0$. If $x \in N(T+D+K-\lambda S)$, then $0 \in Tx+Dx+Kx-\lambda Sx$. Hence, $-Kx \in (T+D-\lambda S)x$. So, $Kx \in R(K) \bigcap R(T+D-\lambda S) = \{0\}$. Therefore, $Kx = 0$ and $0 \in (T+D-\lambda S)x$. This implies that $x \in N(T+D-\lambda S) \bigcap N(K)$. Hence, $x = 0$. Thus, $\alpha(T+D+K-\lambda S) = 0$. In the some way, one proves that $R(T+D+K-\lambda S) = X$. Since $T+D+K-\lambda S$ is closed by Proposition 2.7.2, then $\lambda \in \rho_S(T+D+K)$ for all continuous single valued operators $D \in LR(X)$ such that $\mathcal{D}(D) \supset \mathcal{D}(T)$ and $\|D\| < \varepsilon$. But from the proof of Theorem 4.12.1 $((i) \implies (ii))$, if $\lambda \in \sigma_{\varepsilon,S}(T+K)$, then there exists a continuous single valued operator $D \in LR(X)$ satisfying $\mathcal{D}(D) \supset \mathcal{D}(T+K) = \mathcal{D}(T)$ and $\|D\| < \varepsilon$ such that $\lambda \in \sigma_S(T+K+D)$. Hence, $\lambda \notin \sigma_{\varepsilon,S}(T+K)$. Since $K \in \mathcal{K}_T(X)$, then

$$\lambda \notin \bigcap_{K \in \mathcal{K}_T(X)} \sigma_{\varepsilon,S}(T+K).$$

So, $\lambda \notin \sigma_{e5,\varepsilon,S}(T)$.   Q.E.D.

Remark 4.12.2 *It follow immediately, from Theorem 4.12.3 that for* $T \in CR(X)$ *and* $\varepsilon > 0$,

$$\sigma_{e5,\varepsilon,S}(T) = \bigcup_{\substack{\|D\| < \varepsilon \\ \mathcal{D}(D) \supset \mathcal{D}(T)}} \sigma_{e5,S}(T+D) = \bigcup_{\substack{\|B\| < \varepsilon \\ B(0) \subset T(0) \\ \mathcal{D}(B) \supset \mathcal{D}(T)}} \sigma_{e5,S}(T+B).$$

◇

Proposition 4.12.3 *Let $T \in C\mathcal{R}(X)$. Then,*

(i) If $0 < \varepsilon_1 < \varepsilon_2$, then $\sigma_{e5,S}(T) \subset \sigma_{e5,\varepsilon_1,S}(T) \subset \sigma_{e5,\varepsilon_2,S}(T)$.
(ii) For $\varepsilon > 0$, $\sigma_{e5,\varepsilon,S}(T) \subset \sigma_{\varepsilon,S}(T)$.
(iii) $\bigcap_{\varepsilon>0} \sigma_{e5,\varepsilon,S}(T) = \sigma_{e5,S}(T)$. ◇

Proof. (*i*) If $\lambda \notin \sigma_{e5,\varepsilon_2,S}(T)$, then by Theorem 4.12.3 for all continuous linear relations $B \in LR(X)$ such that $\mathcal{D}(B) \supset \mathcal{D}(T)$, $B(0) \subset T(0)$ and $\|B\| < \varepsilon_2$, we have $T + B - \lambda S \in \Phi(X)$ and $i(T + B - \lambda S) = 0$. Hence, for all continuous linear relations $B \in LR(X)$ such that $\mathcal{D}(B) \supset \mathcal{D}(T)$, $B(0) \subset T(0)$ and $\|B\| < \varepsilon_1$, we have $T + B - \lambda S \in \Phi(X)$ and $i(T + B - \lambda S) = 0$. Then, $\lambda \notin \sigma_{\varepsilon_1,S}(T)$. On other hand, if $\lambda \notin \sigma_{e5,\varepsilon_1,S}(T)$, then for all continuous linear relations $B \in LR(X)$ such that $\mathcal{D}(B) \supset \mathcal{D}(T)$, $B(0) \subset T(0)$ and $\|B\| < \varepsilon_1$, we have $T + B - \lambda S \in \Phi(X)$ and $i(T + B - \lambda S) = 0$. In particular $B = 0$, then $\lambda \notin \sigma_{e5,S}(T)$.

(*ii*) $\sigma_{e5,\varepsilon,S}(T) = \bigcap_{K \in \mathcal{K}_T(X)} \sigma_{\varepsilon,S}(T + K) \subset \sigma_{\varepsilon,S}(T + K)$ for all $K \in \mathcal{K}_T(X)$. In particular, $K = 0$.

(*iii*) From (*i*), $\sigma_{e5,S}(T) \subset \sigma_{e5,\varepsilon,S}(T)$ for all $\varepsilon > 0$, then

$$\sigma_{e5,S}(T) \subset \bigcap_{\varepsilon>0} \sigma_{e5,\varepsilon,S}(T).$$

Conversely, if $\lambda \notin \sigma_{e5,S}(T)$, then $T - \lambda S \in \Phi(X)$ and $i(T - \lambda S) = 0$. Therefore, $R(T - \lambda S)$ is closed. Hence, $T - \lambda S$ is open. So, $\gamma(T - \lambda S) > 0$. Let ε such that $0 < \varepsilon \leq \gamma(T - \lambda S)$ and let D be a single valued linear relation in $LR(X)$ such that $\mathcal{D}(D) \supset \mathcal{D}(T)$ and $\|D\| < \varepsilon \leq \gamma(T - \lambda S)$. By $\gamma(T - \lambda S) = \gamma((T - \lambda S)^*)$, then $\|D\| < \varepsilon \leq \gamma((T - \lambda S)^*)$. So, $T + D - \lambda S \in \Phi_-(X)$. Hence, $\alpha(T + D - \lambda S) \leq \alpha(T - \lambda S) < \infty$. Thus, $T + D - \lambda S \in \Phi_+(X)$ and $T + D - \lambda S \in \Phi(X)$. Furthermore, we have $i(T + D - \lambda S) = i(T - \lambda S) = 0$. Hence, by using Theorem 4.12.3, we have $\lambda \notin \sigma_{e5,\varepsilon,S}(T)$. So, $\lambda \notin \bigcap_{\varepsilon>0} \sigma_{e5,\varepsilon,S}(T)$. It follows $\bigcap_{\varepsilon>0} \sigma_{e5,\varepsilon,S}(T) \subset \sigma_{e5,S}(T)$.   Q.E.D.

Theorem 4.12.4 *Let X be a Banach space, $\varepsilon > 0$, $T \in C\mathcal{R}(X)$ and let $S \in LR(X)$ such that $S(0) \subset T(0)$ and $\mathcal{D}(S) \supset \mathcal{D}(T)$. Then, $\sigma_{e5,\varepsilon,S}(T) = \sigma_{e5,\varepsilon,S^*}(T^*)$.* ◇

Proof. Let $K \in \mathcal{K}_T(X)$, then K is compact. Therefore, it is continuous by Lemma 2.12.1 and from Proposition 2.11.2, K^* is compact. Moreover, by Lemma 2.8.3, $K^*(0) \subset T^*(0)$ and $\mathcal{D}(K^*) \supset \mathcal{D}(T^*)$. Hence,

$$\mathcal{K}_T(X) \subset \{K \in LR(X) \text{ such that } K \text{ is continuous and } K^* \in \mathcal{K}_{T^*}(X^*)\}.$$

Now, let $K \in \{K \in LR(X) \text{ such that } K \text{ is continuous and } K^* \in \mathcal{K}_{T^*}(X^*)\}$, then K is compact and $K^*(0) = \mathcal{D}(K)^\perp \subset T^*(0) = \mathcal{D}(T)^\perp$. So, $\mathcal{D}(K) \supset \mathcal{D}(T)$, also $K(0) \subset \overline{K}(0) = \mathcal{D}(K^*)^\top \subset \mathcal{D}(T^*)^\top = T(0)$ as T is closed. Then, $K \in \mathcal{K}_T(X)$. Hence,

$$\mathcal{K}_T(X) = \{K \in LR(X) \text{ such that } K \text{ is continuous and } K^* \in \mathcal{K}_{T^*}(X^*)\}. \tag{4.76}$$

On the other hand for $K \in \mathcal{K}_T(X)$, $T + K$ is closed (by Proposition 2.7.2). By using Theorem 4.12.2, we have $\sigma_{\varepsilon,S}(T + K) = \sigma_{\varepsilon,S^*}((T + K)^*)$. But $\mathcal{D}(K) \supset \mathcal{D}(T)$ and K is continuous, then $(T + K)^* = T^* + K^*$. Hence, $\sigma_{\varepsilon,S}(T + K) = \sigma_{\varepsilon,S^*}(T^* + K^*)$ for all $K \in \mathcal{K}_T(X)$. Therefore, by using (4.76), we have

$$
\begin{aligned}
\sigma_{e5,\varepsilon,S}(T) &= \bigcap_{K \in \mathcal{K}_T(X)} \sigma_{\varepsilon,S}(T + K) \\
&= \bigcap_{\substack{K^* \in \mathcal{K}_{T^*}(X^*) \\ K \in LR(X) \\ K \, continuous}} \sigma_{\varepsilon,S^*}(T^* + K^*) \\
&\supset \bigcap_{\substack{K^* \in \mathcal{K}_{T^*}(X^*) \\ K \in LR(X)}} \sigma_{\varepsilon,S^*}(T^* + K^*) \\
&= \sigma_{e5,\varepsilon,S^*}(T^*).
\end{aligned}
$$

Let

$$\mathcal{O} := \bigcap_{\substack{K^* \in \mathcal{K}_{T^*}(X^*) \\ K \in LR(X) \\ K \, continuous}} \sigma_{\varepsilon,S^*}(T^* + K^*),$$

then

$$\mathcal{O} = \sigma_{e5,\varepsilon,S}(T) \supset \sigma_{e5,\varepsilon,S^*}(T^*). \tag{4.77}$$

On the other hand, let $\lambda \in \mathcal{O}$, then for all $K \in LR(X)$, K is continuous such that $K^* \in \mathcal{K}_{T^*}(X^*)$, $\lambda \in \sigma_{\varepsilon,S^*}(T^* + K^*)$. Let $K \in LR(X)$ such that $K^* \in \mathcal{K}_{T^*}(X^*)$. Hence, $\widetilde{K}$ is compact, where $\widetilde{T}$ is given in (2.4). Since $G(\widetilde{K})$ is the completion of $G(K)$, then $G(\widetilde{K}) = \overline{G(K)} = G(\overline{K})$. Hence, $\widetilde{K} = \overline{K}$ is compact and hence, continuous. Furthermore, $G(\overline{K}^*) =$

$G(-\overline{K}^{-1})^{\perp} = \overline{G(-K^{-1})}^{\perp} = G(-K^{-1})^{\perp} = G(K^*)$. Hence, $\overline{K}^* = K^*$. Thus, $\overline{K}$ in $LR(X)$, $\overline{K}$ is continuous and $\overline{K}^* = K^* \in \mathcal{K}_{T^*}(X^*)$. Since $\lambda \in \mathcal{O}$, then $\lambda \in \sigma_{\varepsilon,S^*}(T^* + \overline{K}^*) = \sigma_{\varepsilon,S^*}(T^* + K^*)$. We conclude that if $\lambda \in \mathcal{O}$, for all $K \in LR(X)$ such that $K^* \in \mathcal{K}_{T^*}(X^*)$, then $\lambda \in \sigma_{\varepsilon,S^*}(T^* + K^*)$. So, $\lambda \in \bigcap_{\substack{K^* \in \mathcal{K}_{T^*}(X^*) \\ K \in LR(X)}} \sigma_{\varepsilon,S^*}(T^* + K^*) = \sigma_{e5,\varepsilon,S^*}(T^*)$. Hence, $\mathcal{O} \subset \sigma_{e5,\varepsilon,S^*}(T^*)$. By using (4.77), we have $\mathcal{O} = \sigma_{e5,\varepsilon,S}(T) = \sigma_{e5,\varepsilon,S^*}(T^*)$. Q.E.D.

Theorem 4.12.5 *Let X be a Banach space, $\varepsilon > 0$, $T \in CR(X)$ and let $S \in LR(X)$ such that $S(0) \subset T(0)$ and $\mathcal{D}(S) \supset \mathcal{D}(T)$. Then,*

$$\sigma_{e5,\varepsilon,S}(T) = \bigcap_{P \in \mathcal{PR}(\Phi(X))} \sigma_{\varepsilon,S}(T + P). \qquad \diamond$$

Proof. Let $\mathcal{O} := \bigcap_{P \in \mathcal{PR}(\Phi(X))} \sigma_{\varepsilon,S}(T + P)$. Since $\mathcal{K}_T(X) \subset \mathcal{PR}(\Phi(X))$, we infer that $\mathcal{O} \subset \sigma_{e5,\varepsilon,S}(T)$. Conversely, let $\lambda \notin \mathcal{O}$, then there exist $P \in \mathcal{PR}(\Phi(X))$ such that $\lambda \notin \sigma_{\varepsilon,S}(T + P)$. But P is continuous, then by Proposition 2.7.2, $T + P$ is closed. Thus, by Theorem 4.12.1, we can see that $\lambda \in \rho_S(T + B + P)$ for all continuous linear relations $B \in LR(X)$ such that $B(0) \subset (T + P)(0) = T(0)$, $\mathcal{D}(B) \supset \mathcal{D}(T + P) = \mathcal{D}(T) \bigcap \mathcal{D}(P) = \mathcal{D}(T)$ and $\|B\| < \varepsilon$. But $(B + P - \lambda S)(0) \subset T(0)$, $\mathcal{D}(B + P - \lambda S) = \mathcal{D}(B) \bigcap \mathcal{D}(P) \supset \mathcal{D}(T)$ and T is closed, by Proposition 2.7.2, $T + B + P - \lambda S$ is closed. Hence, $T + B + P - \lambda S$ is injective and surjective. Then,

$$T + B + P - \lambda S \in \Phi(X) \quad \text{and} \quad i(T + B + P - \lambda S) = 0.$$

Since $-P \in \mathcal{PR}(\Phi(X))$, $-P(0) = P(0) \subset (T + B + P - \lambda S)(0)$ and $\mathcal{D}(-P) = \mathcal{D}(P) \supset \mathcal{D}(T + B + P - \lambda S) = \mathcal{D}(T) \bigcap \mathcal{D}(B) \bigcap \mathcal{D}(P) \bigcap \mathcal{D}(S)$, then $-P$ is in $\mathcal{PR}(\Phi(X))$. Using Theorem 3.3.2, we conclude that for all continuous linear relations $B \in LR(X)$ such that $B(0) \subset T(0)$, $\mathcal{D}(B) \supset \mathcal{D}(T)$ and $\|B\| < \varepsilon$,

$$T + B - \lambda S \in \Phi(X) \quad \text{and} \quad i(T + B - \lambda S) = 0.$$

Finally, Theorem 4.12.3 shows that $\lambda \notin \sigma_{e5,\varepsilon,S}(T)$. Q.E.D.

Corollary 4.12.1 *Let $\mathcal{J}(X)$ be a subset of $LR(X)$. If $\mathcal{K}_T(X) \subset \mathcal{J}(X) \subset \mathcal{PR}(\Phi(X))$, then*

$$\sigma_{e5,\varepsilon,S}(T) = \bigcap_{J \in \mathcal{J}(X)} \sigma_{\varepsilon,S}(T + J). \qquad \diamond$$

Proof. It is easy to see

$$\sigma_{e5,\varepsilon,S}(T) = \bigcap_{P\in\mathcal{PR}(\Phi(X))} \sigma_{\varepsilon,S}(T+P) \subset \bigcap_{J\in\mathcal{J}(X)} \sigma_{\varepsilon,S}(T+J) \subset \bigcap_{K\in\mathcal{K}_T(X)} \sigma_{\varepsilon,S}(T+K).$$

This completes the proof.   Q.E.D.

Remark 4.12.3 *It follows, from the definition of S-essential pseudospectra, Theorem 4.12.5 and Corollary 4.12.1 that*

(i) $\sigma_{e5,\varepsilon,S}(T+K) = \sigma_{e5,\varepsilon,S}(T)$ *for all* $K \in \mathcal{K}_T(X)$.
(ii) $\sigma_{e5,\varepsilon,S}(T+P) = \sigma_{e5,\varepsilon,S}(T)$ *for all* $P \in \mathcal{PR}(\Phi(X))$.
(iii) If for all $J_1, J_2 \in \mathcal{J}(X)$, $J_1 \pm J_2 \in \mathcal{J}(X)$, *then* $\sigma_{e5,\varepsilon,S}(T+J) = \sigma_{e5,\varepsilon,S}(T)$
for all $J \in \mathcal{J}(X)$ *such that* $\mathcal{K}_T(X) \subset \mathcal{J}(X) \subset \mathcal{PR}(\Phi(X))$.   ◇

Bibliography

[1] Abdmouleh, F., Álvarez, T., Ammar, A., & Jeribi, A., (2015). Spectral mapping theorem for Rakočević and Schmoeger essential spectra of a multivalued linear operator. *Mediterr. J. Math., 12*(3), 1019–1031.

[2] Abdmouleh, F., Ammar, A., & Jeribi, A., (2013). Stability of the *S*-essential spectra on a Banach space. *Math. Slovaca, 63*(2), 299–320.

[3] Abdmouleh, F., Jeribi, A., & Álvarez, T., (2020). *On a Characterization of the Essential Spectra of Linear Relation.* Preprint.

[4] Aiena, P., (2004). *Fredholm and Local Spectral Theory, with Applications to Multipliers.* Kluwer Academic Publishers, Dordrecht.

[5] Aiena, P., (2007). *Semi-Fredholm Operators, Perturbation Theory, and Localized SVEP.* Caracas, Venezuela.

[6] Alaarabiou, E. H., & Benilan, P., (1996). Sur le carre dâun operateur non lineaire. *Arch. Math., 66*, 335–343.

[7] Albrecht, E., & Vasilescu, F. H., (1986). Stability of the index of a semi-Fredholm complex of Banach spaces. *J. Funct. Anal., 66*(2), 141–172.

[8] Álvarez, T., (2004). Perturbation theorems for upper and lower semi-Fredholm linear relations. *Publ. Math. Debrecen, 65*(1/2), 179–191.

[9] Álvarez, T., (2012). On regular linear relations. *Acta. Math. Sin., 28*, 183–194.

[10] Álvarez, T., Ammar, A., & Jeribi, A., (2014). A Characterization of some subsets of *S*-essential spectra of a multivalued linear operator. *Colloq. Math., 135*(2), 171–186.

[11] Álvarez, T., Ammar, A., & Jeribi, A., (2014). On the essential spectra of some matrix of linear relations. *Math. Methods Appl. Sci., 37*(5), 620–644.

[12] T. Álvarez, T., & Benharrat, M., (2015). Relationship between the Kato spectrum and the Goldberg spectrum of linear relations. *Mediterr. J. Math., 13*, 365–378.

[13] Álvarez, T., Cross, R. W., & Wilcox, D., (2001). Multivalued Fredholm type operators with abstract generalized inverses. *J. Math. Anal. Appl., 261*(1), 403–417.

[14] Álvarez, T., & Wilcox, D., (2007). Perturbation theory of multivalued Atkinson operators in normed spaces. *Bull. Austral. Math. Soc., 76*(2), 195–204.

[15] Ambrozie, C. G., & Vasilescu, F. H., (1995). Banach Space Complexes. *Mathematics and Its Applications* (p. 334). Kluwer Academic Publishers Group, Dordrecht.

[16] Ammar, A., (2017). A characterization of some subsets of essential spectra of a multivalued linear operator. *Complex Anal. Oper. Theory, 11*(1), 175–196.

[17] Ammar, A., (2018). Some results on semi-Fredholm perturbations of multivalued linear operators. *Linear Multilinear Algebra, 66*(7), 1311–1332.

[18] Ammar, A., Benharrat, M., Jeribi, A., & Messirdi, B., (2014). On the Kato, semi regular and essentially semi regular spectra. *Funct. Analy. Approx. Comput., 6*, 9–22.

[19] Ammar, A., Bouchekoua, A., & Jeribi, A., (2019). The ε-pseudospectra and the essential ε-pseudospectra of linear relations. *Journal of Pseudo-Differential Operators and Applications*, 1–37.

[20] Ammar, A., Boukettaya, B., & Jeribi, A., (2016). A note on the essential pseudospectra and application. *Linear Multilinear Algebra, 64*(8), 1474–1483.

[21] Ammar, A., Boukettaya, B., & Jeribi, A., (2017). Essentially semi regular in linear relations *Facta Universitatis (Nis) Ser. Math. Inform., 31*.

[22] Ammar, A., Daoud, H., & Jeribi, A., (2015). Pseudospectra, and essential pseudospectra of multivalued linear relations. *Mediterr. J. Math., 12*(4), 1377–1379.

[23] Ammar, A., Daoud, H., & Jeribi, A., (2016). Stability of pseudospectra and essential pseudospectra of linear relations. *J. Pseudo-Differ. Oper. Appl.*, *7*(4), 473–491.

[24] Ammar, A., Daoud, H., & Jeribi, A., (2018). Demicompact and k-d-set contractive multivalued linear operators. *Mediterr. J. Math.*, *15*(2), Paper No. 41, 18.

[25] Ammar, A., Daoud, H., & Jeribi, A., (2020). S-Pseudospectra and S-essential pseudospectra of multivalued linear relations. *Matematicki Vesnik*, *72*(2), 95–105.

[26] Ammar, A., Dhahri, M. Z., & Jeribi, A., (2015). Some properties of upper triangular 3×3-block matrices of linear relations. *Boll. Unione Mat. Ital.*, *8*(3), 189–204.

[27] Ammar, A., Diagana, T., & Jeribi, A., (2016). Perturbations of Fredholm linear relations in Banach spaces with application to 3×3-block matrices of linear relations. *Arab J. Math. Sci.*, *22*(1), 59–76.

[28] Ammar, A., Fakhfakh, S., & Jeribi, A., (2016). Stability of the essential spectrum of the diagonally and off-diagonally dominant block matrix linear relations. *J. Pseudo-Differ. Oper. Appl.*, *7*(4), 493–509.

[29] Ammar, A., Fakhfakh, S., & Jeribi, A., (2017). A New stability of the S essential spectrum of multivalued linear operators. *International Journal of Analysis and Applications*, *14*, 1–8.

[30] Ammar, A., Fakhfakh, S., & Jeribi, A., (2020). Shechter spectra and relatively demicompact linear relations. *Commun. Korean Math. Soc.* *35*(2), 499–516.

[31] Ammar, A., & Jeribi, A., (2013). A characterization of the essential pseudospectra and application to a transport equation. *Extracta Math.*, *28*, 95–112.

[32] Ammar, A., & Jeribi, A., (2013). A characterization of the essential pseudospectra on a Banach space. *J. Arab. Math.*, *2*, 139–145.

[33] Ammar, A., & Jeribi, A., (2014). Measures of noncompactness and essential pseudospectra on Banach space. *Math. Meth. Appl. Sci.*, *37*, 447–452.

[34] Ammar, A., & Jeribi, A., (2016). The Weyl essential spectrum of a sequence of linear operators in Banach spaces. *Indag. Math. (N.S.)*, *27*(1), 282–295.

[35] Ammar, A., Jeribi, A., & Lazrag, N., (2019). Sequence of linear operators in non-Archimedean Banach spaces. *Mediterr. J. Math.*, *16*(5), Paper No. 130.

[36] Ammar, A., Jeribi, A., & Lazrag, N., (2020). Sequence of multivalued linear operators converging in the generalized sense. *Bull. Iranian Math. Soc.*, *46*(6), 1697–1729.

[37] Ammar, A., Jeribi, A., & Saadaoui, B., (2017). Frobenius-Schur factorization for multivalued 2×2 matrices linear operator. *Mediterr. J. Math.*, *14*(1), *29*, Art. 29.

[38] Ammar, A., Jeribi, A., & Saadaoui, B., (2018). A characterization of essential pseudospectra of the multivalued operator matrix. *Anal. Math. Phys.*, *8*(3), 325–350.

[39] Ammar, A., Jeribi, A., & Saadaoui, B., (2019). On some classes of demicompact linear relation and some results of essential pseudospectra. *Mat. Stud.*, *52*, 195–210.

[40] Ammar, A., Jeribi, A., & Saadaoui, B., (2021). *Demicompactness, Selection of Lineat Relation and Application to Multivalued Matrix.* Filomat.

[41] Ammar, A., Jeribi, A., & Saadaoui, B. (2021). *S-Resolvent and S-spectra of Multivalued Linear Operatior.* Preprint.

[42] Arens, R., (1961). Operational calculus of linear relations. *Pacific J. Math.*, *11*, 9–23.

[43] Artstein, Z., (1977). Continuous dependence of solutions of operator equations. I. *Trans. Amer. Math. Soc.*, *231*(1), 143–166.

[44] Baskakov, A. G., & Chernyshov, K. I., (2002). Spectral analysis of linear relations and degenerate operator semigroups. *Sbornik Math.*, *193*(11), 1573–1610.

[45] Berkani, M., (2000). Restriction of an operator to the range of its powers. *Studia Math.*, *140*(2), 163–175.

[46] Chaitin-Chetelin, F., & Harrabi, A., (1998). *About Definitions of Pseudospectra of Closed Operators in Banach Spaces.* Technical Report TR/PA/98/08, CERFACS, Toulouse, France.

[47] Chaker, W., Jeribi, A., & Krichen, B., (2015). Demicompact linear operators, essential spectrum and some perturbation results. *Math. Nachr., 288*(13), 1476–1486.

[48] Coddington, E. A., (1971). *Multivalued Operators and Boundary Value Problems: Analytic Theory of Differential Equations (Proc. Conf., Western Michigan Univ., Kalamazoo, Mich., 1970)*, pp. 2–8. Lecture Notes in Math. 183, Springer, Berlin.

[49] Coddington, E. A., (1973). *Extension Theory of Formally Normal and Symmetric Subspaces.* Memoirs of the American Mathematical Society, no. 134. American Mathematical Society, Providence, R.I.

[50] Cross, R. W., (1998). *Multivalued Linear Operators, Monographs, and Textbooks in Pure and Applied Mathematics*, p. 213. Marcel Dekker, Inc., New York.

[51] Cross, R., Favini, A., & Yakubov, Y., (2011). Perturbation results for multivalued linear operators. Birkhéser/Springer Basel AG, Basel. *Parabolic Problems, Progr. Nonlinear Differential Equations Appl., 80*, 111–130.

[52] Cvetković, D., (2002). On gaps between bounded operators. *Publ. Inst. Math. (Beograd) (N.S.), 72*(86), 49–54.

[53] Davies, E. B., (2007). *Linear Operators and Their Spectra.* The United States of America by Cambridge University Press, New York.

[54] Dunford, N., (1952). Spectral theory. II. Resolutions of the identity. *Pacific J. Math., 2*, 559–614.

[55] Edmunds, D. E., & Evans, W. D., (1987). *Spectral Theory and Differential Operators.* Oxford Mathematical Monographs, Oxford Science Publications. The Clarendon Press, Oxford University Press, New York.

[56] Favini, A., (1979). Laplace transforms method for a class of degenerate evolution problems. *Rend. Mat., 12*, 511–536.

[57] Favini, A., & Yagi, A., (1993). Multivalued linear operators and degenerate evolution equations. *Ann. Mat. Pura Appl., 163*(4), 353–384.

[58] Gohberg, I., Markus, A., & Feldman, I. A., (1967). Normally solvable operators and ideals associated with them. *Amer. Math. Soc. Tran. Ser., 2*(61), 63–84.

[59] Goldberg, S., (1966). *Unbounded Linear Operators: Theory and Applications.* McGraw-Hill Book Co., New York–Toronto, Ontario–London.

[60] Hinrichsen, D., & Pritchard, A. J., (1994). Robust stability of linear operators on Banach spaces. *J. Cont. Opt., 32*, 1503–1541.

[61] Jeribi, A., (2000). Une nouvelle characterization du spectre essentiel et application. *C. R. Acad. Sci. Paris, Serie I, 331*, 525–530.

[62] Jeribi, A., (2002). A characterization of the essential spectrum and applications. *J. Boll. Dell. Union. Mat. Ital., 5-B*(8), 805–825.

[63] Jeribi, A., (2002). A characterization of the Schechter essential spectrum on Banach spaces and applications. *J. Math. Anal. Appl., 271*, 343–358.

[64] Jeribi, A., (2002). On the Schechter essential spectrum on Banach spaces and application. *Facta Univ. Ser. Math. Inform., 17*, 35–55.

[65] Jeribi, A., (2015). *Spectral Theory and Applications of Linear Operators and Block Operator Matrices.* Springer-Verlag, New York.

[66] Jeribi, A., (2018). *Linear Operators and Their Essential Pseudospectra.* CRC Press, Boca Raton.

[67] Kato, T., (1976). *Perturbation Theory for Linear Operators* (2nd edn.). In: Grundlehren der Mathematischen Wissenschaften, Band 132, Springer-Verlag, Berlin–Heidelberg–New York.

[68] Knapp, A. W., (2006). *Basic Algebra.* Along with a companion volume Advanced algebra. Cornerstones. Birkhéser Boston, Inc., Boston, MA.

[69] Krein, M. G., & Krasnoselski, M. A., (1947). Theoremes fundamentaux sur lâextension dâoperateurs Hermitiens. *Uspekhi Matematicheskikh Nauk, 2*(3(19)), 60–106.

[70] Krichen, B., (2014). Relative essential spectra involving relative demicompact unbounded linear operators. *Acta Math. Sci. Ser. B Engl.*, *34*, 546–556.

[71] Krichen, B., ÓRegan, D., (2018). On the class of relatively weakly demicompact nonlinear operators. *Fixed Point Theory, 19*(2), 625–630.

[72] Labrousse, J. P., (1980). Les operateurs quasi Fredholm: Une generalization des operateurs semi-Fredholm. (French) [[Quasi-Fredholm operators: A generalization of semi-Fredholm operators]] *Rend. Circ. Mat. Palermo, 29*(2), 161–258.

[73] Labrousse, J. P., Sandovici, A., De Snoo, H., & Winkler, H., (2006). Quasi-Fredholm relations in Hilbert spaces. *Stud. Cercet. Stiint. Ser. Mat. Univ. Bacau, 16*, 93–105.

[74] Labrousse, J. P., Sandovici, A., De Snoo, H., & Winkler, H., (2010). The Kato decomposition of quasi-Fredholm relations. *Oper. Matrices, 4*, 1–51.

[75] Labuschagne, L. E., (1992). Certain norm related quantities of unbounded linear operators. *Math. Nachr., 157*, 137–162.

[76] Landau, H. J., (1975). On Szegoâs eigenvalue distribution theorem and non-Hermitian kernels. *J. Analyze Math., 28*, 335–357.

[77] Lebow, A., & Schechter, M., (1971). Semigroups of linear operators and measures of non-compactness. *J. Funct. Anal., 7*, 1–26.

[78] Mbekhta, M., & Muller, V., (1996). On the axiomatic theory of spectrum. II. *Studia Math., 119*(2), 129–147.

[79] Muller-Horrig, V., (1980). Zur Theori der Semi-Fredholm-operatoren mit stetig projizierten Kern und Bild. *Math. Nachr., 99*, 185–197.

[80] Muller, V., (2003). *Spectral Theory of Linear Operator and Spectral Systems in Banach Algebras.* Birkhauser Verlag.

[81] Murray, F. J., (1937). On complementary manifolds and projections in spaces L_p and l_p. *Trans. Amer. Math. Soc., 41*, 138–152.

[82] Opial, Z., (1967). *Nonexpansive and Monotone Mappings in Banach Spaces.* Brown Univ Providence Ri Center for Dynamical Systems, no. CDS-LN-67-1.

[83] Petryshyn, W. V., (1972). Remarks on condensing and k-set-contractive mappings. *J. Math. Anal. Appl., 39*, 717–741.

[84] Sandovici, A., & De Snoo, H., (2009). An index formula for the product of linear relations. *Linear Algebra Appl., 431*(11), 2160–2171.

[85] Sandovici, A., De Snoo, H., & Winkler, H., (2007). Ascent, descent, nullity, defect, and related notions for linear relations in linear spaces. *Linear Algebra Appl., 423*(2/3), 456–497.

[86] Shargorodsky, E., (2009). On the definition of pseudospectra. *Bull. Lond. Math. Soc., 41*(3), 524–534.

[87] Showalter, R. E., (1975). A nonlinear Parabolic-Sobolev equation. *J. Math. Anal. Appl., 50*, 183–190.

[88] Trefethen, L. N., (1992). Pseudospectra of matrices. Numerical analysis 1991 (Dundee, 1991). Harlow. *Pitman Res. Notes Math. Ser., 260, Longman Sci. Tech.*, 234–266.

[89] Trefethen, L. N., (1997). Pseudospectra of linear operators. *Siam Rev., 39*(3), 383–406.

[90] Varah, J. M., (1967). *The Computation of Bounds for the Invariant Subspaces of a General Matrix Operator.* Thesis (PhD), Stanford University. ProQuest LLC, Ann Arbor, MI.

[91] Wilcox, D., (2001). *Multivalued semi-Fredholm Operators in Normed Linear Spaces.* PhD Thesis, University of Cape Town.

[92] Wilcox, D., (2002). *Multivalued Semi-Fredholm Operators in Normed Linear Spaces.* Diss. Thesis (PhD), University of Cape Town.

[93] Wilcox, D., (2014). Essential spectra of linear relations. *Linear Algebra Appl., 462*, 110–125.

[94] Wolff, M. P. H., (2001). Discrete approximation of unbounded operators and approximation of their spectra. *J. Approx. Theory, 113*(2), 229–244.

[95] Zeng, Q. P., Zhong, H., & Wu, Z. Y., (2013). Samuel multiplicity and the structure of essentially semi regular operators: A note on a paper of fang. *Sci. China Math., 56*, 1213–1231.

Index

Printed and bound by CPI Group (UK) Ltd, Croydon, CR0 4YY

23/10/2024

01777675-0009